Quasi-Uniform Spaces

PURE AND APPLIED MATHEMATICS

A Program of Monographs, Textbooks, and Lecture Notes

Contributions to *Lecture Notes in Pure and Applied Mathematics* are reproduced by direct photography of the author's typewritten manuscript. Potential authors are advised to submit preliminary manuscripts for review purposes. After acceptance, the author is responsible for preparing the final manuscript in camera-ready form, suitable for direct reproduction. Marcel Dekker, Inc. will furnish instructions to authors and special typing paper. Sample pages are reviewed and returned with our suggestions to assure quality control and the most attractive rendering of your manuscript. The publisher will also be happy to supervise and assist in all stages of the preparation of your camera-ready manuscript.

LECTURE NOTES
IN PURE AND APPLIED MATHEMATICS

1. *N. Jacobson,* Exceptional Lie Algebras
2. *L.-Å. Lindahl and F. Poulsen,* Thin Sets in Harmonic Analysis
3. *I. Satake,* Classification Theory of Semi-Simple Algebraic Groups
4. *F. Hirzebruch, W. D. Newmann, and S. S. Koh,* Differentiable Manifolds and Quadratic Forms (out of print)
5. *I. Chavel,* Riemannian Symmetric Spaces of Rank One (out of print)
6. *R. B. Burckel,* Characterization of C(X) Among Its Subalgebras
7. *B. R. McDonald, A. R. Magid, and K. C. Smith,* Ring Theory: Proceedings of the Oklahoma Conference
8. *Y.-T. Siu,* Techniques of Extension of Analytic Objects
9. *S. R. Caradus, W. E. Pfaffenberger, and B. Yood,* Calkin Algebras and Algebras of Operators on Banach Spaces
10. *E. O. Roxin, P.-T. Liu, and R. L. Sternberg,* Differential Games and Control Theory
11. *M. Orzech and C. Small,* The Brauer Group of Commutative Rings
12. *S. Thomeier,* Topology and Its Applications
13. *J. M. López and K. A. Ross,* Sidon Sets
14. *W. W. Comfort and S. Negrepontis,* Continuous Pseudometrics
15. *K. McKennon and J. M. Robertson,* Locally Convex Spaces
16. *M. Carmeli and S. Malin,* Representations of the Rotation and Lorentz Groups: An Introduction
17. *G. B. Seligman,* Rational Methods in Lie Algebras
18. *D. G. de Figueiredo,* Functional Analysis: Proceedings of the Brazilian Mathematical Society Symposium
19. *L. Cesari, R. Kannan, and J. D. Schuur,* Nonlinear Functional Analysis and Differential Equations: Proceedings of the Michigan State University Conference
20. *J. J. Schäffer,* Geometry of Spheres in Normed Spaces
21. *K. Yano and M. Kon,* Anti-Invariant Submanifolds
22. *W. V. Vasconcelos,* The Rings of Dimension Two
23. *R. E. Chandler,* Hausdorff Compactifications
24. *S. P. Franklin and B. V. S. Thomas,* Topology: Proceedings of the Memphis State University Conference
25. *S. K. Jain,* Ring Theory: Proceedings of the Ohio University Conference
26. *B. R. McDonald and R. A. Morris,* Ring Theory II: Proceedings of the Second Oklahoma Conference
27. *R. B. Mura and A. Rhemtulla,* Orderable Groups
28. *J. R. Graef,* Stability of Dynamical Systems: Theory and Applications
29. *H.-C. Wang,* Homogeneous Banach Algebras
30. *E. O. Roxin, P.-T. Liu, and R. L. Sternberg,* Differential Games and Control Theory II
31. *R. D. Porter,* Introduction to Fibre Bundles
32. *M. Altman,* Contractors and Contractor Directions Theory and Applications
33. *J. S. Golan,* Decomposition and Dimension in Module Categories
34. *G. Fairweather,* Finite Element Galerkin Methods for Differential Equations
35. *J. D. Sally,* Numbers of Generators of Ideals in Local Rings
36. *S. S. Miller,* Complex Analysis: Proceedings of the S.U.N.Y. Brockport Conference
37. *R. Gordon,* Representation Theory of Algebras: Proceedings of the Philadelphia Conference
38. *M. Goto and F. D. Grosshans,* Semisimple Lie Algebras
39. *A. I. Arruda, N. C. A. da Costa, and R. Chuaqui,* Mathematical Logic: Proceedings of the First Brazilian Conference
40. *F. Van Oystaeyen,* Ring Theory: Proceedings of the 1977 Antwerp Conference

41. *F. Van Oystaeyen and A. Verschoren,* Reflectors and Localization: Application to Sheaf Theory
42. *M. Satyanarayana,* Positively Ordered Semigroups
43. *D. L. Russell,* Mathematics of Finite-Dimensional Control Systems
44. *P.-T. Liu and E. Roxin,* Differential Games and Control Theory III: Proceedings of the Third Kingston Conference, Part A
45. *A. Geramita and J. Seberry,* Orthogonal Designs: Quadratic Forms and Hadamard Matrices
46. *J. Cigler, V. Losert, and P. Michor,* Banach Modules and Functors on Categories of Banach Spaces
47. *P.-T. Liu and J. G. Sutinen,* Control Theory in Mathematical Economics: Proceedings of the Third Kingston Conference, Part B
48. *C. Byrnes,* Partial Differential Equations and Geometry
49. *G. Klambauer,* Problems and Propositions in Analysis
50. *J. Knopfmacher,* Analytic Arithmetic of Algebraic Function Fields
51. *F. Van Oystaeyen,* Ring Theory: Proceedings of the 1978 Antwerp Conference
52. *B. Kedem,* Binary Time Series
53. *J. Barros-Neto and R. A. Artino,* Hypoelliptic Boundary-Value Problems
54. *R. L. Sternberg, A. J. Kalinowski, and J. S. Papadakis,* Nonlinear Partial Differential Equations in Engineering and Applied Science
55. *B. R. McDonald,* Ring Theory and Algebra III: Proceedings of The Third Oklahoma Conference
56. *J. S. Golan,* Structure Sheaves over a Noncommutative Ring
57. *T. V. Narayana, J. G. Williams, and R. M. Mathsen,* Combinatorics, Representation Theory and Statistical Methods in Groups: YOUNG DAY Proceedings
58. *T. A. Burton,* Modeling and Differential Equations in Biology
59. *K. H. Kim and F. W. Roush,* Introduction to Mathematical Consensus Theory
60. *J. Banas and K. Goebel,* Measures of Noncompactness in Banach Spaces
61. *O. A. Nielsen,* Direct Integral Theory
62. *J. E. Smith, G. O. Kenny, and R. N. Ball,* Ordered Groups: Proceedings of the Boise State Conference
63. *J. Cronin,* Mathematics of Cell Electrophysiology
64. *J. W. Brewer,* Power Series Over Commutative Rings
65. *P. K. Kamthan and M. Gupta,* Sequence Spaces and Series
66. *T. G. McLaughlin,* Regressive Sets and the Theory of Isols (in press)
67. *T. L. Herdman, S. M. Rankin, III, and H. W. Stech,* Integral and Functional Differential Equations
68. *R. Draper,* Commutative Algebra: Analytic Methods
69. *W. G. McKay and J. Patera,* Tables of Dimensions, Indices, and Branching Rules for Representations of Simple Lie Algebras
70. *R. L. Devaney and Z. H. Nitecki,* Classical Mechanics and Dynamical Systems
71. *J. Van Geel,* Places and Valuations in Noncommutative Ring Theory
72. *C. Faith,* Injective Modules and Injective Quotient Rings
73. *A. Fiacco,* Mathematical Programming with Data Perturbations I
74. *P. Schultz, C. Praeger, and R. Sullivan,* Algebraic Structures and Applications: Proceedings of the First Western Australian Conference on Algebra
75. *L. Bican, T. Kepka, and P. Němec,* Rings, Modules, and Preradicals
76. *D. C. Kay and M. Breen,* Convexity and Related Combinatorial Geometry: Proceedings of the Second University of Oklahoma Conference
77. *P. Fletcher and W. F. Lindgren,* Quasi-Uniform Spaces

Other Volumes in Preparation

Quasi-Uniform Spaces

Peter Fletcher

Department of Mathematics
Virginia Polytechnic Institute
and State University
Blacksburg, Virginia

William F. Lindgren

Department of Mathematics
Slippery Rock State College
Slippery Rock, Pennsylvania

MARCEL DEKKER, INC. New York and Basel

Library of Congress Cataloging in Publication Data

Fletcher, Peter, [date]
Quasi-uniform spaces.

(Lecture notes in pure and applied mathematics; v. 77)
Bibliography: p. 195
Includes index.
1. Quasi-uniform spaces. I. Lindgren, William F. II. Title. III. Series.
QA611.25.F53 514'.3202 82-1500
ISBN 0-8247-1839-9 AACR2

MARCEL DEKKER, INC.
270 Madison Avenue, New York, New York 10016

Current printing (last digit):
10 9 8 7 6 5 4 3 2 1

PRINTED IN THE UNITED STATES OF AMERICA

6/17/85

To the memory of Bill

Preface

A diverse and widely dispersed literature concerning quasi-uniformities has emerged since 1948 when L. Nachbin began the study of these structures. Since the appearance of the monograph by Murdeshwar and Naimpally [113], there has been a spate of research in quasi-uniform spaces, which can be divided into 5 broad classes:

(1) The work of Brümmer, Carlson and Hicks, Császár, Dowker, Hunsaker, Pervin, Salbany, Sieber, and Ward, who have investigated relationships between quasi-uniformities and quasi-proximities, have established the existence of completions and compactifications and have constructed a theory of quasi-uniformities that parallels the well-established theory for uniformities.

(2) The work of Burke, Fleischman, Heath, Junnila, Perugudov, and Scott on covering properties associated with quasi-uniformities.

(3) The study by Fox, Junnila, and Kofner of transitive neighbornets.

(4) The investigation by Choban, Fox, Heath, Hodel, Junnila, Kofner, Nedev, Stoltenberg, and Veličko of quasi-metrizable spaces and their relation to developable spaces and other generalized metric spaces.

(5) The work of Singal and Lal, Blatter, Redfield, and Seever in furthering the use of quasi-uniformities in the study of Nachbin's uniform order spaces.

We believe that the usefulness of quasi-uniformities in the study of general topology has not yet been fully appreciated and that the work cited above deserves to be more widely known. The purpose of our book, therefore, is to collect and organize work in quasi-uniformities and quasi-proximities in order to encourage the use of these structures in general topology.

We have presumed no mathematical background beyond the contents of Bushaw [21] or Kelley [85]; and, although some knowledge of uniform spaces would be helpful, this knowledge is not required. Throughout the text we have presented twenty-two unsolved problems with which we have tried to indicate the variety and extent of our subject. In this respect we have left unheeded the advice of Mary Ellen Rudin: "As I know from experience, it is dangerous to list open problems in a book: people keep solving them . . ." Perhaps we have left some windfalls and easy pickings; we are confident that each reader will find some of our problems sufficiently challenging and that taken as a whole these problems will provide the basis for much future research.

For the most part our notation is straightforward and a little old fashioned. For example, when we index a countable set by the natural numbers, we write $\{A_n | n \in \mathbb{N}\}$ in room of the current $\{A_\alpha\}_{\alpha<\omega}$. We write $<a_n>$ to indicate the sequence whose nth term is a_n. When it comes to denoting composition we waffle. If U and V are relations that are not known to be functions, $U \circ V = \{(x,z) | \text{there is a } y \text{ so that } (x,y) \in U \text{ and } (y,z) \in V\}$, that is $U \circ V$ means first U and then V. If f and g are functions $f \circ g$ has its usual backward interpretation, first g then f.

Helpful comments were given by G. Brümmer, R. Engelking, R. Heath, T. Phillips, and the members of Math 6321. We profited from reading A. H. Stone's unpublished notes on proximity spaces. We are grateful to W. Hunsaker, who read the manuscript and contributed a number of important criticisms and suggestions. We are deeply indebted to R. Fox, H. Junnila, and J. Kofner, whose contributions profoundly influenced the last three chapters.

We thank Elizabeth Kessler and Roxan Rickborn for their excellent and patient typing of several versions of this manuscript.

Peter Fletcher
William F. Lindgren

Contents

Preface	v
Chapter 1 ELEMENTARY PROPERTIES OF QUASI-UNIFORMITIES AND QUASI-PROXIMITIES	
1. Quasi-uniformities	1
2. Quasi-proximities	10
Chapter 2 APPROXIMATIONS OF SYMMETRY	
1. Transitive quasi-uniformities	27
2. Point-symmetry and local symmetry	36
Chapter 3 COMPLETENESS	
1. Cauchy filters	47
2. Bicompleteness	60
3. Completions and compactifications	67
Chapter 4 TOPOLOGICAL ORDERED SPACES	
1. Topological ordered spaces	73
2. Uniform ordered spaces	77
3. Quasi-proximities and order compactifications	81
4. Linear orders	90
Chapter 5 COVERING PROPERTIES OF QUASI-UNIFORM SPACES	
1. Orthocompact and preorthocompact spaces	97
2. Open-finite covers	104
3. Quasi-uniform covers	115

4. Preorthocompactness of $X \times \beta X$ 120

Chapter 6
TRANSITIVE SPACES

1. Transitive uniformities 125
2. Subspaces and mappings 130
3. Properties that imply transitivity 135

Chapter 7
QUASI-METRIZABLE SPACES

1. The quasi-metrization problem 151
2. Products and mappings 156
3. γ-spaces 161
4. Quasi-metrizable subtopologies 172
5. Complete quasi-metric spaces 176

Notes 183

Bibliography 195

List of symbols 207

Index 209

Quasi-Uniform Spaces

Chapter 1

ELEMENTARY PROPERTIES OF QUASI-UNIFORMITIES AND QUASI-PROXIMITIES

1. QUASI-UNIFORMITIES

Many of the results of §§1.1 through 1.4 are so fundamental that they are obliged to be written but not read, and the reader may wish to consult these sections only as needed.

1.1 A *quasi-uniformity* on a set X is a filter $\mathcal{U}$ on $X \times X$ such that (a) each member of $\mathcal{U}$ contains Δ and (b) if $U \in \mathcal{U}$, then $V \circ V \subset U$ for some $V \in \mathcal{U}$. The pair $(X,\mathcal{U})$ is called a *quasi-uniform space* and the members of $\mathcal{U}$ are called *entourages*. If $\mathcal{U}$ is a quasi-uniformity on X, then $\mathcal{U}^{-1} = \{U^{-1} | U \in \mathcal{U}\}$ is also a quasi-uniformity on X and is called the *conjugate* of $\mathcal{U}$. A quasi-uniformity $\mathcal{U}$ is a *uniformity* provided $\mathcal{U} = \mathcal{U}^{-1}$.

Let $\mathbb{R}$ denote the set of all real numbers. For each $\epsilon > 0$ we define relations U_ϵ, Q_ϵ, Z_ϵ and M_ϵ on $\mathbb{R}$ as follows: $U_\epsilon = \{(x,y) | \, |x - y| < \epsilon\}$, $Q_\epsilon = \{(x,y) | x - y < \epsilon\}$, $Z_\epsilon = \{(x,y) | x \le y < x + \epsilon\}$ and $M_\epsilon = \Delta \cup \{(x,y) | x$ is rational and $|x - y| < \epsilon\}$. Let $\mathcal{E}$ be the filter on $\mathbb{R} \times \mathbb{R}$ for which $\{U_\epsilon | \epsilon > 0\}$ is a base and define filters $\mathcal{Q}$, $\mathcal{Z}$ and $\mathcal{M}$ similarly. Then $\mathcal{E}$ is a uniformity on $\mathbb{R}$ whereas $\mathcal{Q}$, $\mathcal{Z}$ and $\mathcal{M}$ are quasi-uniformities on $\mathbb{R}$ that are not uniformities.

1.2 A subfamily $\mathcal{B}$ of a quasi-uniformity $\mathcal{U}$ is a *base* for $\mathcal{U}$ if each member of $\mathcal{U}$ contains a member of $\mathcal{B}$. Axiom 1.1(b) shows that if $\mathcal{B}$ is a base for $\mathcal{U}$ and n is a positive integer, then $\{B^n | B \in \mathcal{B}\}$ is a base for $\mathcal{U}$. A subfamily $\mathcal{S}$ of $\mathcal{U}$ is a *subbase* for $\mathcal{U}$ if the family of finite intersections of members of $\mathcal{S}$ is a base for $\mathcal{U}$.

PROPOSITION. A family $\mathcal{B}$ of subsets of $X \times X$ is a base for a quasi-uniformity on X if and only if $\mathcal{B}$ is a filter base on $X \times X$ that satisfies (a) and (b) of 1.1.

If $\mathcal{B}_1$ and $\mathcal{B}_2$ are bases for quasi-uniformities on a set X, $\mathcal{B}_1$ is *finer* than $\mathcal{B}_2$ (and $\mathcal{B}_2$ *coarser* than $\mathcal{B}_1$) if each member of $\mathcal{B}_2$ contains a member of $\mathcal{B}_1$. In particular, a quasi-uniformity $\mathcal{U}_1$ is finer than $\mathcal{U}_2$ if and only if $\mathcal{U}_2 \subset \mathcal{U}_1$. Two (sub)bases that determine the same quasi-uniformity are said to be *equivalent*.

In practice, a quasi-uniformity is often most easily described by defining a base for it. For example, all the quasi-uniformities on $\mathbb{R}$ given above were defined in terms of bases. No two of the bases $\{U_\epsilon | \epsilon > 0\}$, $\{Q_\epsilon | \epsilon > 0\}$, $\{Z_\epsilon | \epsilon > 0\}$ and $\{M_\epsilon | \epsilon > 0\}$ are equivalent. It is possible, however, for two equivalent bases to appear dissimilar. For example, if for each $n \in \mathbb{N}$ we set $W_n = \{(x,y) | -2/n \leq x - y \leq 1/n\}$, then $\{W_n | n \in \mathbb{N}\}$ and $\{U_\epsilon | \epsilon > 0\}$ are equivalent bases.

1.3 Let $\{\mathcal{U}_i | i \in I\}$ be a family of quasi-uniformities on a set X. The *supremum* of $\{\mathcal{U}_i | i \in I\}$ is the coarsest quasi-uniformity on X that is finer than every $\mathcal{U}_i$. The supremum always exists and is the coarsest filter finer than every $\mathcal{U}_i$. The *infimum* of $\{\mathcal{U}_i | i \in I\}$ is the finest quasi-uniformity that is coarser than every $\mathcal{U}_i$. The infimum always exists; it is the supremum of the family of all quasi-uniformities on X that are coarser than every $\mathcal{U}_i$.

If $\mathcal{U}$ is a quasi-uniformity on X, then the family $\{U \cap U^{-1} | U \in \mathcal{U}\}$ is a base for a quasi-uniformity $\mathcal{U}^*$, which is the coarsest uniformity containing $\mathcal{U}$. The uniformity that has $\{\Delta\}$ as a base is called the *discrete uniformity*. The discrete uniformity on a set X is the supremum of all quasi-uniformities on X. Each nonempty set X has a coarsest quasi-uniformity consisting only of $X \times X$. In the examples first discussed in §1.1, $\mathcal{Z}^*$ is the discrete uniformity and $\mathcal{Q}^* = \mathcal{E} = \mathcal{E}^*$.

1.4 PROPOSITION. Let $\mathcal{B}$ be a base for a quasi-uniformity $\mathcal{U}$ on X and for each $x \in X$ let $\mathcal{B}(x) = \{B(x) | B \in \mathcal{B}\}$. Then there is a unique topology on X such that, for each $x \in X$, $\mathcal{B}(x)$ is a base for the neighborhood filter of x in this topology.

If $(X,\mathcal{U})$ is a quasi-uniform space, the topology *induced* by $\mathcal{U}$ (or simply the topology $\mathcal{T}(\mathcal{U})$ of $\mathcal{U}$) is the topology defined in Proposition 1.4. If $\mathcal{T}(\mathcal{U}) = \mathcal{T}$, then $\mathcal{U}$ is said to be *compatible* with $\mathcal{T}$ (and $(X,\mathcal{T})$ is said to *admit* $\mathcal{U}$).

A subset A of X belongs to $\mathcal{T}(\mathcal{U})$ if and only if for each $x \in A$, there is an entourage $U \in \mathcal{U}$ such that $U(x) \subset A$. By definition for each $x \in X$ and $U \in \mathcal{U}$, $U(x)$ is a $\mathcal{T}(\mathcal{U})$-neighborhood of x; and, although $U(x)$ need not be a $\mathcal{T}(\mathcal{U})$-open set, there is always a base $\mathcal{B}$ for $\mathcal{U}$ such that for each $B \in \mathcal{B}$ and $x \in X$, $B(x) \in \mathcal{T}(\mathcal{U})$ (as is established in Corollary 1.17). A subset B of X is said to be a *$\mathcal{U}$-neighborhood* of A provided there is an entourage $U \in \mathcal{U}$ such that $U(A) \subset B$. Every neighborhood of a compact set is a $\mathcal{U}$-neighborhood; if A is not a compact set then a neighborhood of A is not necessarily a $\mathcal{U}$-neighborhood of A.

The topology induced by the supremum of a family $\{\mathcal{U}_i | i \in I\}$ of quasi-uniformities on a set X is the supremum of the topologies induced by the $\mathcal{U}_i$. The topology induced by the discrete uniformity is the discrete topology.

1.5 A *quasi-pseudo-metric* for a set X is a function d on $X \times X$ to the non-negative real numbers such that for all x, y, $z \in X$, $d(x,z) \leq d(x,y) + d(y,z)$ and $d(x,x) = 0$. A quasi-pseudo-metric d such that $x = y$ whenever $d(x,y) = 0$ is a *quasi-metric*. Let d be a quasi-pseudo-metric for a set X. Then the family of all sets of the form $\{(x,y) | d(x,y) < 1/2^n\}$ for $n \in \mathbb{N}$ is a base for a quasi-uniformity called *the quasi-uniformity generated by* d. The *conjugate of* d, d^{-1} is given by $d^{-1}(x,y) = d(y,x)$. If d generates $\mathcal{U}$, then d^{-1} generates $\mathcal{U}^{-1}$.

We omit the proof of the following lemma on the assumption that the elegant presentation of this result as given by J. L. Kelley is one argument concerning quasi-uniform spaces that is easily accessible and well known.

LEMMA. Let $\langle U_n \rangle$ be a sequence of reflexive relations on a set X such that for each $n \in \mathbb{N}$, $U_{n+1} \circ U_{n+1} \circ U_{n+1} \subset U_n$. Then there is a quasi-pseudo-metric d for X such that $U_{n+1} \subset \{(x,y) | d(x,y) < 1/2^n\} \subset U_n$ for each $n \in \mathbb{N}$. If each U_n is symmetric, d can be taken to be a pseudo-metric.

THEOREM. Let $(X,\mathcal{U})$ be a quasi-uniform space. Then there is a quasi-pseudo-metric d such that $\mathcal{U}$ is the quasi-uniformity generated by d if and only if $\mathcal{U}$ has a countable base.

Let X be a set and let d be a quasi-pseudo-metric for X. The set $\{y|d(x,y) < \epsilon\}$ is called the open ϵ-sphere about x and is denoted by $S_d(x,\epsilon)$. We often omit the distinguishing d when the quasi-pseudo-metric is clear from context. The topology that has the family $\{S_d(x,\epsilon)|x \in X, \epsilon > 0\}$ as a base is called the topology *induced* by d and is denoted by $\mathcal{T}(d)$. The topology induced by d is identical with the topology induced by the quasi-uniformity generated by d. A topological space $(X,\mathcal{T})$ is *quasi-metrizable* provided there is a quasi-metric d such that $\mathcal{T} = \mathcal{T}(d)$.

The quasi-uniformities $\mathcal{E}$, $\mathcal{Z}$ and $\mathcal{M}$ discussed previously are compatible respectively with the Euclidean topology, the Sorgenfrey topology and the topology of the Michael line. That is, $\mathcal{T}(\mathcal{E})$ has as a basis all open intervals (a,b), $\mathcal{T}(\mathcal{Z})$ has as a basis all intervals of the form [a,b), a < b and $\mathcal{T}(\mathcal{M}) = \{G \cup H|G \in \mathcal{T}(\mathcal{E})$ and H is a subset of the set of all irrational numbers$\}$. The topology $\mathcal{T}(\mathcal{Q})$ is the "lower topology" for the real line and has a basis $\{(x,\infty)|x \in \mathbb{R}\}$. The quasi-uniformities $\mathcal{E}$, $\mathcal{Q}$, $\mathcal{Z}$ and $\mathcal{M}$ are all generated by quasi-metrics.

1.6 Let $\mathcal{U}$ be a quasi-uniformity on a set X. In order to define $\mathcal{T}(\mathcal{U})$ no use is made of condition 1.1(b). On the other hand this condition is required to show that for each entourage U and each $x \in X$, U(x) is a $\mathcal{T}(\mathcal{U})$-neighborhood of x. Let W be a reflexive relation on X and suppose that for each $x \in X$, W(x) is a $\mathcal{T}(\mathcal{U})$-neighborhood of x. In general, W may fail to be a member of any quasi-uniformity compatible with $\mathcal{T}(\mathcal{U})$. Nonetheless, relations such as W are so closely tied to the entourages of a quasi-uniformity that it is worth studying these relations explicitly.

If $(X,\mathcal{T})$ is a topological space and W is a relation on X such that for each $x \in X$, W(x) is a neighborhood of x, then W is called a *neighbornet* of $(X,\mathcal{T})$. Let W be a neighbornet of $(X,\mathcal{T})$. If W is a symmetric (transitive) relation on X, then W is a *symmetric (transitive) neighbornet*. If for each $x \in X$, W(x) is an open set, then W is an *open neighbornet*; similarly if for each $x \in X$, W(x) is a closed set, then W is a *closed neighbornet*. If W(x) = W(y) for each $y \in W(x) \cap W^{-1}(x)$, then W is an *unsymmetric neighbornet*. Equivalently W is an unsymmetric neighbornet provided for all x, $y \in X$ if

$x \in W(y)$ and $y \in W(x)$, then $W(x) = W(y)$. A neighbornet is an equivalence relation if and only if it is both symmetric and unsymmetric.

A sequence $\langle U_n \rangle$ of neighbornets of a space $(X,\mathcal{T})$ is a *normal sequence* of neighbornets if $U_{n+1}^2 \subset U_n$ for each $n \in \mathbb{N}$, and a neighbornet U of $(X,\mathcal{T})$ is a *normal neighbornet* provided that U is a member of some normal sequence of neighbornets of $(X,\mathcal{T})$. Every transitive neighbornet is a normal neighbornet. If $\langle U_n \rangle$ is a normal sequence of neighbornets, then for each $k \in \mathbb{N}$, $U_1 \circ U_2 \circ \ldots \circ U_k \subset U_1^2$ and $\cap_{n=1}^{\infty} U_n$ is a transitive relation.

For any neighbornet U we set $U^{\infty} = \cup\{U^n | n \in \mathbb{N}\}$; the relation U^{∞} is always a transitive neighbornet. If $(X,\mathcal{U})$ is a quasi-uniform space and $U \in \mathcal{U}$, then U is a normal neighbornet of $(X,\mathcal{T}(\mathcal{U}))$. As we note in §2.3, it follows from Proposition 2.1 that every normal neighbornet of $(X,\mathcal{T})$ belongs to a quasi-uniformity compatible with $\mathcal{T}$. Although a neighbornet may have a "square root" without being a normal neighbornet, we note the following proposition, which is of use in Chapter 6.

PROPOSITION. Let n be a natural number and let $(X,\mathcal{T})$ be a topological space. If for each neighbornet U of X there exists a neighbornet V of X such that $V^{2n} \subset U^n$, then for each neighbornet W of X, W^n is a normal neighbornet.

1.7 PROPOSITION. Let $(X,\mathcal{T})$ be a topological space and let V be a neighbornet of $(X,\mathcal{T})$. Then for each $A \subset X$, $\overline{A} \subset V^{-1}(A)$, and the following statements are equivalent.

(a) V^{-1} is a neighbornet.

(b) V contains a symmetric neighbornet.

(c) For each subset A of X, $\overline{A} \subset V(A)$.

Proof. Let A be a subset of X. If $p \in \overline{A}$, then $V(p) \cap A \neq \emptyset$ so that $p \in V^{-1}(A)$. Thus $\overline{A} \subset V^{-1}(A)$. The implications (a) $\Rightarrow$ (b) and (b) $\Rightarrow$ (c) are evident. To see that (c) $\Rightarrow$ (a), suppose that for each subset A of X, $\overline{A} \subset V(A)$ and let $x \in X$. Evidently $x \notin V(X - V^{-1}(x))$ and so $x \notin \overline{X - V^{-1}(x)}$. Hence $V^{-1}(x)$ is a neighborhood of x. ■

1.8 PROPOSITION. Let $\mathcal{B}$ be a base for a quasi-uniformity $\mathcal{U}$ on a set X and let $A \subset X$. The closure of A in $\mathcal{T}(\mathcal{U})$ is given by $\overline{A} = \cap\{U^{-1}(A) | U \in \mathcal{B}\}$.

1.9 Throughout this text, we accept the synonymity of the following two statements:

(a) $(X,\mathcal{U})$ is a quasi-uniform space that has property P.

(b) $\mathcal{U}$ is a quasi-uniformity on X and $\mathcal{U}$ has property P.

Moreover, if P is a topological property, we often write "$(X,\mathcal{U})$ has property P" instead of the more precise, but certainly more cumbersome, "$(X,\mathcal{U})$ is a quasi-uniform space and $(X,\mathcal{T}(\mathcal{U}))$ has property P."

A topological space $(X,\mathcal{T})$ is an R_0-*space* provided that if $G \in \mathcal{T}$ and $x \in G$, then $\overline{\{x\}} \subset G$.

PROPOSITION. Let $(X,\mathcal{U})$ be a quasi-uniform space.

(a) $(X,\mathcal{U})$ is R_0 if and only if $\cap\mathcal{U}$ is symmetric.

(b) $(X,\mathcal{U})$ is T_0 if and only if $\cap\mathcal{U}$ is a partial order.

(c) $(X,\mathcal{U})$ is T_0 if and only if $(X,\mathcal{U}^*)$ is T_2.

(d) $(X,\mathcal{U})$ is T_1 if and only if $\cap\mathcal{U} = \Delta$.

(e) $(X,\mathcal{U})$ is T_2 if and only if $\bigcap_{U\in\mathcal{U}} U \circ U^{-1} = \Delta$.

1.10 The following characterization of T_1 quasi-uniform spaces, though less elegant than the characterization given in Proposition 1.9, proves useful.

PROPOSITION. Let $(X,\mathcal{U})$ be a quasi-uniform space. Then $(X,\mathcal{U})$ is a T_1 space if and only if the following condition holds.

If a filter $\mathcal{F}$ converges in $(X,\mathcal{T}(\mathcal{U}^{-1}))$ to x, then $\mathcal{F}$ has no $\mathcal{T}(\mathcal{U})$ cluster point different from x.

1.11 Let $(X,\mathcal{U})$ and $(Y,\mathcal{V})$ be quasi-uniform spaces. A function $f : X \to Y$ is *quasi-uniformly continuous* if for each $V \in \mathcal{V}$ there exists $U \in \mathcal{U}$ such that $(f(x), f(y)) \in V$ whenever $(x,y) \in U$; that is for each $V \in \mathcal{V}$, $f_2^{-1}[V] \in \mathcal{U}$. Let $f : X \to Y$ and $g : Y \to Z$. Since $(g \circ f)_2 = g_2 \circ f_2$, the composition of two quasi-uniformly continuous functions is quasi-uniformly continuous. A bijection $f : X \to Y$ is a *quasi-unimorphism* if f and f^{-1} are quasi-uniformly continuous.

PROPOSITION. Let $\{\mathcal{U}_i | i \in I\}$ and $\{\mathcal{V}_i | i \in I\}$ be families of quasi-uniformities on X and Y respectively, $\mathcal{U} = \sup\{\mathcal{U}_i\}$ and $\mathcal{V} = \sup\{\mathcal{V}_i\}$. If for each

$i \in I$, $f : (X,\mathcal{U}_i) \to (Y,\mathcal{V}_i)$ is quasi-uniformly continuous, then $f : (X,\mathcal{U}) \to (Y,\mathcal{V})$ is quasi-uniformly continuous.

1.12 PROPOSITION. If $f : (X,\mathcal{U}) \to (Y,\mathcal{V})$ is quasi-uniformly continuous, so are $f : (X,\mathcal{U}^{-1}) \to (Y,\mathcal{V}^{-1})$ and $f : (X,\mathcal{U}^*) \to (Y,\mathcal{V}^*)$.

1.13 PROPOSITION. Let X be a set. For each $i \in I$ let $(Y_i,\mathcal{V}_i)$ be a quasi-uniform space and let $f_i : X \to Y_i$. The family of all sets of the form $\{(x,y) | (f_i(x),f_i(y)) \in V\}$, for $i \in I$ and $V \in \mathcal{V}_i$ is a subbase for a quasi-uniformity $\mathcal{U}$ on X; moreover $\mathcal{U}$ is the coarsest quasi-uniformity for which each f_i is quasi-uniformly continuous.

1.14 PROPOSITION. Let $f : (X,\mathcal{U}) \to (Y,\mathcal{V})$ be a quasi-uniformly continuous function. Then f is $\mathcal{T}(\mathcal{U}) - \mathcal{T}(\mathcal{V})$ continuous, $\mathcal{T}(\mathcal{U}^{-1}) - \mathcal{T}(\mathcal{V}^{-1})$ continuous and $\mathcal{T}(\mathcal{U}^*) - \mathcal{T}(\mathcal{V}^*)$ continuous.

1.15 Let $(X,\mathcal{U})$ be a quasi-uniform space and E be a subset of X. The quasi-uniformity on E whose set of entourages is the trace on $E \times E$ of $\mathcal{U}$ is called the quasi-uniformity *induced* by $\mathcal{U}$ (or the *relative* quasi-uniformity) and is denoted by $\mathcal{U}|E \times E$. Equivalently, $\mathcal{U}|E \times E$ is the coarsest quasi-uniformity on E such that the inclusion map from E into X is quasi-uniformly continuous. The topology $\mathcal{T}(\mathcal{U}|E \times E)$ is the same as $\mathcal{T}(\mathcal{U})|E$. If $f : (X,\mathcal{U}) \to (Y,\mathcal{V})$ is quasi-uniformly continuous, then the restriction $f|E$ is a quasi-uniformly continuous mapping of E into Y. If $F \subset E \subset X$, then $\mathcal{U}$ and $\mathcal{U}|E \times E$ induce the same quasi-uniformity on F. A quasi-uniform space $(X,\mathcal{U})$ is *embedded* in a quasi-uniform space $(Y,\mathcal{V})$ by an *embedding* f provided that f is a quasi-uniform isomorphism from X onto some subspace of Y with the quasi-uniformity induced by $\mathcal{V}$.

1.16 Let $\{(X_i,\mathcal{U}_i) | i \in I\}$ be a family of quasi-uniform spaces and let $X = \Pi_{i\in I} X_i$. The *product quasi-uniformity* is the coarsest quasi-uniformity on X for which all the projections $\pi_i : X \to X_i$ are quasi-uniformly continuous. The family of all sets of the form $\{(x,y) | (\pi_i(x), \pi_i(y)) \in U\}$ for $i \in I$ and $U \in \mathcal{U}_i$ is a subbase for the product quasi-uniformity. In particular if $(X,\mathcal{U})$ and $(Y,\mathcal{V})$ are quasi-uniform spaces, a base for the product quasi-uniformity on $X \times Y$ consists of the collection of relations on $X \times Y$ to which B belongs provided there is a $U \in \mathcal{U}$ and a $V \in \mathcal{V}$ such that for each

$(x,y) \in X \times Y$, $B((x,y)) = U(x) \times V(y)$. The topology induced by the product quasi-uniformity is the product topology.

PROPOSITION. Let $f : (X,\mathcal{U}) \to \Pi_{i \in I} (Y_i,\mathcal{V}_i)$. Then f is quasi-uniformly continuous if and only if for each $i \in I$, $\pi_i \circ f$ is quasi-uniformly continuous.

1.17 PROPOSITION. Let $\mathcal{U}$ and $\mathcal{V}$ be quasi-uniformities on X, let $U \in \mathcal{U}$, $V \in \mathcal{V}$ and $M \subset X \times X$. Then $U \circ M \circ V$ is a neighborhood of M in the topology of $\mathcal{U}^{-1} \times \mathcal{V}$.

COROLLARY. Let $(X,\mathcal{U})$ be a quasi-uniform space. Then $\{U | U \in \mathcal{U}$ and U is $\mathcal{T}(\mathcal{U}^{-1} \times \mathcal{U})$ open$\}$ is a base for $\mathcal{U}$.

We note that the basic entourages Q_ϵ for $\mathcal{Q}$ defined in §1.1 are not $\mathcal{T}(\mathcal{Q} \times \mathcal{Q})$ neighborhoods of Δ, although the previous corollary asserts that they are $\mathcal{T}(\mathcal{Q}^{-1} \times \mathcal{Q})$ neighborhoods of Δ.

1.18 Although we provide a simple proof of the following proposition, this proposition is also a consequence of the preceding corollary.

PROPOSITION. Let $\mathcal{T}_1$ and $\mathcal{T}_2$ be topologies on a set X, let $(Y,\mathcal{V})$ be a quasi-uniform space and let $f : X \to Y$ be a $\mathcal{T}_1$ - $\mathcal{T}(\mathcal{V})$ and $\mathcal{T}_2$ - $\mathcal{T}(\mathcal{V}^{-1})$ continuous function. Then for each $V \in \mathcal{V}$, $f_2^{-1}[V]$ is a $\mathcal{T}_2 \times \mathcal{T}_1$ neighborhood of Δ.

Proof. Let $V \in \mathcal{V}$ and let $W \in \mathcal{V}$ such that $W^2 \subset V$. Let $x \in X$. There exist $G_1 \in \mathcal{T}_1$ and $G_2 \in \mathcal{T}_2$ such that $x \in G_1 \cap G_2$, $f(G_1) \subset W(f(x))$ and $f(G_2) \subset W^{-1}(f(x))$. It is easily verified that $(x,x) \in G_2 \times G_1 \subset f_2^{-1}[V]$. ∎

1.19 PROPOSITION. Let $\mathcal{U}$ and $\mathcal{V}$ be quasi-uniformities on X and let $M \subset X \times X$. The closure of M in the topology of $\mathcal{U} \times \mathcal{V}$ is given by $\overline{M} = \cap\{U \circ M \circ V^{-1} | U \in \mathcal{U}$ and $V \in \mathcal{V}\}$.

COROLLARY. Let $\mathcal{U}$ be a quasi-uniformity on X. Then $\{U | U \in \mathcal{U}$ and U is $\mathcal{T}(\mathcal{U} \times \mathcal{U}^{-1})$-closed$\}$ is a base for $\mathcal{U}$.

1.20 THEOREM. Let $(X,\mathcal{T})$ be a compact Hausdorff space and let G be a closed partial order on X. There is exactly one quasi-uniformity $\mathcal{U}$ on X such that $\cap\mathcal{U} = G$ and $\mathcal{T}(\mathcal{U}^*) = \mathcal{T}$.

Proof. Suppose that $\mathcal{U}$ is a quasi-uniformity on X such that $\cap\mathcal{U} = G$ and $\mathcal{T}(\mathcal{U}^*) = \mathcal{T}$. We show that $\mathcal{U}$ consists of all $\mathcal{T} \times \mathcal{T}$ neighborhoods of G. By Corollary 1.17, all members of $\mathcal{U}$ are $\mathcal{T} \times \mathcal{T}$ neighborhoods of G. Suppose that there exists a $\mathcal{T} \times \mathcal{T}$ neighborhood V of G that is not a member of $\mathcal{U}$. Then $\{U - V \mid U \in \mathcal{U}\}$ is a base for a filter $\mathcal{V}$ on $X \times X$. Since $(X,\mathcal{T})$ is compact, $\mathcal{V}$ has a $\mathcal{T} \times \mathcal{T}$ cluster point (x,y) that does not belong to G. Since $\mathcal{U}$ is coarser than $\mathcal{V}$, (x,y) is a cluster point of $\mathcal{U}$. It follows from the preceding corollary that the intersection of the $\mathcal{T} \times \mathcal{T}$ closures of members of $\mathcal{U}$ is G - a contradiction.

The proof may be completed by establishing that the family $\mathcal{U}$ of all $\mathcal{T} \times \mathcal{T}$ neighborhoods of G is a quasi-uniformity on X such that $\cap\mathcal{U} = G$ and $\mathcal{T}(\mathcal{U}^*) = \mathcal{T}$. Clearly $\mathcal{U}$ is a filter on $X \times X$, and $\cap\mathcal{U} = G$. Suppose that 1.1(b) is not satisfied; then there exists a $\mathcal{T} \times \mathcal{T}$ open set $U \in \mathcal{U}$ such that for all $V \in \mathcal{U}$, $V^2 - U \neq \emptyset$. For each $V \in \mathcal{U}$ let $V' = \{((x,y),z) \in X^2 \times X \mid (x,y) \notin U, (x,z) \in V, (z,y) \in V\}$. It follows that $\mathcal{B} = \{V' \mid V \in \mathcal{U}\}$ is a filter base on $(X \times X - U) \times X$. Since $(X \times X - U) \times X$ is compact, $\mathcal{B}$ has a cluster point $((a,b),c)$. We assert that $(a,c) \in G$. Suppose that $(a,c) \notin G$; as G is compact, there exist disjoint open sets V and H such that $G \subset V$ and $(a,c) \in H$. Set $W = \{((x,y),z) \mid (x,z) \in H\}$. Then $((a,b),c) \in W$ and $W \cap V' = \emptyset$ - a contradiction. Thus $(a,c) \in G$ and it follows as above that $(c,b) \in G$. Since G is transitive, $(a,b) \in G \subset U$ - a contradiction. Finally it is evident that $\mathcal{T}(\mathcal{U}^*) \subset \mathcal{T}$; as $\cap\mathcal{U}$ is a partial order $\mathcal{T}(\mathcal{U}^*)$ is a Hausdorff topology. Hence $\mathcal{T}(\mathcal{U}^*) = \mathcal{T}$. ∎

COROLLARY. Let $(X,\mathcal{T})$ be a compact Hausdorff space. The unique uniformity compatible with $\mathcal{T}$ is the family of all neighborhoods of Δ.

PROBLEM A. For which quasi-uniform spaces $(X,\mathcal{U})$ is the family of all $\mathcal{T}(\mathcal{U}^* \times \mathcal{U}^*)$ neighborhoods of $\cap\mathcal{U}$ a quasi-uniformity compatible with $\mathcal{T}(\mathcal{U})$?

1.21 THEOREM. Let $(X,\mathcal{U})$ and $(Y,\mathcal{V})$ be quasi-uniform spaces and suppose that $(X,\mathcal{U}^*)$ is a compact Hausdorff space. If $f : X \to Y$ is $\mathcal{T}(\mathcal{U})$ - $\mathcal{T}(\mathcal{V})$ continuous and $\mathcal{T}(\mathcal{U}^{-1})$ - $\mathcal{T}(\mathcal{V}^{-1})$ continuous, then $f : (X,\mathcal{U}) \to (Y,\mathcal{V})$ is quasi-uniformly continuous.

Proof. Let $V \in \mathcal{V}$. By Proposition 1.18, $f_2^{-1}[V]$ is a $\mathcal{T}(\mathcal{U}^{-1} \times \mathcal{U})$ neighborhood of Δ and hence of $\cap\mathcal{U}$. By the proof of Theorem 1.20, $f_2^{-1}[V] \in \mathcal{U}$. ∎

EXAMPLE. It is not sufficient in the previous theorem that $(X,\mathcal{U})$ and $(X,\mathcal{U}^{-1})$ be compact. Let X denote the set of all *nonzero* real numbers in $[-1,1]$. Let $\mathcal{U}$ denote the restriction to $X \times X$ of the quasi-uniformity $\mathcal{Q}$ defined in §1.1. Then $(X,\mathcal{T}(\mathcal{U}))$ and $(X,\mathcal{T}(\mathcal{U}^{-1}))$ are compact spaces. Define $f : X \to X$ by $f(x) = -1$ if $x < 0$ and $f(x) = 1$ if $x > 0$. Then f is a $\mathcal{T}(\mathcal{U})$ - $\mathcal{T}(\mathcal{U})$ and $\mathcal{T}(\mathcal{U}^{-1})$ - $\mathcal{T}(\mathcal{U}^{-1})$ continuous function, but $f : (X,\mathcal{U}) \to (X,\mathcal{U})$ is not quasi-uniformly continuous.

2. QUASI-PROXIMITIES

1.22 The notion that two sets can be near is fundamentally sound and although the notion is often obscured by ϵ's and δ's it remains the basis of much mathematical intuition. We now list a set of axioms for a relation δ in $\mathcal{P}(X)$ that makes precise the appropriate behavior of nearness. We write $A \,\delta\, B$ for $(A,B) \in \delta$ and $A \not\delta B$ instead of $(A,B) \notin \delta$.

A relation δ in $\mathcal{P}(X)$ is a *quasi-proximity* for a set X if it satisfies the following conditions:

(a) $X \not\delta \emptyset$ and $\emptyset \not\delta X$.

(b) $C \,\delta\, A \cup B$ if and only if $C \,\delta\, A$ or $C \,\delta\, B$.
$A \cup B \,\delta\, C$ if and only if $A \,\delta\, C$ or $B \,\delta\, C$.

(c) $\{x\} \,\delta\, \{x\}$ for each $x \in X$.

(d) If $A \not\delta B$, there exists $C \subset X$ such that $A \not\delta C$ and $X - C \not\delta B$.

The pair (X,δ) is called a *quasi-proximity space*. If δ is a quasi-proximity on X, then so is δ^{-1}. A quasi-proximity δ is a *proximity* if $\delta = \delta^{-1}$. Let A and B be subsets of a quasi-proximity space (X,δ). If $A \,\delta\, B$, then A is said to be *near* B and if $A \not\delta B$, then A is said to be *far from* B.

Axioms (a), (b) and (c) are natural and axiom (d) will be seen to correspond to the property of quasi-uniformities given by axiom 1.1(b).

1.23 PROPOSITION. Let (X,δ) be a quasi-proximity space. (a) If $A \subset A'$, $B \subset B'$ and $A \,\delta\, B$, then $A' \,\delta\, B'$. (b) If $A \cap B \neq \emptyset$, then $A \,\delta\, B$.

1.24 A set B is said to be a δ-*neighborhood* of a set A if $A \not\delta X - B$.

PROPOSITION. Let (X,δ) be a quasi-proximity space and let $\ll$ be the relation on $\mathcal{P}(X)$ defined by $A \ll B$ if and only if B is a δ-neighborhood of A.

Then $<<$ satisfies the following conditions.

(a) $X << X$ and $\emptyset << \emptyset$.
(b) If $A << B$, then $A \subset B$.
(c) If $A \subset B << C \subset D$, then $A << D$.
(d) If $A << B_1$ and $A << B_2$, then $A << B_1 \cap B_2$.
(e) If $A_1 << B$ and $A_2 << B$, then $A_1 \cup A_2 << B$.
(f) If $A << B$, then there exists $C \subset X$ such that $A << C << B$.

The conditions of the previous proposition provide a notion of "strong inclusion" in a quasi-proximity space. The next proposition, however, establishes that one could just as well start with the notion of strong inclusion and derive the notion of a quasi-proximity. Since the two ideas are basically the same, it might well seem redundant to have introduced them both. While "nearness" is undeniably intuitive, in many instances "strong inclusion" is the more straightforward way of expressing a given relationship between sets. For example, condition (f) of strong inclusion is more expressive than axiom (d) of nearness, which it replaces.

1.25 PROPOSITION. Let a relation $<<$ satisfying conditions 1.24 (a)-(f) be defined on $\mathcal{P}(X)$. The relation δ defined by $A \not\delta B$ provided $A << X - B$ is a quasi-proximity on X. Further B is a δ-neighborhood of A if and only if $A << B$.

1.26 PROPOSITION. If (X,δ) is a quasi-proximity space, the function $cl_\delta:\mathcal{P}(X) \to \mathcal{P}(X)$ defined by $cl_\delta(A) = \{x | \{x\}\ \delta\ A\}$ is a closure operator on X.

1.27 If (X,δ) is a quasi-proximity space, then the topology *induced* by δ (or simply the topology $\mathcal{T}(\delta)$ of δ) is the topology generated by the closure operator defined in Proposition 1.26. A topological space $(X,\mathcal{T})$ is said to *admit* a quasi-proximity δ (and δ is said to be *compatible* with $\mathcal{T}$) provided δ induces $\mathcal{T}$. If $x \in X$, the $\mathcal{T}(\delta)$-neighborhoods of x are precisely the δ-neighborhoods of x. Every neighborhood of a compact set is a δ-neighborhood; if A is not a compact set, a neighborhood of A need not be a δ-neighborhood of A. We note that if $A \not\delta B$, then $cl_\delta B \subset X - A$. Moreover since $int(X - B) = X - cl_\delta B$, it also follows that if $A \not\delta B$ then $A \subset int(X - B)$.

PROPOSITION. Let (X,δ) be a quasi-proximity space. If $\{x\} \not\delta A$, there is a $\mathcal{T}(\delta)$-neighborhood G of x such that $G \not\delta A$.

Proof. Since $\{x\} \not\delta A$, there is a set C such that $\{x\} \not\delta C$ and $X - C \not\delta A$. Set $G = \text{int}(X - C)$. As noted above, $x \in G$ and $G \not\delta A$. ∎

1.28 PROPOSITION. Let $\mathcal{U}$ be a quasi-uniformity on X and let $\delta_{\mathcal{U}}$ denote the relation in $P(X)$ defined by $A \,\delta_{\mathcal{U}}\, B$ provided that for each $U \in \mathcal{U}$, $A \times B \cap U \neq \emptyset$. Then $\delta_{\mathcal{U}}$ is a quasi-proximity on X and $\mathcal{T}(\mathcal{U}) = \mathcal{T}(\delta_{\mathcal{U}})$. Moreover the relation $\ll_{\mathcal{U}}$ corresponding to $\delta_{\mathcal{U}}$ is given by $A \ll_{\mathcal{U}} B$ if and only if there is a $U \in \mathcal{U}$ such that $U(A) \subset B$.

If $(X,\mathcal{U})$ is a quasi-uniform space, the quasi-proximity *induced* by $\mathcal{U}$ is the quasi-proximity of Proposition 1.28 defined by $A \,\delta_{\mathcal{U}}\, B$ if and only if for each $U \in \mathcal{U}$, $A \times B \cap U \neq \emptyset$. A quasi-uniformity $\mathcal{U}$ is said to be *compatible* with δ if $\delta_{\mathcal{U}} = \delta$. If δ is a quasi-proximity on X, then $\pi(\delta)$ denotes the class of all quasi-uniformities compatible with δ. Two quasi-uniformities that belong to the same quasi-proximity class are said to be *qp-equivalent*.

1.29 PROPOSITION. Let $\mathcal{U}$ and $\mathcal{V}$ be quasi-uniformities on a set X. If $\mathcal{U} \subset \mathcal{V}$, then $\mathcal{T}(\mathcal{U}) \subset \mathcal{T}(\mathcal{U})$ and $\delta_{\mathcal{V}} \subset \delta_{\mathcal{U}}$.

1.30 PROPOSITION. Let δ be a quasi-proximity on a set X and E be a subset of X. Then $\delta_E = \delta \cap (P(E) \times P(E))$ is a quasi-proximity on E and $\mathcal{T}(\delta_E) = \mathcal{T}(\delta)|E$. Further, if $\mathcal{U}$ induces δ, then $\mathcal{U}|E \times E$ induces δ_E.

1.31 For all subsets A and B of $\mathbb{R}$ define $d(A,B) = \inf\{|a - b| \mid a \in A, b \in B\}$. Then $A \,\delta_{\mathcal{E}}\, B$ if and only if $d(A,B) = 0$. Define the relation $\ll$ on $P(\mathbb{R})$ by $A \ll B$ provided $A \subset B$ and $d(Q \cap A, X - B) > 0$. Then $\ll$ is a strong inclusion relation and the corresponding quasi-proximity is $\delta_{\mathcal{M}}$. Finally we note that for nonempty sets A and B, $A \,\delta_{\mathcal{Q}}\, B$ if and only if $\inf A \leq \sup B$.

1.32 A quasi-uniform space $(X,\mathcal{U})$ is *totally bounded* provided that for each (subbasic) entourage U there is a finite cover $\mathcal{A}$ of X such that for each $A \in \mathcal{A}$, $A \times A \subset U$. Evidently, if $\mathcal{U}$ is a quasi-uniformity on a set X each of $\mathcal{U}$, $\mathcal{U}^{-1}$ and $\mathcal{U}^*$ is totally bounded or none $\mathcal{U}$, $\mathcal{U}^{-1}$, and $\mathcal{U}^*$ is totally bounded. Let E and F be subsets of a quasi-uniform space $(X,\mathcal{U})$. If $\mathcal{U}$ is totally bounded, so is $\mathcal{U}|E \times E$; and if $\mathcal{U}|E \times E$ and $\mathcal{U}|F \times F$ are totally bounded so is $\mathcal{U}|(E \cup F) \times (E \cup F)$. A product of totally bounded quasi-uniformities is totally bounded, and since total boundedness is preserved by quasi-uniformly continuous functions each factor of a totally bounded product quasi-uniformity is totally bounded. The inverse image of a totally bounded quasi-uniformity is totally bounded, and any quasi-uniformity coarser than a totally bounded quasi-uniformity is totally bounded. The supremum of a family of totally bounded quasi-uniformities on a set X is totally bounded.

None of the quasi-uniformities $\mathcal{E}$, $\mathcal{Z}$, $\mathcal{M}$, or $\mathcal{Q}$ is totally bounded; however it is instructive to verify that $\mathcal{Q}$ (equivalently $\mathcal{Q}^* = \mathcal{E}$) is totally bounded when restricted to any bounded subset of $\mathbb{R}$.

1.33 Let (X,δ) be a quasi-proximity space. We proceed to show that there is a unique totally bounded quasi-uniformity compatible with δ and that this quasi-uniformity is the coarsest quasi-uniformity compatible with δ. The result is fundamental. It establishes a one-to-one correspondence between the quasi-proximities and the totally bounded quasi-uniformities that are compatible with a given topology. One reason for an interest in quasi-proximities is that some important properties of quasi-uniform spaces are qp-invariant in the sense that if $\mathcal{U}$ and $\mathcal{V}$ are qp-equivalent quasi-uniformities then $\mathcal{U}$ has the given property if and only if $\mathcal{V}$ does.

LEMMA. Let δ be a quasi-proximity on X and suppose that $A \delta B$ and $E \not\delta F$. Then $(A - E) \delta B$ or $A \delta (B - F)$.

Proof. Suppose $(A - E) \not\delta B$. Since $A = (A \cap E) \cup (A - E)$ and $A \delta B$, $A \cap E \delta B$. Since $B = (B \cap F) \cup (B - F)$ and $E \not\delta F$, $A \cap E \delta B - F$ so that $A \delta B - F$. ■

Let X be a set and let $(A,B) \in \mathcal{P}(X) \times \mathcal{P}(X)$. We let $T(A,B)$ denote $X \times X - A \times B$.

THEOREM. Let (X,δ) be a quasi-proximity space. The collection $\mathcal{S}$ of all sets of the form $T(A,B)$, where $A \not\delta B$, is a subbase for a totally bounded quasi-uniformity, $\mathcal{U}_\delta$, which is compatible with δ. Moreover, $\mathcal{U}_\delta$ is the coarsest quasi-uniformity in $\pi(\delta)$ and is the only totally bounded member of $\pi(\delta)$.

Proof. <u>$\mathcal{S}$ is a subbase for a quasi-uniformity, $\mathcal{U}_\delta$.</u> Let $T(A,B) \in \mathcal{S}$. Since $A \not\delta B$, there is a subset C of X such that $A \not\delta C$ and $X - C \not\delta B$. It follows that $[T(A,C) \cap T(X - C, B)]^2 \subset T(A,B)$. Each $T(A,B)$ in $\mathcal{S}$ is reflexive, and the result follows from Proposition 1.2.

$\underline{\mathcal{U}_\delta \text{ is totally bounded}}$. Let $T(A,B) \in \mathcal{S}$. Since $X - A \cup X - B = X$ and $(X - A) \times (X - A) \cup (X - B) \times (X - B) \subset T(A,B)$, $\mathcal{U}_\delta$ is totally bounded.

$\underline{\mathcal{U}_\delta \text{ is compatible with } \delta}$. Let α denote the quasi-proximity induced by $\mathcal{U}_\delta$. If $A \not\delta B$, then $A \times B \cap T(A,B) = \emptyset$ and $A \not\alpha B$. We show that $A \alpha B$ whenever $A \alpha B$ by establishing the following:

For each $n \in \mathbb{N}$ and each (A,B) with $A \delta B$ and each subfamily $\{T(E_i,F_i) \mid 1 \le i \le n\}$ of $\mathcal{S}$,

$A \times B \cap [\cap\{T(E_i,F_i) \mid 1 \le i \le n\}] \neq \emptyset$

The proof is by induction. The case $n = 1$ follows easily from the preceding lemma. Let $n \in \mathbb{N}$ and suppose that the statement holds for all $k < n$. Let $A \delta B$ and let $\{T(E_i,F_i) \mid 1 \le i \le n\}$ be a subfamily of $\mathcal{S}$. For each $k \le n$ set $G_k = \cap\{T(E_i,F_i) \mid 1 \le i \le k\}$. Then $A \times B \cap G_n = (A \times B \cap (X - E_n) \times X) \cap G_{n-1} \cup (A \times B \cap X \times (X - F_n)) \cap G_{n-1} = ((A - E_n) \times B) \cap G_{n-1} \cup (A \times (B - F_n)) \cap G_{n-1}$. By the preceding lemma and the inductive hypothesis, one of the terms in the union given above is nonempty.

$\underline{\mathcal{U}_\delta \text{ is the coarsest quasi-uniformity compatible with } \delta}$. Let $\mathcal{U} \in \pi(\delta)$. If $A \not\delta B$, there exists $U \in \mathcal{U}$ such that $A \times B \cap U = \emptyset$. Thus $U \subset T(A,B)$, so that $\mathcal{U}_\delta \subset \mathcal{U}$.

$\underline{\mathcal{U}_\delta \text{ is the unique totally bounded quasi-uniformity compatible with } \delta}$. Let $\mathcal{U}$ be a totally bounded quasi-uniformly compatible with δ. Let $U \in \mathcal{U}$ and choose $W \in \mathcal{U}$ such that $W^2 \subset U$. Since $\mathcal{U}$ is totally bounded, there exists a finite cover $\{A_i \mid 1 \le i \le n\}$ of X such that $A_i \times A_i \subset W$, whenever $1 \le i \le n$. Since $A_i \times (X - W(A_i)) \cap W = \emptyset$, we have $A_i \not\delta (X - W(A_i))$. Let $V = \cap\{T(A_i, X - W(A_i)) \mid 1 \le i \le n\}$. Then $V \in \mathcal{U}_\delta$ and it follows that $V \subset \cup\{A_i \times W(A_i) \mid 1 \le i \le n\} \subset W^2 \subset U$. Therefore $U \in \mathcal{U}_\delta$ and $\mathcal{U} \subset \mathcal{U}_\delta$. ■

1.34 COROLLARY. Let (X,δ) be a proximity space. The collection of all sets of the form $T(A,B)$, where $A \not\delta B$, is a subbase for a totally bounded uniformity, $\mathcal{U}_\delta$, which is compatible with δ. Moreover $\mathcal{U}_\delta$ is the coarsest quasi-uniformity in $\pi(\delta)$ and is the only totally bounded member of $\pi(\delta)$.

1.35 COROLLARY. If δ and ρ are quasi-proximities on X such that $\delta \subset \rho$, then $\mathcal{U}_\rho \subset \mathcal{U}_\delta$.

1.36 COROLLARY. Let (X,δ) be a quasi-proximity space. A base for $\mathcal{U}_\delta$ is the collection $\mathcal{V}$ of all sets of the form $\cup\{A_i \times B_i \mid 1 \le i \le n\}$, where

$\{A_i | 1 \le i \le n\}$ is a finite cover of X and $A_i << B_i$ for every i, $1 \le i \le n$.

Proof. Let $V = \cup\{A_i \times B_i | 1 \le i \le n\}$ be an element of $\mathcal{V}$. Let $U = \cap\{T(A_i, X - B_i) | 1 \le i \le n\}$; then $U \in \mathcal{U}_\delta$ and $U(A_i) \subset B_i$. Therefore $U \subset V$ and $\mathcal{V} \subset \mathcal{U}_\delta$.

Let $A \not\delta B$. Then $X - A << X$ and $A << X - B$. As $T(A,B) = X - A \times X \cup A \times X - B$, $T(A,B) \in \mathcal{V}$. Hence $\mathcal{V}$ is finer than $\mathcal{U}_\delta$. ∎

1.37 If $\mathcal{U}$ is a quasi-uniformity on X, then $\mathcal{U}_\omega$ denotes the totally bounded member of $\pi(\delta_{\mathcal{U}})$. The notation $\mathcal{U}_\omega$ is standard and, as we will not use ω to denote a quasi-proximity, no ambiguity will arise. It follows from Theorem 1.33 that $\mathcal{U}_\omega$ is the finest totally bounded quasi-uniformity on X that is coarser than $\mathcal{U}$ and that two quasi-uniformities $\mathcal{U}$ and $\mathcal{V}$ are qp-equivalent if and only if $\mathcal{U}_\omega = \mathcal{V}_\omega$.

1.38 We consider the family of all quasi-proximities on a set X to be partially ordered by *reverse* set inclusion and we say that δ is *coarser* than ρ (and ρ is *finer* than δ) provided $\rho \subset \delta$. In light of Proposition 1.29, this partial order squares with the usual partial order on the family of all quasi-uniformities on a set X. Every set X has a finest quasi-proximity: the discrete quasi-proximity given by $A \delta B$ if and only if $A \cap B \ne \emptyset$. It also has a coarsest quasi-proximity given by $A \delta B$ if and only if $A \ne \emptyset$ and $B \ne \emptyset$.

Let X be a set. Since the union of a family of quasi-uniformities on X is a subbase for the supremum of that family, it appears at first glance that the supremum of a family of quasi-proximities on X might be the intersection of all the quasi-proximities of the family. Unfortunately, it is unlikely even for the family $\{\delta, \delta^{-1}\}$ that $\delta \cap \delta^{-1}$ is a quasi-proximity; condition (b) of the definition of a quasi-proximity is nearly always violated. Nonetheless every family of quasi-proximities on X has a supremum and it will follow from Proposition 1.39 that the supremum of a nonempty family of quasi-proximities compatible with a given topology is a quasi-proximity that is compatible with that topology. The infimum of a family $\{\delta_i | i \in I\}$ of quasi-proximities on X always exists; it is the supremum of the family of all quasi-proximities that are coarser than each δ_i.

PROPOSITION. Let $\{\delta_i | i \in I\}$ be a nonempty family of quasi-proximities on a set X and let δ_0 be defined by $A \delta_0 B$ if and only if for each finite cover $\mathcal{A}$ of A and each finite cover $\mathcal{B}$ of B there exist A' $\in \mathcal{A}$ and B' $\in \mathcal{B}$ such that for each i $\in$ I, A' δ_i B'. Then δ_0 is a quasi-proximity on X and δ_0 is the coarsest quasi-proximity that is finer than δ_i for each i $\in$ I.

As a special case, the preceding proposition establishes that for each quasi-proximity δ, there is a coarsest proximity that is finer than both δ and δ^{-1}. This proximity is denoted by δ^*. Evidently $\delta^* = \sup \{\delta, \delta^{-1}\}$. Thus A δ^* B, provided that if $\mathcal{A}$ is a finite cover of A and $\mathcal{B}$ is a finite cover of B then there exist A' $\in \mathcal{A}$ and B' $\in \mathcal{B}$ such that A' δ B' and B' δ A'.

EXAMPLE. $\delta_{\mathcal{Q}^*} \neq (\delta_{\mathcal{Q}})^*$. Let A be the even integers and B be the odd integers. Then A $(\delta_{\mathcal{Q}})^*$ B; and A $\delta_{\mathcal{Q}^*}$ B is false since $\mathcal{Q}^* = \mathcal{E}$.

1.39 Let $\{\mathcal{U}_i | i \in I\}$ be a collection of quasi-uniformities on a set X and let $\mathcal{U} = \sup \{\mathcal{U}_i | i \in I\}$. It might appear reasonable to hope that $\sup \{\delta_{\mathcal{U}_i} | i \in I\} = \delta_{\mathcal{U}}$. While this hope is unfounded, we do have the following proposition, which relies upon Theorem 1.33.

PROPOSITION. Let X be a set, let $\{\mathcal{U}_i | i \in I\}$ be a collection of totally bounded quasi-uniformities on X, let $\delta_0 = \sup \{\delta_{\mathcal{U}_i} | i \in I\}$ and let $\mathcal{U} = \sup \{\mathcal{U}_i | i \in I\}$. Then $\delta_0 = \delta_{\mathcal{U}}$.

Proof. For each i $\in$ I, $\mathcal{U}_i \subset \mathcal{U}$ and $\delta_{\mathcal{U}} \subset \delta_{\mathcal{U}_i}$. Thus $\delta_{\mathcal{U}} \subset \delta_0$. Note that if A $\not\delta_{\mathcal{U}_i}$ B, then A $\not\delta_0$ B and that if T(A,B) $\in \mathcal{U}_i$, then T(A,B) $\in \mathcal{U}_{\delta_0}$. By Theorem 1.33, for each i $\in$ I, $\mathcal{U}_i \subset \mathcal{U}_{\delta_0}$. Thus $\mathcal{U} = \sup \{\mathcal{U}_i | i \in I\} \subset \mathcal{U}_{\delta_0}$ so that $\delta_0 \subset \delta_{\mathcal{U}}$. ■

COROLLARY. If $\{\delta_i | i \in I\}$ is a family of quasi-proximities on a set X, then $\sup \{\delta_i | i \in I\}$ induces $\sup \{\mathcal{T}(\delta_i) | i \in I\}$.

Theorem 1.46 establishes that $\inf \{\delta_i | i \in I\}$ need not induce $\inf \{\mathcal{T}(\delta_i) | i \in I\}$.

1.40 LEMMA. Let (X,δ) be a quasi-proximity space and let V be a subset of X × X such that for each subset C of X, C << V(C). Then C << $(U \cap V)(C)$ whenever C $\subset$ X and U $\in \mathcal{U}_\delta$.

Proof. Let C $\subset$ X and let U $\in \mathcal{U}_\delta$. In view of Theorem 1.33 we may assume that $U = \cap_{i=1}^{n} T(A_i,B_i)$ where for each i, $A_i \not\delta B_i$. We first verify that $C \ll (U \cap V)(C)$ for the case $n = 1$, that is for the case that $U = T(A,B) = (X - A \times X) \cup (A \times X - B)$. In this case, $(U \cap V)(C) = (U \cap V)(C \cap A) \cup (U \cap V)(C - A) = [(X - B) \cap V(C \cap A)] \cup V(C - A)$. By hypothesis $C \cap A \ll V(C \cap A)$ and $C - A \ll V(C - A)$, and since $A \not\delta B$, $C \cap A \ll X - B$. Thus $C \ll (U \cap V)(C)$. The result now follows by induction on n. ■

PROPOSITION. Let $\mathcal{U}$ and $\mathcal{V}$ be quasi-uniformities on a set X. If $\mathcal{U}$ is totally bounded, then $\delta_{\mathcal{U} \vee \mathcal{V}} = \delta_{\mathcal{U}} \vee \delta_{\mathcal{V}}$.

Proof. It is clear that $\delta_{\mathcal{U} \vee \mathcal{V}} \subset \delta_{\mathcal{U}} \vee \delta_{\mathcal{V}}$. For notational convenience we let $\ll$ denote the strong inclusion determined by $\delta_{\mathcal{U}} \vee \delta_{\mathcal{V}}$ and $\ll'$ denote the strong inclusion determined by $\delta_{\mathcal{U} \vee \mathcal{V}}$. We show that $\delta_{\mathcal{U}} \vee \delta_{\mathcal{V}} \subset \delta_{\mathcal{U} \vee \mathcal{V}}$ by establishing that $A \ll B$ whenever $A \ll' B$. Suppose that $A \ll' B$. Then there exist $U \in \mathcal{U}$ and $V \in \mathcal{V}$ such that $(U \cap V)(A) \subset B$. For each subset C of X, $C \ll V(C)$; and $\mathcal{U}$ is coarser tnan the totally bounded member of $\pi(\delta_{\mathcal{U}} \vee \delta_{\mathcal{V}})$. By the preceding lemma $A \ll (U \cap V)(A) \subset B$. ■

COROLLARY. Let X be a set and let $\{\mathcal{U}_i | i \in I\}$ be a collection of quasi-uniformities on X and suppose that for at most one $i \in I$, $\mathcal{U}_i$ is not totally bounded. Then the quasi-proximity induced by sup $\{\mathcal{U}_i | i \in I\}$ is the supremum of $\{\delta_{\mathcal{U}_i} | i \in I\}$.

An interesting consequence of the previous corollary is that if $\mathcal{U}$ and $\mathcal{V}$ are quasi-uniformities on a set X, then $\mathcal{U}_\omega \vee \mathcal{V}$ and $\mathcal{V}_\omega \vee \mathcal{U}$ are qp-equivalent. It is also a consequence of the previous proposition that if $\mathcal{U}$ is totally bounded then $\delta_{\mathcal{U}^*} = (\delta_{\mathcal{U}})^*$. In section 1.59 we show that $(\mathcal{U}^*)_\omega$ is not necessarily equal to $(\mathcal{U}_\omega)^*$ so that unless $\mathcal{U}$ is totally bounded $\delta_{\mathcal{U}^*}$ is not necessarily equal to $(\delta_{\mathcal{U}})^*$. The following equalities, however, are easily verified.

(a) For each quasi-uniformity $\mathcal{U}$, $(\delta_{\mathcal{U}})^{-1} = \delta_{(\mathcal{U}^{-1})}$.

(b) For each quasi-uniformity $\mathcal{U}$, $(\mathcal{U}_\omega)^{-1} = (\mathcal{U}^{-1})_\omega$.

(c) For each quasi-proximity δ, $(\mathcal{U}_\delta)^{-1} = \mathcal{U}_{(\delta^{-1})}$.

(d) For each quasi-proximity δ, $(\mathcal{U}_\delta)^* = \mathcal{U}_{(\delta^*)}$.

1.41 PROPOSITION. Let (X,δ) be a quasi-proximity space. The following statements are equivalent.

(a) $A \delta B$.

(b) $cl_{\delta^*}A \, \delta \, cl_{\delta^*}B$.

(c) $cl_{\delta^{-1}}A \, \delta \, cl_{\delta}B$.

Proof. It follows from Proposition 1.23(a) that (a) ⇒ (b) ⇒ (c). Suppose that $cl_{\delta^{-1}}A \, \delta \, cl_{\delta}B$ and that $A \not\delta B$. Then there is a subset C of X such that $A \not\delta C$ and $X - C \not\delta B$. It follows that $cl_{\delta}B \subset C$ so that $A \not\delta cl_{\delta}B$. Consequently, $cl_{\delta}B \not\delta^{-1} A$ and there is a subset D of X such that $cl_{\delta}B \not\delta^{-1} D$ and $X - D \not\delta^{-1} A$. As $cl_{\delta^{-1}} A \subset D$ we have that $cl_{\delta}B \not\delta^{-1} cl_{\delta^{-1}} A$ so that $cl_{\delta^{-1}} A \not\delta cl_{\delta}B$ - a contradiction. ■

1.42 PROPOSITION. Let $(X,\mathcal{T})$ be a normal Hausdorff space. The relation ρ defined by $A \rho B$ if and only if $\overline{A} \cap \overline{B} \neq \emptyset$ is the finest proximity compatible with $\mathcal{T}$.

The following proposition is easily obtained from results of §1.27.

1.43 PROPOSITION. Let (X,δ) be a quasi-proximity space. If A is a $\mathcal{T}(\delta)$-compact subset of X and $B \subset X$, then the following statements are equivalent.

(a) There is an $x \in A$ such that $\{x\} \delta B$.

(b) $A \cap cl_{\delta}B \neq \emptyset$.

(c) $A \delta B$.

1.44 PROPOSITION. Let $(X,\mathcal{T})$ be a compact topological space, let δ be a quasi-proximity on X such that $\mathcal{T}(\delta^*) \subset \mathcal{T}$, and let A and B be subsets of X. Then the following statements are equivalent.

(a) $A \delta B$.

(b) There exist $x \in cl_{\mathcal{T}}A$ and $y \in cl_{\mathcal{T}}B$ such that $\{x\} \delta \{y\}$.

(c) $cl_{\delta^{-1}}A \cap cl_{\delta}B \neq \emptyset$.

Proof. (a) ⇒ (b): Let $\overline{A}$ denote $cl_{\mathcal{T}}A$ and $\overline{B}$ denote $cl_{\mathcal{T}}B$. Since $\mathcal{T}(\delta) \subset \mathcal{T}$, $\overline{A}$ is $\mathcal{T}(\delta)$-compact so that by Proposition 1.43 there is an $x \in \overline{A}$ such that $\{x\} \delta \overline{B}$. Since $\overline{B}$ is $\mathcal{T}(\delta^{-1})$-compact, by Proposition 1.43, there is a $y \in \overline{B}$ such that $\{y\} \delta^{-1} \{x\}$. Thus there exist $x \in \overline{A}$ and $y \in \overline{B}$ such that $\{x\} \delta \{y\}$.

(b) ⇒ (c): Let $x \in cl_{\mathcal{T}}A$ and $y \in cl_{\mathcal{T}}B$ such that $\{x\} \delta \{y\}$. Then $\{x\} \delta cl_{\delta}B$ and since $\{x\}$ is compact it follows from Proposition 1.43 that $x \in cl_{\delta^{-1}}A \cap cl_{\delta}B$.

(c) ⇒ (a): This implication is a consequence of Proposition 1.41. ■

COROLLARY. The only proximity compatible with the topology of a compact Hausdorff space is the one defined in Proposition 1.42.

1.45 PROPOSITION. Let $(X,\mathcal{T})$ be a compact Hausdorff space. Then the proximity ρ defined in Proposition 1.42 is the coarsest quasi-proximity compatible with $\mathcal{T}$, and $\mathcal{U}_\rho$ is the coarsest quasi-uniformity compatible with $\mathcal{T}$.

Proof. Let δ be any quasi-proximity compatible with $\mathcal{T}$. If $A \not\delta B$, then by Proposition 1.41, $\overline{A} \not\delta \overline{B}$. By the preceding proposition $\overline{A} \not\rho \overline{B}$ and so $A \not\rho B$. By Theorem 1.33 and Corollary 1.35, $\mathcal{U}_\rho$ is the coarsest quasi-uniformity compatible with $\mathcal{T}$. ∎

COROLLARY. Let $(X,\mathcal{U})$ be a compact Hausdorff space, let $(Y,\mathcal{V})$ be a uniform space, and let $f : (X,\mathcal{T}(\mathcal{U})) \to (Y,\mathcal{T}(\mathcal{V}))$ be a continuous function, then $f : (X,\mathcal{U}) \to (Y,\mathcal{V})$ is quasi-uniformly continuous.

Proof. As in the previous proposition, let ρ denote the coarsest quasi-proximity compatible with $\mathcal{T}(\mathcal{U})$. By Theorem 1.21, $f : (X,\mathcal{U}_\rho) \to (Y,\mathcal{V})$ is uniformly continuous. By the previous proposition, $\mathcal{U}_\rho \subset \mathcal{U}$ and so the result follows. ∎

1.46 The results of Corollary 1.44 and Proposition 1.45 naturally suggest two questions: Which topological spaces admit only one quasi-proximity and which spaces admit a coarsest quasi-proximity? We do not know the complete answer to either of these questions and we postpone considering the first until the next chapter. The following theorem provides an answer to the second question for Tychonoff spaces.

LEMMA. Let $(X,\mathcal{T})$ be a locally compact Hausdorff space and let $\ll$ denote the relation defined by $A \ll B$ provided that $B = X$ or there exists an open set G and a compact set K such that $A \subset K \subset G \subset B$. Then $\ll$ satisfies conditions 1.24 (a)-(f) and the associated quasi-proximity is compatible with $\mathcal{T}$.

Proof. The verification that $\ll$ satisfies conditions (a)-(e) is straightforward. To see that $\ll$ satisfies condition (f) suppose that $A \ll B$ and $B \neq X$. Then there is a compact set K and an open set G such that $A \subset K \subset G \subset B$. For each $x \in K$ let C_x be an open set containing x such that $\overline{C}_x$ is compact and $\overline{C}_x \subset G$. There is a finite subset I of K such that $K \subset \cup\{C_x | x \in I\}$. Set $C = \cup\{C_x | x \in I\}$. Then $A \subset K \subset C \subset \overline{C} \subset G \subset B$ so that $A << C$ and $C << B$. If $B = X$, choose $C = X$. ■

THEOREM. Let $(X,\mathcal{T})$ be a Tychonoff space. The following statements are equivalent.

(a) $(X,\mathcal{T})$ admits a coarsest quasi-proximity.
(a') $(X,\mathcal{T})$ admits a coarsest quasi-uniformity.
(b) $(X,\mathcal{T})$ admits a coarsest proximity.
(b') $(X,\mathcal{T})$ admits a coarsest uniformity.
(c) $(X,\mathcal{T})$ is locally compact.

Proof. By Theorem 1.33 and Corollary 1.35, conditions (a) and (a') are equivalent and conditions (b) and (b') are equivalent. It therefore suffices to prove that conditions (a), (b) and (c) are equivalent.

(a) ⇒ (b): Let δ be the coarsest quasi-proximity compatible with $\mathcal{T}$ and let ρ denote any proximity compatible with $\mathcal{T}$. Then $\rho \subset \delta$ so that $\rho \subset \delta^*$. Thus δ^* is the coarsest proximity compatible with $\mathcal{T}$.

(b) ⇒ (c): This implication is a well-known result from the theory of proximities. (See the notes.)

(c) ⇒ (a): Let δ be the quasi-proximity determined by the strong inclusion $<<$ of the preceding lemma and let ρ be any quasi-proximity compatible with $\mathcal{T}$. Suppose that $A \not\delta B$. Then $B = \emptyset$ or there is a compact set K and a closed set F such that $A \subset K$, $B \subset F$, and $K \cap F = \emptyset$. By Proposition 1.43, $A \not\rho B$. Thus $\rho \subset \delta$. ■

1.47 PROPOSITION. Let $(X,\mathcal{T})$ be a Hausdorff topological space. Then $(X,\mathcal{T})$ is compact if and only if $(X,\mathcal{T})$ is locally compact and the coarsest quasi-proximity compatible with $\mathcal{T}$ is a proximity.

Proof. Let δ be the coarsest quasi-proximity compatible with $\mathcal{T}$. In light of Proposition 1.45 it suffices to show that if δ is a proximity, then $(X,\mathcal{T})$ is compact. Let $\mathcal{C}$ be an open cover of X and let C be a nonempty member of $\mathcal{C}$ other than X. Let $x \in C$ and set $F = X - C$. Since $\{x\} \cap F = \emptyset$, $\{x\} \not\delta F$ and so $F \not\delta \{x\}$. It follows from the previous lemma that there is a compact set containing F. In particular, F is compact and so $\mathcal{C}$ has a finite subcover. ■

1.48 Let (X,δ) and (Y,ρ) be quasi-proximity spaces and let $f : X \to Y$. The relation δ' given by $A \,\delta'\, B$ if and only if $f(A) \,\rho\, f(B)$ is a quasi-proximity on X. The function $f : (X,\delta) \to (Y,\rho)$ is said to be *qp-continuous* provided that $\delta \subset \delta'$ (that is, $f(A)\ \rho\ f(B)$ whenever $A\ \delta\ B$). Evidently δ' is the coarsest quasi-proximity for which $f : X \to (Y,\rho)$ is qp-continuous. If $\mathcal{V} \in \pi(\rho)$ and $\mathcal{U}$ denotes the quasi-uniformity for which $\{f_2^{-1}[V] \mid V \in \mathcal{V}\}$ is a base, then $\mathcal{U} \in \pi(\delta')$. Thus the following proposition holds.

PROPOSITION. Let $f : X \to Y$ and let $\mathcal{V}$ be a quasi-uniformity on Y. Let $\mathcal{U}$ be the quasi-uniformity for which $\{f_2^{-1}[V] \mid V \in \mathcal{V}\}$ is a base. Then $\{f_2^{-1}[V] \mid V \in \mathcal{V}_\omega\}$ is a base for $\mathcal{U}_\omega$.

1.49 The composition of two qp-continuous functions is qp-continuous. A bijection f such that f and f^{-1} are qp-continuous is a *qp-isomorphism*. A quasi-proximity space (X,δ) is *embedded* in a quasi-proximity space (Y,ρ) by a *qp-embedding* f provided that f is a qp-isomorphism from X onto some subspace of Y.

PROPOSITION. If $f : (X,\delta) \to (Y,\rho)$ is qp-continuous, then so are $f : (X,\delta^{-1}) \to (Y,\rho^{-1})$ and $f : (X,\delta^*) \to (Y,\rho^*)$.

1.50 PROPOSITION. Let $f : (X,\delta) \to (Y,\rho)$ be a qp-continuous function. Then f is $\mathcal{T}(\delta)$ - $\mathcal{T}(\rho)$ continuous, $\mathcal{T}(\delta^{-1})$ - $\mathcal{T}(\rho^{-1})$ continuous, and $\mathcal{T}(\delta^*)$ - $\mathcal{T}(\rho^*)$ continuous.

1.51 PROPOSITION. If $f : (X,\mathcal{U}) \to (Y,\mathcal{V})$ is quasi-uniformly continuous, then $f : (X,\delta_{\mathcal{U}}) \to (Y,\delta_{\mathcal{V}})$ is qp-continuous.

1.52 THEOREM. Let (X,δ) and (Y,ρ) be quasi-proximity spaces and suppose that (X,δ^*) is a compact Hausdorff space. If $f : X \to Y$ is $\mathcal{T}(\delta)$ - $\mathcal{T}(\rho)$ continuous and $\mathcal{T}(\delta^{-1})$ - $\mathcal{T}(\rho^{-1})$ continuous, then $f : (X,\delta) \to (Y,\rho)$ is qp-continuous.

Proof. By Theorem 1.21, $f : (X,\mathcal{U}_\delta) \to (Y,\mathcal{U}_\rho)$ is quasi-uniformly continuous. Thus by Proposition 1.51, $f : (X,\delta) \to (Y,\rho)$ is qp-continuous. ■

1.53 Let $\{(X_i,\delta_i) \mid i \in I\}$ be a nonempty collection of quasi-proximity spaces and let $X = \Pi\{X_i \mid i \in I\}$. The *product quasi-proximity* is the coarsest quasi-proximity on X such that π_i is qp-continuous for each $i \in I$. For each $i \in I$, let δ_i' denote the coarsest quasi-proximity for which π_i is qp-continuous.

Then sup $\{\delta_i' | i \in I\}$ is the product quasi-proximity and in light of Proposition 1.38, the product quasi-proximity δ may be characterized explicitly as follows. If A and B are subsets of X, then A δ B provided that if $\mathcal{A}$ and $\mathcal{B}$ are finite covers of A and B then there exist $A' \in \mathcal{A}$ and $B' \in \mathcal{B}$ so that $\pi_i(A')\ \delta_i\ \pi_i(B')$ for each $i \in I$. It is false in general that the product quasi-uniformity induces the product quasi-proximity even though both do induce the product topology. (See Example 1.60.) We have, however, the following proposition as a consequence of Corollary 1.40.

PROPOSITION. For each $i \in I$ let $\mathcal{U}_i$ be a quasi-uniformity that induces a quasi-proximity δ_i and suppose that at most one $\mathcal{U}_i$ fails to be totally bounded. Then $\Pi\{\mathcal{U}_i | i \in I\}$ induces $\Pi\{\delta_i | i \in I\}$.

1.54 Let $(X,\mathcal{U})$ be a quasi-uniform space. A subset A of X is *uniformly discrete* provided that $\mathcal{U}|A \times A$ is the discrete uniformity on A. One obvious consequence of the following proposition is that a uniform space is totally bounded if and only if every uniformly discrete subset is finite.

PROPOSITION. Let $(X,\mathcal{U})$ and $(Y,\mathcal{V})$ be quasi-uniform spaces. Each of the statements (a) and (b) implies its successor, and if $(X,\mathcal{U})$ and $(Y,\mathcal{V})$ are uniform spaces, then the following statements are equivalent.

(a) One of $\mathcal{U}$ and $\mathcal{V}$ is totally bounded.

(b) $\delta_{\mathcal{U}} \times \delta_{\mathcal{V}} = \delta_{\mathcal{U}\times\mathcal{V}}$.

(c) One of $(X,\mathcal{U})$ and $(Y,\mathcal{V})$ has no infinite uniformly discrete subset.

Proof. We noted in Proposition 1.53 that (a) $\Rightarrow$ (b). (b) $\Rightarrow$ (c): Suppose that $A = \{a_n | n \in \mathbb{N}\}$ and $B = \{b_n | n \in \mathbb{N}\}$ are uniformly discrete subsets of X and Y such that $a_m \neq a_n$ and $b_m \neq b_n$ unless $m = n$. Set $C = \{(a_m,b_n) | m \neq n\}$ and set $D = \{(a_n,b_n) | n \in \mathbb{N}\}$. Then C $\delta_{\mathcal{U}} \times \delta_{\mathcal{V}}$ D whereas C $\delta_{\mathcal{U}\times\mathcal{V}}$ D is false.

(c) $\Rightarrow$ (a). Suppose $\mathcal{U}$ is not totally bounded. By Corollary 3.14 there is a symmetric entourage $U \in \mathcal{U}$ such that for each finite set F, $U(F) \neq X$. Choose a sequence $\langle x_n \rangle$ in X such that for each $i < n$, $x_n \notin U(x_i)$. Then $\{x_n | n \in \mathbb{N}\}$ is an infinite uniformly discrete subset of X. ∎

EXAMPLE. (c) $\not\Rightarrow$ (b). Let $\mathcal{W}$ be a totally bounded uniformity on $\mathbb{R}$ that is compatible with the Euclidean topology and let $\mathcal{V}$ be the quasi-uniformity on $\mathbb{R}$ for which $\{\{(x,y) \mid (x,y) \in W \text{ and } x \leq y\} \mid W \in \mathcal{W}\}$ is a base. Note that $\mathcal{T}(\mathcal{V})$ is the Sorgenfrey topology on $\mathbb{R}$ and that every uniformly discrete subset of $(X,\mathcal{V})$ is finite. Set $A = \{(x,y) \mid x > -y\}$ and set $B = \{(x,-x) \mid x \in \mathbb{R}\}$. Then $A \times B \cap \{((x,y),(u,v)) \mid x \leq u \text{ and } y \leq v\} = \emptyset$ so that $A \not\delta_{\mathcal{V}\times\mathcal{V}} B$. To show that $A \ \delta_{\mathcal{V}} \times \delta_{\mathcal{V}} \ B$ let $\mathcal{A}$ and $\mathcal{B}$ be finite covers of A and B. Since $\mathbb{R} = \cup\{\pi_1(B') \mid B' \in \mathcal{B}\}$ it follows from the Baire Category Theorem that there exists $B' \in \mathcal{B}$ and an interval $(a,b) \subset \mathbb{R}$ such that (a,b) is contained in the Euclidean closure of $\pi_1(B')$. There is a point $(x,y) \in A$ so that $(-y,y) \in B'$ and $a < x < b$. Let $A' \in \mathcal{A}$ be chosen so that $(x,y) \in A'$. Then $y \in \pi_2(A') \cap \pi_2(B')$ and $x \in \pi_1(A') \cap \overline{\pi_1(B')}$, where the closure of the last term is taken with respect to the Sorgenfrey topology. An example to show that (b) $\not\Rightarrow$ (a) is given in Example 2.38.

1.55 PROPOSITION. Let $f : (X,\delta) \to (Y,\rho)$ be a qp-continuous function and let $\mathcal{V} \in \pi(\rho)$. Then there exists $\mathcal{U} \in \pi(\delta)$ such that $f : (X,\mathcal{U}) \to (Y,\mathcal{V})$ is quasi-uniformly continuous.

Proof. Let $\mathcal{W}$ denote the quasi-uniformity for which $\{f_2^{-1}[V] \mid V \in \mathcal{V}\}$ is a base. As is noted in §1.48, $\delta_{\mathcal{W}}$ is the coarsest quasi-proximity on X for which $f : X \to (Y,\rho)$ is qp-continuous. Thus $\delta \subset \delta_{\mathcal{W}}$. Let $\mathcal{U} = \mathcal{U}_\delta \vee \mathcal{W}$. By Proposition 1.40, $\mathcal{U} \in \pi(\delta)$. Clearly $f : (X,\mathcal{U}) \to (Y,\mathcal{V})$ is quasi-uniformly continuous. ∎

If in the preceding proposition $\mathcal{V}$ is assumed to be totally bounded, then $\mathcal{W}$ is totally bounded. Thus by Theorem 1.33 the quasi-uniformity $\mathcal{U}$ obtained in the preceding proposition is $\mathcal{U}_\delta$ and we have the following corollary.

COROLLARY. Let $(X,\mathcal{U})$ be a quasi-uniform space and let $(Y,\mathcal{V})$ be a totally bounded quasi-uniform space. Then $f : (X,\mathcal{U}) \to (Y,\mathcal{V})$ is quasi-uniformly continuous if and only if $f : (X,\delta_{\mathcal{U}}) \to (Y,\delta_{\mathcal{V}})$ is qp-continuous.

1.56 For each $\epsilon > 0$ and $f : X \to \mathbb{R}$, we let $U_{(\epsilon,f)} = f_2^{-1}[Q_\epsilon] = \{(x,y) \mid f(x) - f(y) < \epsilon\}$. Let $\mathcal{U}$ be a quasi-uniformity on X. Then $f : (X,\mathcal{U}) \to (\mathbb{R},\mathcal{Q})$ is quasi-uniformly continuous if and only if for each $\epsilon > 0$, $U_{(\epsilon,f)} \in \mathcal{U}$.

Let $(X,\mathcal{U})$ be a quasi-uniform space. Then $Q(\mathcal{U})$ $(QB(\mathcal{U}))$ denotes the set of all (bounded) quasi-uniformly continuous functions from $(X,\mathcal{U})$ to $(\mathbb{R},\mathcal{Q})$. Similarly, if (X,δ) is a quasi-proximity space, $Q(\delta)$ $(QB(\delta))$ denotes the set of all (bounded) qp-continuous functions from (X,δ) to $(\mathbb{R},\delta_{\mathcal{Q}})$. We could just as well have taken $Q(\mathcal{U})$ to be the set of all quasi-uniformly continuous functions from $(X,\mathcal{U})$ to $(\mathbb{R},\mathcal{Q}^{-1})$; our choice of definition is motivated by certain applications of quasi-uniformities to topological ordered spaces.

Let $\mathcal{U}$ and $\mathcal{V}$ be quasi-uniformities on a set X and let δ be a quasi-proximity on X. We summarize some immediate consequences of previous results in the terminology established above.

(a) If $\mathcal{U} \subset \mathcal{V}$, then $Q(\mathcal{U}) \subset Q(\mathcal{V})$ and $QB(\mathcal{U}) \subset QB(\mathcal{V})$.

(b) $Q(\mathcal{U}) \subset Q(\delta_{\mathcal{U}})$. (Proposition 1.51.)

(c) If $f \in Q(\mathcal{U})$, then f is $\mathcal{T}(\mathcal{U}^*) - \mathcal{T}(\mathcal{E})$ continuous. (Proposition 1.14.)

(d) If $f \in Q(\delta)$, then f is $\mathcal{T}(\delta^*) - \mathcal{T}(\mathcal{E})$ continuous. (Proposition 1.50.)

(e) If $f \in Q(\mathcal{U})$ (or $Q(\delta)$), then f is lower semi-continuous.

(f) $QB(\delta) \subset QB(\mathcal{U}_\delta)$. (Proposition 1.55.)

1.57 We omit the proof of the following lemma for two reasons. First, a proof of the lemma can be given that differs only slightly from the well-known proof of Urysohn's Lemma. Second, for (X,δ) a T_0-space, we show in Section 4.26 that this lemma is a corollary to a generalization of Urysohn's Lemma.

LEMMA. Let (X,δ) be a quasi-proximity space, and let A and B be subsets of X such that $A \not\delta B$. Then there exists $f : X \to [0,1]$ such that $f \in QB(\delta)$, $f(A) = 1$ and $f(B) = 0$.

THEOREM. Let (X,δ) be a quasi-proximity space. Then $\mathcal{U} \in \pi(\delta)$ if and only if $QB(\mathcal{U}) = QB(\delta)$.

Proof. Let $\mathcal{U} \in \pi(\delta)$. We show that $Q(\mathcal{U}_\delta) = QB(\mathcal{U}) = QB(\delta)$. Let $f \in Q(\mathcal{U}_\delta)$. As $\mathcal{U}_\delta$ is totally bounded, there exists a finite cover $\{A_i | 1 \le i \le m\}$ of X such that $\cup\{A_i \times A_i | 1 \le i \le m\} \subset U_{(1,f)}$. For each i with $1 \le i \le m$ choose $a_i \in A_i$ and set $F = \{a_i | 1 \le i \le m\}$. Then $U_{(1,f)}(F) = X$ so that $1 + \max \{f(x) | x \in F\}$ is an upper bound for f and $-1 + \min \{f(x) | x \in F\}$ is a lower bound for f. Thus $Q(\mathcal{U}_\delta) = QB(\mathcal{U}_\delta)$. By 1.56(a), $QB(\mathcal{U}_\delta) \subset QB(\mathcal{U})$; by 1.56(b), $QB(\mathcal{U}) \subset QB(\delta)$; by 1.56(f), $QB(\delta) \subset QB(\mathcal{U}_\delta)$.

Now suppose that $QB(\mathcal{U}) = QB(\delta)$. It follows from the preceding lemma that $\delta_{\mathcal{U}} = \delta$. ∎

COROLLARY. Let (X,δ) be a quasi-proximity space. Then $\{U_{(\epsilon,f)} | \epsilon > 0$ and $f \in QB(\delta)\}$ is a subbase for $\mathcal{U}_\delta$.

Proof. Let $\mathcal{U}$ denote the quasi-uniformity on X for which $\{U_{(\epsilon,f)} | \epsilon > 0, f \in QB(\delta)\}$ is a subbase. As $QB(\delta) = QB(\mathcal{U}_\delta)$, it is clear that $\mathcal{U} \subset \mathcal{U}_\delta$ and that $QB(\mathcal{U}) \subset QB(\delta) \subset QB(\mathcal{U})$. By the previous theorem, $\mathcal{U} \in \pi(\delta)$. It remains to show that $\mathcal{U}$ is totally bounded. Let $f \in QB(\delta)$, let $\epsilon > 0$, and assume without loss of generality that $f(x) \geq 0$ for all $x \in X$. For each nonnegative integer i, let $A_i = \{t | |f(t) - i\epsilon/2| < \epsilon/2\}$. There is a positive integer m such that $\cup_{i=0}^{m} A_i = X$. Let j be an integer with $0 \leq j \leq m$ and let $(x,y) \in A_j \times A_j$. Then $|f(x) - j\epsilon/2| < \epsilon/2$ and $|f(y) - j\epsilon/2| < \epsilon/2$. Consequently, $f(x) - f(y) \leq |f(x) - f(y)| < \epsilon$ so that $(x,y) \in U_{(\epsilon,f)}$. Thus $\mathcal{U}$ is totally bounded and since there is only one totally bounded quasi-uniformity in $\pi(\delta)$, $\mathcal{U} = \mathcal{U}_\delta$. ■

1.58 We have seen that every quasi-proximity class has a coarsest member. The basic information concerning the product quasi-proximity given in §1.54 is all that is required to provide an example of a quasi-proximity class that has no finest member.

EXAMPLE. Let $\mathcal{U}$ be the discrete uniformity on an infinite set X. By Proposition 1.54, $\mathcal{U} \times \mathcal{U}_\omega$ and $\mathcal{U}_\omega \times \mathcal{U}$ both induce $\delta_\mathcal{U} \times \delta_\mathcal{U}$. This proximity is strictly coarser than the discrete proximity on $X \times X$, which is induced by $\mathcal{U} \times \mathcal{U}$. Since $\mathcal{U} \times \mathcal{U}$ is the only quasi-uniformity on $X \times X$ that is finer than both $\mathcal{U} \times \mathcal{U}_\omega$ and $\mathcal{U}_\omega \times \mathcal{U}$, the quasi-proximity class $\pi(\delta_\mathcal{U} \times \delta_\mathcal{U})$ has no finest member. Indeed $\pi(\delta_\mathcal{U} \times \delta_\mathcal{U})$ has no finest uniformity.

1.59 THEOREM. Let (X,δ) be a proximity space and let $\mathcal{U}$ be a quasi-uniformity such that $\mathcal{U}_\delta \subset \mathcal{U}$. If has a countable base, $\mathcal{U}$ contains every uniformity in $\pi(\delta)$.

Proof. Let $\langle U_n \rangle$ be a countable nested base for $\mathcal{U}$. Suppose that there is a uniformity $\mathcal{V}$ in $\pi(\delta)$ and suppose that there exists an entourage $V \in \mathcal{V} - \mathcal{U}$. For each $n \in \mathbb{N}$ choose $(x_n,y_n) \in U_n - V$. For each infinite subset M of $\mathbb{N}$, $\{x_m | m \in M\} \delta \{y_m | m \in M\}$. Let W be a symmetric member of $\mathcal{V}$ such that $W^3 \subset V$. For each $n \in \mathbb{N}$ set $A_n = \{m \in \mathbb{N} | (x_n,y_m) \in W\}$ and set $B_n = \{m \in \mathbb{N} | (x_m,y_n) \in W\}$. We assert that for each $n \in \mathbb{N}$, A_n is finite. Suppose that for some $n \in \mathbb{N}$, A_n is infinite. Then $\{x_m | m \in A_n\} \delta \{y_m | m \in A_n\}$

and there exists s, t εA_n such that $(x_s,y_t) \varepsilon W$. Thus $(y_t,y_s) \varepsilon W^2$ and $(x_s,y_s) \varepsilon V$ - a contradiction. A similar argument establishes that each B_n is finite. Let $n_1 = 1$ and for each $j \varepsilon \mathbb{N}$ choose $n_{j+1} \varepsilon \mathbb{N}$ greater than n_j and every member of $\cup_{i=1}^{n_j} A_i \cup B_i$. Then $\{x_{n_j} | j \varepsilon \mathbb{N}\} \delta \{y_{n_j} | j \varepsilon \mathbb{N}\}$ so that there exist j, k $\varepsilon \mathbb{N}$ such that $(x_{n_j}, y_{n_k}) \varepsilon W$ - a contradiction. ■

COROLLARY. Let $\mathcal{U}$ be a uniformity that has a countable base. Then $\mathcal{U}$ is the finest uniformity in $\pi(\delta_{\mathcal{U}})$.

EXAMPLE. A proximity δ such that $\pi(\delta)$ contains a finest uniformity but contains no finest quasi-uniformity. Let X = (0,1) and let $\mathcal{T}$ be the discrete topology on X. For each n $\varepsilon \mathbb{N}$, let $U_n = \{(x,y) | x = y \text{ or } x \le y < 1/n\}$ and let $\mathcal{U}$ denote the quasi-uniformity for which $\{U_n | n \varepsilon \mathbb{N}\}$ is a base. Then $\mathcal{T} = \mathcal{T}(\mathcal{U}) = \mathcal{T}(\mathcal{U}^{-1})$. The quasi-proximity $\delta_{\mathcal{U}}$ is given by A $\delta_{\mathcal{U}}$ B if and only if $A \cap B \ne \emptyset$ or inf A = 0 = inf B. As $\delta_{\mathcal{U}}$ is a proximity, $\mathcal{U}^{-1} \varepsilon \pi(\delta_{\mathcal{U}})$. Suppose that $\pi(\delta_{\mathcal{U}})$ has a largest member $\mathcal{V}$; then $\mathcal{V}$ contains $\mathcal{U}^*$. It follows that $\delta_{\mathcal{U}}$ is the discrete quasi-proximity - a contradiction. The family $\{U_n \cup U_n^{-1} | n \varepsilon \mathbb{N}\}$ is a base for the finest uniformity that belongs to $\pi(\delta_{\mathcal{U}})$. We note for future reference that $(\mathcal{U}_\omega)^* \ne (\mathcal{U}^*)_\omega$ and $\delta_{\mathcal{U}^*} \ne (\delta_{\mathcal{U}})^*$.

1.60 According to Proposition 1.51 if $f : (X,\mathcal{U}) \to (Y,\mathcal{V})$ is quasi-uniformly continuous, then $f : (X,\delta_{\mathcal{U}}) \to (Y,\delta_{\mathcal{V}})$ is qp-continuous. Although the converse of this proposition fails even in the case that $\mathcal{U}$ and $\mathcal{V}$ are uniformities and f is the identity mapping, Corollary 1.55 provides one instance in which qp-continuity implies quasi-uniform continuity and Theorem 1.59 provides still another. A special case of the following proposition is the important result that a mapping between two metric spaces is uniformly continuous if and only if it is p-continuous.

PROPOSITION. Let $(Y,\mathcal{V})$ be a uniform space, let $(X,\mathcal{U})$ be a quasi-uniform space, and suppose that $\mathcal{U}$ has a countable base. Then $f : (X,\mathcal{U}) \to (Y,\mathcal{V})$ is quasi-uniformly continuous if and only if $f : (X,\delta_{\mathcal{U}}) \to (Y,\delta_{\mathcal{V}})$ is qp-continuous.

Proof. Suppose that f is qp-continuous. By Corollary 1.55, $f : (X,\mathcal{U}) \to (Y,\mathcal{V}_\omega)$ is quasi-uniformly continuous. Let $\mathcal{W}$ be the quasi-uniformity on X for which $\{f_2^{-1}[V] | V \varepsilon \mathcal{V}\}$ is a base. It follows from Proposition 1.48 that $\mathcal{W}_\omega \subset \mathcal{U}$. By Theorem 1.59, $\mathcal{W} \subset \mathcal{U}$. ■

Chapter 2

APPROXIMATIONS OF SYMMETRY

1. TRANSITIVE QUASI-UNIFORMITIES

2.1 A (sub)base $\mathcal{B}$ for a quasi-uniformity is *transitive* provided that each $B \in \mathcal{B}$ is a transitive relation. A quasi-uniformity with a transitive (sub)base is called a *transitive quasi-uniformity*.

The problem of determining which topological spaces admit a quasi-uniformity has as its antitype the classical result of L. Pontrjagin and A. Weil that a topological space admits a uniformity if and only if it is completely regular. Curiously, the problem for quasi-uniform spaces is simply disposed of; all topological spaces admit a quasi-uniformity. Two proofs of this result, which at first appear entirely dissimilar, are given below. We shall see subsequently, however, that both proofs yield transitive quasi-uniformities.

PROPOSITION. Let $(X,\mathcal{T})$ be a topological space. Then $\mathcal{S} = \{T(A, X - A) | A \in \mathcal{T}\}$ is a subbase for a totally bounded transitive quasi-uniformity compatible with $\mathcal{T}$.

Proof. For notational convenience denote $T(A, X - A)$ by $S(A)$. Since for each $A \in \mathcal{T}$, $\Delta \subset S(A)$ and $S(A) \circ S(A) = S(A)$, $\mathcal{S}$ is a subbase for a transitive quasi-uniformity $\mathcal{U}$ on X. Let $U \in \mathcal{U}$ and let $x \in X$. There is a family $\{S(A_i) | 1 \leq i \leq n\}$ so that $\cap_{i=1}^{n} S(A_i) \subset U$. If $x \notin \cup_{i=1}^{n} A_i$, then $X = \cap_{i=1}^{n} S(A_i)(x) \subset U(x)$ so that $U(x) = X \in \mathcal{T}$. If $x \in \cup_{i=1}^{n} A_i$, then

$\cap_{i=1}^{n} S(A_i)(x) = \cap\{A_i | x \in A_i, 1 \le i \le n\}$ is a T-open subset of $U(x)$; hence $T(U) \subset T$. If $A \in T$ and $x \in A$, then $S(A) \in U$ and $S(A)(x) = A$. Thus $T \subset T(U)$ and U is compatible with T.

To see that U is totally bounded it suffices to note that for each $S(A) \in S$, $A \times A \cup (X - A) \times (X - A) \subset S(A)$ and $A \cup (X - A) = X$. ■

2.2 The quasi-uniformity constructed in Proposition 2.1 is called the *Pervin quasi-uniformity for* (X,T) and is denoted by P. It is evident from Propositions 2.1 and 1.28 that every topological space admits a quasi-proximity, namely the *Pervin quasi-proximity* δ_P. Since $A \,\delta_P\, B$ if and only if $A \cap \overline{B} \neq \emptyset$, δ_P is the finest quasi-proximity compatible with T. This observation is comparable to the result stated in Proposition 1.42 that the finest proximity compatible with a normal Hausdorff space is given by $A \,\delta\, B$ if and only if $\overline{A} \cap \overline{B} \neq \emptyset$. Since δ_P is the finest quasi-proximity compatible with T, any totally bounded quasi-uniformity compatible with T is coarser than P.

EXAMPLE. Let T be the upper topology of the real line and let P denote the Pervin quasi-uniformity for $(\mathbb{R}, T)$. Then $T(P^*)$ is the Sorgenfrey topology.

2.3 Let (X,T) be a topological space. The supremum of all quasi-uniformities compatible with T is called the *fine quasi-uniformity* for (X,T) and is denoted $FINE$. Similarly the supremum of all the uniformities compatible with a completely regular space is called the fine uniformity. Since the topology induced by the supremum of a family $\{U_i | i \in I\}$ of quasi-uniformities is the supremum of the topologies induced by the U_i, it follows from Proposition 2.1 that $FINE$ is compatible with T. We note that $FINE$ consists of all normal neighbornets in (X,T). The fine quasi-uniformity can also be characterized as that quasi-uniformity U on X with the property that whenever (Y,V) is a quasi-uniform space and $f : (X,T) \to (Y,T(V))$ is a continuous function then f is a U - V quasi-uniformly continuous function. Finally, we state yet another characterization of $FINE$, which depends upon Lemma 1.5. Let D denote the family of all quasi-pseudo-metrics d for X such that for each $x \in X$, $d(x,\_)$ is upper semi-continuous. The family $\{\{(x,y) | d(x,y) < \epsilon\} | d \in D \text{ and } \epsilon > 0\}$ is a base for $FINE$.

2.4 We omit the proof of the following proposition, whose proof is similar to the proof of Proposition 2.1.

PROPOSITION. Let $(X,\mathcal{T})$ be a noncompact topological space and let $\mathcal{F}$ be a filter on X that has no cluster point. Then $\{T(G, X - G) | G \in \mathcal{T}$ and $X - G \in \mathcal{F}\}$ is a subbase for a totally bounded transitive quasi-uniformity that is compatible with $\mathcal{T}$.

2.5 A collection $\mathcal{C}$ of subsets of a topological space is *interior preserving* provided that if $\mathcal{C}' \subset \mathcal{C}$, then $\text{int} \bigcap\{C | C \in \mathcal{C}'\} = \bigcap\{\text{int } C | C \in \mathcal{C}'\}$; and a collection $\mathcal{C}$ is *closure preserving* provided that if $\mathcal{C}' \subset \mathcal{C}$, then $\overline{\bigcup\{C | C \in \mathcal{C}'\}} = \bigcup\{\overline{C} | C \in \mathcal{C}'\}$. A collection $\mathcal{C}$ of open subsets is interior preserving if and only if for each subcollection $\mathcal{C}'$ of $\mathcal{C}$, $\bigcap\mathcal{C}'$ is open; a collection $\mathcal{C}$ of closed subsets is closure preserving if and only if for each subcollection $\mathcal{C}'$ of $\mathcal{C}$, $\bigcup\mathcal{C}'$ is closed. The following proposition is a straightforward consequence of de Morgan's laws.

PROPOSITION. A collection $\mathcal{C}$ of subsets of a topological space $(X,\mathcal{T})$ is interior preserving if and only if $\{X - C | C \in \mathcal{C}\}$ is a closure-preserving collection.

2.6 If $\mathcal{C}$ is a collection of subsets of a set X and $x \in X$, then $\mathcal{C}_x$ denotes $\{C \in \mathcal{C} | x \in C\}$ so that $\bigcap\mathcal{C}_x = \bigcap\{C \in \mathcal{C} | x \in C\}$. In terms of this notation a collection $\mathcal{C}$ of open subsets of a topological space $(X,\mathcal{T})$ is interior preserving if and only if for each $x \in X$, $\bigcap\mathcal{C}_x \in \mathcal{T}$. If $\mathcal{C}$ is a collection of subsets of a set X, then $U_{\mathcal{C}}$ denotes the reflexive transitive relation $\{(x,y) | x \in X \text{ and } y \in \bigcap\mathcal{C}_x\}$; and if $\mathcal{A}$ is a nonempty family of collections of subsets of X, then $\mathcal{U}_{\mathcal{A}}$ denotes the quasi-uniformity on X for which $\{U_{\mathcal{C}} | \mathcal{C} \in \mathcal{A}\}$ is a subbase. For the nonce we make use of the following notation: If V is a neighbornet of a space $(X,\mathcal{T})$, then $\mathcal{C}_V$ denotes $\{V(x) | x \in X\}$.

THEOREM. Let $(X,\mathcal{T})$ be a topological space and let $\mathcal{A}$ be a family of interior-preserving open collections such that $\bigcup\mathcal{A}$ is a subbase for $\mathcal{T}$. Then $\mathcal{U}_{\mathcal{A}}$ is a transitive quasi-uniformity that is compatible with $\mathcal{T}$. Moreover, if $\mathcal{U}$ is any transitive quasi-uniformity compatible with $\mathcal{T}$, there is a collection $\mathcal{A}$ of interior-preserving open covers of X such that $\bigcup\mathcal{A}$ is a subbase for $\mathcal{T}$ and $\mathcal{U} = \mathcal{U}_{\mathcal{A}}$.

Proof. For each $\mathcal{C} \in \mathcal{A}$, $U_{\mathcal{C}}$ is transitive so that $\mathcal{U}$ is a transitive quasi-uniformity on X. Let $G \in \mathcal{T}(\mathcal{U}_{\mathcal{A}})$ and let $x \in G$. Then there is a finite subcollection $\{\mathcal{C}_i | 1 \leq i \leq n\}$ of $\mathcal{A}$ such that $x \in \cap_{i=1}^{n} [U_{\mathcal{C}_i}(x)] \subset G$. For each i with $1 \leq i \leq n$, $\mathcal{C}_i$ is an interior-preserving open collection so that $U_{\mathcal{C}_i}(x) \in \mathcal{T}$. Thus $\cap_{i=1}^{n} [U_{\mathcal{C}_i}(x)] \in \mathcal{T}$ and so $\mathcal{T}(\mathcal{U}_{\mathcal{A}}) \subset \mathcal{T}$.

Let $G \in \mathcal{T}$ and let $x \in G$. Since $\cup\mathcal{A}$ is a subbase for $\mathcal{T}$, there are A_1, A_2, ... , $A_n \in \cup\mathcal{A}$ such that $x \in \cap_{i=1}^{n} A_i \subset G$. For each i with $1 \leq i \leq n$ let $\mathcal{C}_i \in \mathcal{A}$ such that $A_i \in \mathcal{C}_i$. Then $\cap_{i=1}^{n} U_{\mathcal{C}_i} \in \mathcal{U}_{\mathcal{A}}$ and $\cap_{i=1}^{n} U_{\mathcal{C}_i}(x) \subset G$. Thus $\mathcal{T} \subset \mathcal{T}(\mathcal{U}_{\mathcal{A}})$ so that $\mathcal{U}_{\mathcal{A}}$ is compatible with $\mathcal{T}$.

Now suppose that $\mathcal{U}$ is a quasi-uniformity that is compatible with $\mathcal{T}$ and let $\mathcal{B}$ be a transitive base for $\mathcal{U}$. Let $V \in \mathcal{B}$, let $y \in X$ and let $G = \cap\{V(x) | y \in V(x), x \in X\}$. Let $p \in G$. For each $x \in X$ such that $y \in V(x)$, $V(p) \subset V(x)$. Hence $V(p) \subset G$ and $G \in \mathcal{T}$. Thus for each $V \in \mathcal{B}$, $\mathcal{C}_V$ is an interior-preserving open cover of X. Let $\mathcal{A} = \{\mathcal{C}_V | V \in \mathcal{B}\}$. In order to see that $\mathcal{U} = \mathcal{U}_{\mathcal{A}}$, it suffices to observe that for each $x \in X$ and each $V \in \mathcal{B}$, $V(x) = U_{\mathcal{C}_V}(x)$ so that for each $V \in \mathcal{B}$, $V = U_{\mathcal{C}_V}$. It is apparent that $\cup\mathcal{A}$ is a subbase for $\mathcal{T}$. ■

COROLLARY. Let $(X,\mathcal{T})$ be a topological space and let $\mathcal{A}$ be the collection of all interior-preserving open covers of X. Then $\mathcal{U}_{\mathcal{A}}$ is the finest transitive quasi-uniformity compatible with $\mathcal{T}$.

2.7 The finest transitive quasi-uniformity for a space $(X,\mathcal{T})$, whose existence is guaranteed by Corollary 2.6, is denoted by $\mathcal{FT}$ and is called the *fine transitive quasi-uniformity for* $(X,\mathcal{T})$. We note that the collection of all transitive neighbornets of a space is a base for $\mathcal{FT}$. If $(X,\mathcal{T})$ is a topological space and $\mathcal{A}$ is the collection of all point-finite (locally finite) open covers of X, then $\mathcal{U}_{\mathcal{A}}$ is called the *point-finite (locally finite) covering quasi-uniformity for* $(X,\mathcal{T})$ and is denoted by $\mathcal{PF}(\mathcal{LF})$. The next proposition establishes that $\mathcal{P}$ also can be obtained from the construction given in Theorem 2.6.

PROPOSITION. Let $(X,\mathcal{T})$ be a topological space and let $\mathcal{A}$ be the collection of all finite open covers of X. Then $\mathcal{U}_{\mathcal{A}} = \mathcal{P}$.

2.8 PROPOSITION. Let $(X,\mathcal{T})$ be a topological space and let $\mathcal{A}$ be a family of interior-preserving open collections such that $\cup\mathcal{A}$ is a subbase for $\mathcal{T}$. A necessary and sufficient condition that $\mathcal{U}_{\mathcal{A}}$ be totally bounded is that each member of $\mathcal{A}$ be a finite collection.

Proof. If each member of $\mathcal{A}$ is a finite collection, then by the previous proposition $\mathcal{U}_{\mathcal{A}} \subset \mathcal{P}$. Since $\mathcal{P}$ is totally bounded, $\mathcal{U}_{\mathcal{A}}$ is totally bounded as well.

Now suppose that $\mathcal{U}_{\mathcal{A}}$ is totally bounded and let $\mathcal{C} \in \mathcal{A}$. For each $x \in X$ set $G_x = \cap\mathcal{C}_x$, $U = U_{\mathcal{C}}$, $\mathcal{G} = \{G_x | x \in X\}$, $H_x = \{y | G_x = G_y\}$, and $\mathcal{H} = \{H_x | x \in X\}$. Then $\mathcal{H}$ is a partition of X and card $(\mathcal{H})$ = card $(\mathcal{G})$. Since $\mathcal{U}_{\mathcal{A}}$ is totally bounded, there exists a collection $\{A_i | i = 1, 2, \ldots, n\}$ such that $\cup\{A_i | i = 1, 2, \ldots, n\} = X$ and $\cup\{A_i \times A_i | i = 1, 2, \ldots, n\} \subset U$. If $\mathcal{H}$ (equivalently $\mathcal{G}$) is infinite, there exist A_i, H_x, H_y such that $H_x \neq H_y$, and there exist $w \in A_i \cap H_x$ and $z \in A_i \cap H_y$. Since $(w,z) \in A_i \times A_i \subset U$, $z \in U(w) = G_w$; hence $G_z \subset G_w$. Since $(z,w) \in A_i \times A_i \subset U$, $w \in U(z) = G_z$; hence $G_w \subset G_z$ and $G_w = G_z$. Now $w \in H_x$ so that $G_w = G_x$, and $z \in H_y$ so that $G_z = G_y$. Then $G_x = G_w = G_z = G_y$ and $H_x = H_y$ - a contradiction. We have proved that $\mathcal{H}$ is finite; thus $\mathcal{G}$ is finite. Since for each $G \in \mathcal{C}$, $G = \cup\{G_x | x \in G\}$, $\mathcal{C}$ must be finite as well. ∎

COROLLARY. Let $(X,\mathcal{T})$ be a topological space. Then $\mathcal{P} = \mathcal{FT}$ if and only if every interior-preserving open cover of X is finite.

2.9 Let $(X,\mathcal{T})$ be a topological space. We recall from §1.56 that if f is a real-valued function on X and $\epsilon > 0$ then $U_{(\epsilon,f)} = \{(x,y) \in X \times X | f(x) - f(y) < \epsilon\}$. We adopt the usual notation that C(X) denotes the set of all continuous real-valued functions on X and C*(X) denotes the set of all bounded members of C(X). If $(X,\mathcal{T})$ is a completely regular space, then $\{U_{(\epsilon,f)} | \epsilon > 0, f \in C(X)\}$ is a subbase for a uniformity on X compatible with $\mathcal{T}$. This uniformity, which plays a central role in the study of realcompactness, is denoted by $\mathcal{C}(X)$ or just $\mathcal{C}$ when the topological space $(X,\mathcal{T})$ is understood. It follows from Corollary 1.57 that for every totally bounded quasi-uniformity $\mathcal{U}$ there is a collection of bounded lower semi-continuous functions $\mathbb{F}$ such that $\{U_{(\epsilon,f)} | \epsilon > 0, f \in \mathbb{F}\}$ is a subbase for $\mathcal{U}$ and so the uniformity $\mathcal{C}^*(X)$ determined by all bounded continuous real-valued functions is $(\mathcal{C}(X))_\omega$. Although $\mathcal{C}(X)$ and $\mathcal{C}^*(X)$ can be defined as above for an arbitrary topological space $(X,\mathcal{T})$, these uniformities are compatible with

$\mathcal{T}$ only when $C^*(X)$ separates points and closed sets (that is only when $(X,\mathcal{T})$ is completely regular). For an arbitrary topological space X, the characteristic function of each open set is lower semi-continuous so that natural analogues of $C(X)$ and $C^*(X)$ can be defined for X using lower semi-continuous functions in much the same way that continuous functions are used to define C and C* for completely regular spaces.

PROPOSITION. Let $(X,\mathcal{T})$ be a topological space and let E be a collection of lower semi-continuous functions on X such that if $G \in \mathcal{T}$ and $x \in G$ then there exists $f \in E$ such that $f(x) = 1$ and $f(X - G) = 0$. Let $\mathcal{S} = \{U_{(\epsilon,f)} | f \in E, \epsilon > 0\}$. Then $\mathcal{S}$ is a subbase for a quasi-uniformity that is compatible with $\mathcal{T}$.

Proof. Since for each $f \in E$ and $\epsilon > 0$, $\Delta \subset U_{(\epsilon,f)}$ and $U_{(\epsilon/2,f)} \circ U_{(\epsilon/2,f)} \subset U_{(\epsilon,f)}$, $\mathcal{S}$ is a subbase for a quasi-uniformity $\mathcal{U}$ on X. Let $f \in E$, $\epsilon > 0$ and $x \in X$. Then $U_{(\epsilon,f)}(x) = \{y | f(y) > f(x) - \epsilon\} = f^{-1}(f(x) - \epsilon, \infty)$; and $f^{-1}(f(x) - \epsilon, \infty) \in \mathcal{T}$, since f is lower semi-continuous. Hence $\mathcal{T}(\mathcal{U}) \subset \mathcal{T}$. Let $G \in \mathcal{T}$, let $x \in G$, and let $f \in E$ such that $f(x) = 1$ and $f(X - G) = 0$. Then $x \in U_{(1,f)}(x) \subset G$ so that $\mathcal{T} \subset \mathcal{T}(\mathcal{U})$ and $\mathcal{U}$ is compatible with $\mathcal{T}$. ∎

2.10 The quasi-uniformity generated in Proposition 2.9 by taking E to be the set of all lower semi-continuous functions is called the *semi-continuous quasi-uniformity* for $(X,\mathcal{T})$ and is denoted by $\mathcal{SC}$. This quasi-uniformity is the coarsest quasi-uniformity on X for which each continuous function $f : X \to (\mathbb{R}, \mathcal{Q})$ is quasi-uniformly continuous. It was originally defined in terms of upper semi-continuous functions. (See the notes.)

It follows from Corollary 1.57 that $\{U_{(\epsilon,f)} | \epsilon > 0, f \in QB(\mathcal{P})\}$ is a subbase for $\mathcal{P}$. We can show even more.

PROPOSITION. Let $(X,\mathcal{T})$ be a topological space and let E be the collection of all bounded lower semi-continuous functions on X. Then $\{U_{(\epsilon,f)} | \epsilon > 0, f \in E\}$ is a subbase for $\mathcal{P}$.

Proof. Let $\mathcal{U}$ be the quasi-uniformity for which $\{U_{(\epsilon,f)} | \epsilon > 0, f \in E\}$ is a subbase. Since E contains all characteristic functions of open sets, it follows from Proposition 2.9 that $\mathcal{T}(\mathcal{U}) = \mathcal{T}$. By Corollary 1.57, $\mathcal{P} \subset \mathcal{U}$. As $\mathcal{U}$ is totally bounded, it follows from remarks in §2.2 that $\mathcal{U} = \mathcal{P}$. ∎

2.11 Since δ_P is the finest quasi-proximity compatible with a given topology T, any quasi-uniformity compatible with T that contains P belongs to the quasi-proximity class $\pi(\delta_P)$. By the previous proposition, SC as well as PF, LF, FT, and $FINE$ are members of $\pi(\delta_P)$.

Although Corollary 1.57 shows that not every quasi-uniformity defined as in Proposition 2.9 is transitive, we now establish that SC is transitive.

2.12 An *open spectrum* a in a set X is a family $\{A_n | n \in \mathbb{Z}\}$ of open subsets of X such that for each $n \in \mathbb{Z}$, $A_n \subset A_{n+1}$, $\cap_{n\in\mathbb{Z}} A_n = \emptyset$, and $\cup_{n\in\mathbb{Z}} A_n = X$.

Each open spectrum is an interior-preserving open cover. If a is an open spectrum, there is an integer n such that $A_n = X$ if and only if a is a point-finite open cover. In this case we call a a *point-finite open spectrum*. If a is an open spectrum, it is easily verified that $U_a = \cup_{n\in\mathbb{Z}} (A_n - A_{n-1}) \times A_n$.

THEOREM. Let A be the collection of all open spectra in a space (X,T). Then $U_A = SC$.

Proof. Let E be the collection of all lower semi-continuous functions. It suffices to show that $\{U_a | a \in A\}$ and $\{U_{(\epsilon,f)} | \epsilon > 0,\ f \in E\}$ are equivalent subbases. Let $f \in E$ and $\epsilon > 0$ be given; take $x_0 \in X$. For each $n \in \mathbb{Z}$ set $A_n = \{x | f(x) > f(x_0) - (n + 1)\epsilon\}$ and set $a = \{A_n | n \in \mathbb{Z}\}$. Let $(x,y) \in U_a$. There is an integer n such that $f(x_0) - n\epsilon \geq f(x) > f(x_0) - (n + 1)\epsilon$. Then $y \in U_a(x) = A_n$ so that $f(y) > f(x_0) - (n + 1)\epsilon$. Hence $f(x) - f(y) < f(x_0) + (n\epsilon) - (f(x_0) - (n + 1)\epsilon) = \epsilon$ and $(x,y) \in U_{(\epsilon,f)}$.

Let $a \in A$. For each $x \in X$ define $f(x) = -n$ where n is the integer for which $x \in A_n - A_{n-1}$. Then f is lower semi-continuous. Let $(x,y) \in U_{(1,f)}$ and suppose that $x \in A_n - A_{n-1}$. Then $f(y) > f(x) - 1 = -(n + 1)$. Hence $y \in A_n$ and $(x,y) \in (A_n - A_{n-1}) \times A_n \subset U_a$. ∎

2.13 COROLLARY. Let (X,T) be a topological space. Then SC is a transitive quasi-uniformity.

2.14 COROLLARY. Let (X,T) be a countably compact topological space. Then each open spectrum of X is a point-finite open spectrum and $SC \subset PF$.

2.15 COROLLARY. Let (X,T) be a topological space. For each $U \in SC$ there is a countable subset D of X such that $U(D) = X$.

2.16 We omit the proof of the following theorem, which is similar to the proof of Theorem 2.12.

THEOREM. Let $(X,\mathcal{T})$ be a topological space and let $\mathcal{A}$ be the collection of all point-finite open spectra. For each lower semi-continuous function f that is bounded below, $f : (X,\mathcal{U}_{\mathcal{A}}) \to (\mathbb{R},\mathcal{Q})$ is quasi-uniformly continuous, and $\mathcal{U}_{\mathcal{A}}$ is the coarsest quasi-uniformity that has this property.

2.17 PROPOSITION. Let $f : (X,\mathcal{T}) \to (Y,\mathcal{T}')$ be a continuous function, let $\mathcal{U}$ denote the Pervin (resp. point finite, locally finite, semi-continuous, fine transitive, fine) quasi-uniformity on X and let $\mathcal{V}$ denote the corresponding quasi-uniformity on Y. Then $f : (X,\mathcal{U}) \to (Y,\mathcal{V})$ is quasi-uniformly continuous.

Proof. The proof of the case that $\mathcal{U}$ and $\mathcal{V}$ are the fine transitive quasi-uniformities on X and Y respectively is representative. Let $V \in \mathcal{V}$ be a transitive entourage. Then $\mathcal{C} = \{f^{-1}(V(y)) \mid y \in Y\}$ is an interior-preserving open cover of X and so $U_{\mathcal{C}} \in \mathcal{U}$. If $(a,b) \in U_{\mathcal{C}}$, then $(f(a),f(b)) \in V$; hence f is quasi-uniformly continuous. ■

2.18 Since each of $\mathcal{P}$, $\mathcal{PF}$, $\mathcal{LF}$, $\mathcal{SC}$, $\mathcal{FT}$, and $\mathcal{FINE}$ is defined for every topological space, there is defined for each of these quasi-uniformities a rule that assigns to each topological space the corresponding quasi-uniformity. Let f denote the rule corresponding to any of $\mathcal{P}$, $\mathcal{PF}$, $\mathcal{LF}$, $\mathcal{SC}$, $\mathcal{FT}$, or $\mathcal{FINE}$ and let X and Y be topological spaces. Then $f(X) \times f(Y) \subset f(X \times Y)$ and if A is a subspace of X, $f(X)|A \subset f(A)$. While in general these two comparisons are not equalities, it is easily seen from Definition 2.2 that if f is the rule corresponding to the Pervin quasi-uniformity and A is a subspace of X then $f(X)|A = f(A)$. Even for this rule, however, it is not in general true that $f(X) \times f(Y) = f(X \times Y)$. Indeed the following proposition shows that if $(X,\mathcal{T})$ is a Hausdorff space, f is any one of the rules discussed in this section and if $f(X) \times f(X) = f(X \times X)$, then f(X) is the discrete uniformity. In particular $\mathcal{FINE} \times \mathcal{FINE}$ is the fine quasi-uniformity only when $\mathcal{FINE}$ is the discrete uniformity.

PROPOSITION. Let $(X,\mathcal{T})$ be a Hausdorff space and let $\mathcal{U}$ be a quasi-uniformity compatible with $\mathcal{T}$ such that $\delta_{\mathcal{U}\times\mathcal{U}}$ is the Pervin quasi-proximity on $X \times X$. Then $\mathcal{U}$ is the discrete uniformity.

Proof. As $\delta_{U\times U}$ is the Pervin quasi-proximity on $X \times X$, $(X \times X - \Delta) \not\delta_{U\times U} \Delta$ so that there is an entourage $U \in U$ such that $\{((a,b), (c,d)) | (a,c), (b,d) \in U\} \cap (X \times X - \Delta) \times \Delta = \emptyset$. It follows that for each $x \in X$, $U^{-1}(x) = \{x\}$ and so $U = \Delta$. ■

COROLLARY. Let (X,T) be a Hausdorff space such that $P_X \times P_X = P_{X\times X}$. Then X is a finite set.

2.19 PROPOSITION. Let X be a topological space and let S be an open or closed subspace of X. Every normal neighbornet of S is the restriction to S of a normal neighbornet of X. Moreover every transitive neighbornet of S is the restriction to S of transitive neighbornet of X.

Proof. Let V be a normal neighbornet of S. If S is open, set $V^\dagger = V \cup (X - S \times X)$ and if S is closed, set $V^\dagger = V \cup (X \times X - S)$. Then $V^\dagger$ is the required neighbornet of X and it is easily seen that $V^\dagger$ is transitive whenever V is transitive. ■

COROLLARY. Let X be a topological space and let a subset S of X be the intersection of an open set and a closed set. Then $FINE_X|S \times S = FINE_S$ and $FT_X|S \times S = FT_S$.

2.20 We close this section with a technical theorem, which shows in particular that the restriction of the fine quasi-uniformity on $\mathbb{R}$ to the subset Q is not the fine quasi-uniformity on Q.

THEOREM. Let X be a T_1 space and let $D = \{x_n | n \in \mathbb{N}\}$ be a dense subspace of X such that X - D is of second category in X. Then there is a normal neighbornet U of D so that for each neighbornet V of X, $V^2 \cap D \times D - U \neq \emptyset$.

Proof. For each m, $n \in \mathbb{N}$ set $g(m,x_n) = D - \{x_i | i \leq m, x_i \neq x_n\}$ and let $U = \cup_{n=1}^{\infty}\{x_n\} \times g(n,x_n)$. Since U is a member of the point-finite quasi-uniformity on D, U is a normal neighbornet of D. Suppose that there is a neighbornet V of X such that $V^2 \cap D \times D \subset U$. For each $n \in N$ set $A_n = \{x \in X - D | \text{for each } y \in D, V(x) \cap D - g(n,y) \neq \emptyset\}$. Then $\cup_{n=1}^{\infty} A_n = X - D$ so that there is an $n \in \mathbb{N}$ such that $\text{int}(\bar{A}_n) \neq \emptyset$. There is an $m > n$ so that $x_m \in D \cap \text{int}(\bar{A}_n)$. Let $x \in V(x_m) \cap A_n$. Then $V(x) \cap D \subset V^2(x_m) \cap D \subset U(x_m) = g(m,x_m) \subset g(n,x_m)$ - a contradiction. ■

2. POINT-SYMMETRY AND LOCAL SYMMETRY

2.21 Although the difference between proximity spaces and arbitrary quasi-proximity spaces is only a matter of assuming an axiom of symmetry, the difference between these two classes of spaces is considerable. We now investigate some approximations of symmetry, which serve to narrow the gap between quasi-proximities and proximities.

A quasi-proximity δ on a set X is *point-symmetric* provided that $\{x\} \delta A$ whenever $A \delta \{x\}$.

PROPOSITION. Let $(X,\mathcal{U})$ be a quasi-uniform space. Then the following statements are equivalent.

(a) $(X,\delta_{\mathcal{U}})$ is point-symmetric.

(b) For each $U \in \mathcal{U}$ and $x \in X$, there exists a symmetric $V \in \mathcal{U}$ such that $V(x) \subset U(x)$.

(c) For each $U \in \mathcal{U}$ and $x \in X$, there exists a $V \in \mathcal{U}$ such that $V^{-1}(x) \subset U(x)$.

(d) $\mathcal{T}(\mathcal{U}) \subset \mathcal{T}(\mathcal{U}^{-1})$.

Proof. We prove only that (a) $\Rightarrow$ (b), since the remaining implications are apparent. Let $U \in \mathcal{U}$ and let $x \in X$. Then $\{x\} \not\delta_{\mathcal{U}} X - U(x)$ so that $X - U(x) \not\delta_{\mathcal{U}} \{x\}$. Set $V = T(\{x\}, X - U(x)) \cap T(X - U(x), \{x\})$. By Theorem 1.33, $V \in \mathcal{U}_{\omega} \subset \mathcal{U}$. Further V is symmetric and $V(x) \subset U(x)$. ∎

2.22 We say that a quasi-uniformity is *point-symmetric* provided it satisfies any of the conditions of the previous proposition. We note that if δ is a point-symmetric quasi-proximity, then $\mathcal{T}(\delta^{-1}) = \mathcal{T}(\delta^*)$. Thus $(X,\mathcal{T}(\delta))$ is completely regular if δ^{-1} is point-symmetric. In addition $\mathcal{T}(\delta) = \mathcal{T}(\delta^{-1})$ if and only if both δ and δ^{-1} are point-symmetric. Let $\mathcal{V}$ denote the quasi-uniformity on $Y = (0,1)$ induced by the quasi-uniformity $\mathcal{U}$ of Example 3.8. Although $\delta_{\mathcal{V}}$ is not a proximity, $\mathcal{T}(\delta_{\mathcal{V}}) = \mathcal{T}(\delta_{\mathcal{V}^{-1}})$, which is the discrete topology. It follows from the preceding proposition that a topological space that admits a point-symmetric quasi-proximity is an R_0 space. Moreover if $(X,\mathcal{T})$ is an R_0 space, then $\delta_{\mathcal{P}}$ is a point-symmetric quasi-proximity.

2.23 A quasi-proximity δ and a set X is *locally symmetric* provided that $\{x\} \delta A$ whenever $A \delta G$ for each $\mathcal{T}(\delta)$-neighborhood G of x.

By Proposition 1.27 every proximity is locally symmetric. Another consequence of this proposition is that if δ is locally symmetric and

$\{x\} \not\delta A$, there is an open set G containing x such that $A \not\delta G$ and $G \not\delta A$. If ρ is locally symmetric (point-symmetric) $\delta \subset \rho$ and $\mathcal{T}(\delta) = \mathcal{T}(\rho)$, then δ is locally symmetric (point-symmetric). Both point-symmetry and locally symmetry are hereditary properties.

PROPOSITION. Let $(X,\mathcal{U})$ be a quasi-uniform space. The following statements are equivalent.

(a) $(X,\delta_{\mathcal{U}})$ is locally symmetric.

(b) For each $U \in \mathcal{U}$ and $x \in X$, there exists a symmetric $V \in \mathcal{U}$ such that $V^2(x) \subset U(x)$.

(c) For each $U \in \mathcal{U}$ and $x \in X$, there exists a $V \in \mathcal{U}$ such that $V^{-1}(V(x)) \subset U(x)$.

(d) For each $x \in X$, $\{U^{-1}(U(x)) \mid U \in \mathcal{U}\}$ is a base for the $\mathcal{T}(\mathcal{U})$-neighborhood filter of x.

Proof. We prove only that (a) $\Rightarrow$ (b) since the remaining implications are apparent. Let $U \in \mathcal{U}$ and let $x \in X$. Then $\{x\} \not\delta_{\mathcal{U}} X - U(x)$ so that there is a $G \in \eta_x$ such that $X - U(x) \not\delta_{\mathcal{U}} G$ and $G \not\delta_{\mathcal{U}} X - U(x)$. Since $\{x\} \not\delta_{\mathcal{U}} X - G$, there is an $H \in \eta_x$ such that $X - G \not\delta_{\mathcal{U}} H$ and $H \not\delta_{\mathcal{U}} X - G$. Set $V = T(X - U(x),G) \cap T(G,X - U(x)) \cap T(H,X - G) \cap T(X - G,H)$. Then V is a symmetric member of $\mathcal{U}$ and $V(V(x)) \subset V(G) \subset U(x)$. ∎

2.24 We say that a quasi-uniformity is *locally symmetric* provided it satisfies any of the conditions of the previous proposition. It follows from this proposition that a topological space that admits a locally symmetric quasi-uniformity is a regular space. Furthermore if $(X,\mathcal{T})$ is a regular space, then $\delta_{\mathcal{P}}$ is a locally symmetric quasi-proximity. Manifestly point-symmetry and local symmetry are qp-invariant properties.

We have considered two qp-invariant properties possessed by the Pervin quasi-proximity of any regular topological space and not possessed by any quasi-proximity of a topological space that fails to be an R_0 space. These two properties are related to a third qp-invariant property that is possessed by the Pervin quasi-proximity of any topological space. A quasi-proximity δ on a set X is *equinormal* provided that if A and B are disjoint closed subsets of X then $A \not\delta B$. A quasi-uniformity $\mathcal{U}$ is *equinormal* provided $\delta_{\mathcal{U}}$ is equinormal. A quasi-uniform space $(X,\mathcal{U})$ is equinormal if and only if for each closed subset F of X, and each open set G containing F there is a $U \in \mathcal{U}$ such that $U(F) \subset G$.

PROPOSITION. A quasi-uniform space $(X,\mathcal{U})$ is equinormal if $(X,\mathcal{U})$ is compact or $(X,\mathcal{U})$ is countably compact and $\mathcal{U}$ has a countable base.

Proof. Let A and B be disjoint closed sets of a quasi-uniform space $(X,\mathcal{U})$ satisfying either hypothesis stated above. Suppose that $A \;\delta_{\mathcal{U}}\; B$. For each $U \in \mathcal{U}$, $(A \times B) \cap U \neq \emptyset$ and by hypothesis $\mathrm{fil}(\{A \cap U^{-1}(B) \mid U \in \mathcal{U}\})$ has a cluster point p. Then $p \in A \cap (\cap\{U^{-1}(B) \mid U \in \mathcal{U}\}) = A \cap \overline{B} = A \cap B$ - a contradiction. ■

2.25 We have seen that point symmetry is related to the R_0 separation axiom and that local symmetry is related to regularity; similarly, equinormality and normality are related concepts. Indeed the following proposition shows that for a normal Hausdorff space $(X,\mathcal{T})$ the only equinormal proximity compatible with $\mathcal{T}$ is the fine proximity. Proposition 2.26 provides a more significant comparison of equinormality, point symmetry, and local symmetry.

PROPOSITION. Let $(X,\mathcal{T})$ be a normal Hausdorff space and let β be the finest proximity compatible with $\mathcal{T}$. A quasi-proximity δ compatible with $\mathcal{T}$ is equinormal if and only if $\delta \subset \beta$.

Proof. Suppose first that $\delta \subset \beta$. It follows from Proposition 1.42 that β, and hence δ, is an equinormal quasi-proximity.

Now suppose that δ is equinormal and that A and B are subsets of X such that $A \;\not\beta\; B$. Then $\overline{A} \;\not\beta\; \overline{B}$ so that $\overline{A} \cap \overline{B} = \emptyset$. Since δ is equinormal, $\overline{A} \;\not\delta\; \overline{B}$ and so $A \;\not\delta\; B$. ■

2.26 PROPOSITION. Let (X,δ) be an equinormal quasi-proximity space.

(a) If X is an R_0 space, δ is point symmetric.

(b) If X is a regular space, δ is locally symmetric.

Proof. Let x be an element of X and let A be a subset of X such that $\{x\} \;\not\delta\; A$. By Proposition 1.27 there is an open set G such that $x \in G$ and $G \;\not\delta\; A$.

Suppose first that X is an R_0 space. Then $\overline{\{x\}} \subset G$ so that $\overline{\{x\}} \;\not\delta\; \overline{A}$ and $A \;\not\delta\; \{x\}$.

Now suppose that X is a regular space. There is an open set H such that $x \in H \subset \overline{H} \subset G$. Since $G \;\not\delta\; A$, $\overline{A} \cap \overline{H} = \emptyset$ and so $A \;\not\delta\; H$. ■

2.27 PROPOSITION. Let $(X,\mathcal{T})$ be a topological space for which every quasi-proximity compatible with $\mathcal{T}$ is point-symmetric. Then $(X,\mathcal{T})$ is compact. Moreover if (X,δ) is a subspace of a compact R_0 (regular) quasi-proximity space, then δ is point-(locally) symmetric.

Proof. Suppose that $(X,\mathcal{T})$ is not compact and let $\mathcal{F}$ be a filter on X that has no cluster point. By Proposition 2.4, $\{T(G,X - G) \mid G \in \mathcal{T}$ and $X - G \in \mathcal{F}\}$ is a subbase for a quasi-uniformity $\mathcal{U}$ compatible with $\mathcal{T}$. Let F be a closed set such that $F \in \mathcal{F} - \{X\}$ and let $x \in X - F$. Then $F\ \delta_{\mathcal{U}}\ \{x\}$, and since $\delta_{\mathcal{U}}$ is point-symmetric $\{x\}\ \delta_{\mathcal{U}}\ F$ - a contradiction.

The remaining implications follow from Propositions 2.24 and 2.26. ∎

2.28 It is natural to consider the problems of characterizing those topological spaces that admit a point-symmetric, locally symmetric or equinormal quasi-metric. In Chapter 7 we obtain a complete characterization of those topological spaces that admit a point-symmetric quasi-metric; we note here only that if X admits a point-symmetric quasi-metric then X also admits a structure called a development, which is defined as follows.

A *development* for a topological space $(X,\mathcal{T})$ is a sequence $\langle \mathcal{G}_n \rangle$ of open covers of X such that if $x \in X$ and G is an open set containing x then there is an $n \in \mathbb{N}$ such that $st(x,\mathcal{G}_n) \subset G$. A T_1 space that has a development is called a *developable space* and a regular developable space is called a *Moore space*.

PROPOSITION. A topological space that admits a point-symmetric quasi-metric is a developable space.

Proof. Let X be a topological space that admits a point-symmetric quasi-metric d. Let $\langle U_n \rangle$ be a nested base of open neighbornets for the quasi-uniformity generated by d and for each $n \in \mathbb{N}$ set $\mathcal{G}_n = \{U_n(x) \mid x \in X\}$. Let $x \in X$ and let G be an open set containing x. There is an $n \in \mathbb{N}$ such that $U_n(x) \subset G$, an $m \in \mathbb{N}$ such that $U_m^2 \subset U_n$, and a $k > m$ such that $U_k^{-1}(x) \subset U_m(x)$. It is easily verified that $st(x,\mathcal{G}_k) \subset G$. ∎

2.29 Although Example 5.34 provides a locally compact Moore space that is not quasi-metrizable, we establish in Theorem 7.26 that every quasi-metrizable Moore space admits a point-symmetric quasi-metric. The following proposition is an immediate consequence of Propositions 2.24 and 2.26.

PROPOSITION. Let $(X,\mathcal{T})$ be a countably compact space. Every quasi-metric compatible with $\mathcal{T}$ is a point-symmetric quasi-metric.

COROLLARY. Every countably compact quasi-metrizable space is compact.

Proof. Let X be a countably compact quasi-metrizable space. It follows from the previous two propositions that X is developable, and every countably compact developable space is compact. ■

2.30 Let X be a countably infinite set and let $\mathcal{T}$ be the cofinite topology on X. Then $(X,\mathcal{T})$ is a compact quasi-metrizable space that is not a Hausdorff space. It is well known and easy to prove, however, that every compact Moore space is metrizable. Thus, since every locally metrizable paracompact space is metrizable, the previous two propositions yield the following simple metrization theorem.

COROLLARY. Every locally compact paracompact Hausdorff quasi-metrizable space is a metrizable space.

2.31 In order to characterize those topological spaces that admit a locally symmetric quasi-metric it is convenient to state the following consequence of Theorem 1.5 in terms of a development.

PROPOSITION. Let $(X,\mathcal{T})$ be a T_1 space and let $\langle \mathcal{G}_n \rangle$ be a development for X. Suppose that for each $n \in \mathbb{N}$ if two members of $\mathcal{G}_{n+1}$ have nonempty intersection then some member of $\mathcal{G}_n$ contains their union. Then $(X,\mathcal{T})$ is metrizable.

Proof. For each $n \in \mathbb{N}$ set $U_n = \{(x,y) | y \in st(x,\mathcal{G}_n)\}$. It is easily verified that $\langle U_n \rangle$ is a base for a uniformity compatible with $\mathcal{T}$. ■

2.32 THEOREM. A topological space that admits a locally symmetric quasi-metric is metrizable.

Proof. Let $(X,\mathcal{T})$ be a topological space that admits a locally symmetric quasi-metric and let $\langle U_n \rangle$ be a decreasing sequence of open neighbornets of X such that $\langle U_n \rangle$ is a base for a locally symmetric quasi-uniformity compatible with $\mathcal{T}$. For each $x \in X$ and $n \in \mathbb{N}$ let g[n,x] be the least natural number $j > n$ such that $U_j \circ U_j^{-1} \circ U_j(x) \subset U_n(x)$. For each

$x \in X$ choose a sequence of natural numbers as follows. Set $f(1,x) = 1$, $f(2,x) = g[f(1,x),x] \ldots f(r+1,x) = g[f(r,x),x]$. It follows easily by induction that for each $x \in X$ and $n \in \mathbb{N}$, $f(n,x) \geq n$. For each $n \in \mathbb{N}$ set $V_n = \{(x,y) \mid y \in U_{f(n,x)}(x)\}$ and set $\mathcal{G}_n = \{V_n(x) \mid x \in X\}$.

We show first that the sequence $\langle \mathcal{G}_n \rangle$ of open covers of X is a development for X. Let $G \in \mathcal{T}$, let $a \in G$, and let $r \in \mathbb{N}$ such that $U_r(a) \subset G$. Set $m = g[r,a]$ and let $b \in st(a,\mathcal{G}_m)$. There is an $x \in X$ such that $\{a,b\} \subset V_m(x) = U_{f(m,x)}(x) \subset U_m(x)$. Thus $b \in U_m(x) \subset U_m^{-1} \circ U_m(a) \subset U_r(a)$ so that $st(a,\mathcal{G}_m) \subset G$.

To see that $\langle \mathcal{G}_n \rangle$ satisfies the conditions of Proposition 2.31 let $m \in \mathbb{N}$ and let $p, q \in X$ such that $V_{m+1}(p) \cap V_{m+1}(q) \neq \emptyset$. For notational convenience we set $f(m+1,p) = j$ and $f(m+1,q) = k$ and we assume without loss of generality that $j \geq k$. Then $V_{m+1}(p) \cup V_{m+1}(q) = U_j(p) \cup U_k(q) \subset U_k \circ U_k^{-1} \circ U_k(q)$ and as $k = f(m+1,q) = g[f(m,q),q]$, $U_k \circ U_k^{-1} \circ U_k(q) \subset U_{f(m,q)}(q) = V_m(q) \in \mathcal{G}_m$. ∎

2.33 In §2.30 we pointed out a simple example of a topological space that admits an equinormal quasi-metric and fails to be a Hausdorff space. If $(X,\mathcal{T})$ is a regular space that admits an equinormal quasi-metric, then by Proposition 2.26, $(X,\mathcal{T})$ admits a locally symmetric quasi-metric and so is metrizable.

THEOREM. For a Tychonoff space X the following statements are equivalent.

(a) X admits an equinormal quasi-metric.

(b) X is metrizable and the set of all nonisolated points of X is a compact subset of X.

(c) The fine uniformity of X has a countable base.

(d) X admits an equinormal metric.

Proof. (a) ⇒ (b) : Let d be an equinormal quasi-metric and let $\langle U_n \rangle$ be a nested base for the quasi-metric generated by d. Let I denote the set of all isolated points of X. In light of the previous theorem it suffices to show that $X - I$ is countably compact. Suppose that $X - I$ is not countably compact and let D be a closed infinite discrete subset of $X - I$. We may assume that $D = \{x_n \mid n \in \mathbb{N}\}$ where $x_n = x_m$ only if $n = m$. There is a family $\{G_n \mid n \in \mathbb{N}\}$ of pairwise disjoint open sets of X such that for each $n \in \mathbb{N}$, $x_n \in G_n$. Since no point of D is isolated, for each $n \in \mathbb{N}$ there is a $y_n \neq x_n$ such that $y_n \in U_n(D) \cap G_n$. Set $F = \overline{\{y_n | n \in \mathbb{N}\}}$. Since D and F are disjoint closed sets, there is an $n \in \mathbb{N}$ such that $U_n(D) \cap F = \emptyset$ - a contradiction.

(b) ⇒ (c) : Let $K = \{(x,x) | x$ is not an isolated point of $X\}$, let d be a metric compatible with the topology of $X \times X$ and for each $n \in \mathbb{N}$ set $G_n = \{y \in X \times X | d(y,K) < 1/n\}$. For each $n \in \mathbb{N}$ set $U_n = G_n \cup \{(x,x) | x \in I\}$. Evidently $\{U_n | n \in \mathbb{N}\}$ is a base for the fine uniformity, which consists of all neighborhoods of the diagonal. (See the notes to §5.21.)

(c) ⇒ (d) : Since X is metrizable, it is an immediate consequence of Proposition 2.25 that the fine uniformity is equinormal.

(d) ⇒ (a) : This implication is evident. ∎

2.34 We now characterize those regular T_1 spaces for which the fine quasi-uniformity has a countable base. Our characterization relies upon Theorem 2.33(b), to which it should be compared.

PROPOSITION. Let $(X,\mathcal{T})$ be a regular T_1 space. The fine quasi-uniformity for X has a countable base if and only if X is a metric space with only finitely many nonisolated points.

Proof. Suppose that $(X,\mathcal{T})$ is a metric space with only finitely many nonisolated points $x_1, x_2, \ldots, x_n$ and for $1 \leq i \leq n$ let $\{U_j(x_i) | j \in \mathbb{N}\}$ be a base for η_{x_i} with the property that $x_i \in U_j(x_k)$ only if $i = k$. Then $\{\Delta \cup [\cup_{i=1}^n \{x_i\} \times U_j(x_i)] | j \in \mathbb{N}\}$ is a subbase for $\mathcal{FINE}$.

Now suppose that $\mathcal{FINE}$ has a countable base. Since this quasi-uniformity belongs to the Pervin quasi-proximity class, it is equinormal and so by Theorem 2.33 the set X' of all nonisolated points of X is a compact metric space. Let G' be an open subset of X' and let $G \in \mathcal{T}$ such that $G' = X' \cap G$. Let I denote the set of isolated points of G and let $K = G - I$. By Corollary 2.19, $\mathcal{FINE}|G \times G = \mathcal{FINE}_G$ so that by Theorem 2.33 K is compact. Consequently $G' = X' \cap G = X' \cap (K \cup I) = X' \cap K$, which is compact. Thus every open subset of X' is compact and so X' is finite. ∎

2.35 It is natural to pose the problem of determining those topological spaces for which $\mathcal{P}$ is the only compatible quasi-uniformity. We have postponed considering this problem until now, since it is inextricably tied to the study of those topological spaces for which every compatible quasi-uniformity is (locally) symmetric.

PROPOSITION. Let $(X,\mathcal{T})$ be a topological space. If $\mathcal{P}$ is a uniformity, then for each $x \in X$, $\overline{\{x\}}$ is the smallest open set containing x.

Proof. Suppose that $\mathcal{P}$ is a uniformity, let $x \in X$ and let $U = T(X - \overline{\{x\}}, \overline{\{x\}})$. Then $U^{-1} \in \mathcal{P}$; hence $\overline{\{x\}} = U^{-1}(x)$ is a neighborhood of x. Since $(X,\mathcal{T})$ is completely regular, $\overline{\{x\}}$ is a subset of every neighborhood of x. It follows that $\overline{\{x\}}$ is the smallest open set containing x. ■

COROLLARY. Let $(X,\mathcal{T})$ be a T_0 space. Then $\mathcal{P}$ is a uniformity if and only if $\mathcal{T}$ is the discrete topology.

2.36 A topological space X is an R_1 *space* provided that if x and y are points of X such that $\overline{\{x\}} \neq \overline{\{y\}}$ then x and y have disjoint neighborhoods. Evidently every R_1 space is an R_0 space and a topological space is a Hausdorff space if and only if it is both an R_1 space and a T_1 space.

THEOREM. Let $(X,\mathcal{T})$ be a topological space. Each of the statements (a) through (d) implies its successor, and if $(X,\mathcal{T})$ is an R_1 space then the following statements are equivalent.

(a) $\mathcal{T}$ is finite.

(b) $\mathcal{P}$ is the only quasi-uniformity compatible with $\mathcal{T}$.

(c) Every interior-preserving open collection is finite.

(d) $(X,\mathcal{T})$ is hereditarily compact.

(e) $\delta_{\mathcal{P}}$ is the only quasi-proximity compatible with $\mathcal{T}$.

Proof. We first prove that (d) ⇒ (c) , since this implication expedites the proof that (a) ⇒ (b) . Suppose that $(X,\mathcal{T})$ is hereditarily compact and let δ be a quasi-proximity that is compatible with $\mathcal{T}$. Then $\delta_{\mathcal{P}} \subset \delta$. Suppose that $A \not\delta_{\mathcal{P}} B$. Then $A \cap \overline{B} \neq \emptyset$ and since A is compact it follows from Proposition 1.43 that $A \not\delta B$.

(a) ⇒ (b) : Let $\mathcal{U}$ be a quasi-uniformity compatible with $\mathcal{T}$. As $(X,\mathcal{T})$ is hereditarily compact, $\delta_{\mathcal{U}} = \delta_{\mathcal{P}}$ so that $\mathcal{P} \subset \mathcal{U}$. Since $\mathcal{T}$ is finite, $U_{\mathcal{T}} \in \mathcal{P}$. Furthermore every neighbornet of X contains $U_{\mathcal{T}}$; hence $\mathcal{U} \subset \mathcal{P}$.

(b) ⇒ (c) : Since $\mathcal{FT} = \mathcal{P}$, we have from Corollary 2.8 that every interior-preserving open collection is finite.

(c) ⇒ (d) : It is well known that a topological space is hereditarily compact if and only if every strictly increasing sequence of open sets is finite. As every increasing sequence of open sets is interior preserving, $(X,\mathcal{T})$ is hereditarily compact.

(e) $\Rightarrow$ (a) : Now suppose that (X,T) is an R_1 space for which δ_P is the only quasi-proximity compatible with T. By Proposition 2.27 we have that (X,T) is compact. As every compact R_1 space is completely regular, P is a uniformity. Thus by Proposition 2.35, $C = \{\overline{\{x\}} \mid x \in X\}$ is an open cover; (X,T) is compact and R_1 so C is finite. For each $G \in T$, $G = \bigcup\{\overline{\{x\}} \mid x \in G\}$. Hence T is finite. ■

COROLLARY. Let (X,T) be a Hausdorff topological space. Then the following statements are equivalent.

(a) X is finite.

(b) P is the only quasi-uniformity compatible with T.

(c) δ_P is the only quasi-proximity compatible with T.

2.37 If the R_1 separation axiom is not assumed, then it is possible for a T_1 topological space with an infinite topology to admit only one quasi-uniformity.

EXAMPLE. Let T be the cofinite topology on the set $\mathbb{R}$ of all real numbers. Since $(\mathbb{R},T)$ is hereditarily compact, it follows from Theorem 2.36 that P is the only totally bounded quasi-uniformity compatible with T. Suppose that there exists a quasi-uniformity U compatible with T that is not totally bounded. Let $U \in U$ be an entourage such that if $\{A_i\}_{i=1}^{n}$ is a finite cover of $\mathbb{R}$, then there is j with $1 \leq j \leq n$ such that $A_j \times A_j - U \neq \emptyset$. Let $V \in U$ such that $V^2 \subset U$. Then $A = \bigcup\{\mathbb{R} - U(x) \mid x \in \mathbb{R}\}$ is infinite and we may select an infinite subset $B = \{y_i \mid i \in \mathbb{N}\}$ of A such that for each $y_i \in B$, there exists $x_i \in \mathbb{R}$ with $(x_i,y_i) \notin U$. For each $i \in \mathbb{N}$, $V(x_i) \cap V^{-1}(y_i) = \emptyset$ so that $V^{-1}(y_i)$ is finite; hence $V^{-1}(B)$ is countable. Thus there is an $x \in \mathbb{R} - V^{-1}(B)$. Consequently $B \cap V(x) = \emptyset$ and B is finite. This contradiction shows that every quasi-uniformity compatible with T is totally bounded.

2.38 EXAMPLE. A topological space that admits more than one quasi-uniformity, for which δ_P is the only compatible quasi-proximity.

Let $(\mathbb{N},T)$ be the set of positive integers with the cofinite topology. Since $(\mathbb{N},T)$ is hereditarily compact, by Theorem 2.36 δ_P is the only compatible quasi-proximity. Set $G_1 = \mathbb{N}$ and for each $n > 1$ set $G_n = \mathbb{N} - \{1, 2, \ldots, n-1\}$. Then $G = \{G_n \mid n \in \mathbb{N}\}$ is an infinite interior-preserving open cover of $\mathbb{N}$. By Proposition 2.8, FT is not totally bounded and so P is not the only quasi-uniformity compatible with T. Since the product of two hereditarily compact spaces is hereditarily compact, (see the notes) $\delta_{FT} \times \delta_{FT} = \delta_{FT}$ and so the present example shows that conditions (a) and (b) of Proposition 1.54 are not in general equivalent.

PROBLEM B. The previous two examples leave unanswered two questions: (a): If δ_P is the only quasi-proximity compatible with T, is (X,T) hereditarily compact?; (b): If every interior-preserving open cover of X is finite, is P the only quasi-uniformity compatible with T? It is interesting to note that this question has an affirmative answer in the class of transitive spaces (see §6.10).

Chapter 3

COMPLETENESS

1. CAUCHY FILTERS

3.1 We now consider the theory of completeness and completions for quasi-uniform spaces. In Section 2 of this chapter we establish a satisfactory analogue of the completion theory of uniform spaces; indeed the results of Section 2 extend the usual theory. In this analogue, as in its uniform space counterpart, we have an ideal economy; every space has a completion and no space has two essentially different completions. The theory of completeness that we first consider has an oversupply of complete spaces and the reader who has become accustomed to the ideal economy of uniform completions may be dismayed by problems inherent in an imbalance of supply and demand. Some of the problems caused by the over-abundance of complete spaces are of interest (for example Problem C) and while the concept of completeness that we first consider does not allow an entirely adequate theory of completion it does provide a valuable link between topological properties and quasi-uniform properties.

A filter base $\mathcal{B}$ on a quasi-uniform space $(X,\mathcal{U})$ is a *$\mathcal{U}$-Cauchy filter base* (or simply a Cauchy filter base when the quasi-uniformity $\mathcal{U}$ is apparent) provided that for each $U \in \mathcal{U}$ there exist $x \in X$ and $B \in \mathcal{B}$ such that $B \subset U(x)$. It follows that a filter $\mathcal{F}$ on $(X,\mathcal{U})$ is a $\mathcal{U}$-Cauchy filter provided that for each $U \in \mathcal{U}$ there exists $x \in X$ such that $U(x) \in \mathcal{F}$. Evidently every $\mathcal{T}(\mathcal{U})$-convergent filter base is a $\mathcal{U}$-Cauchy filter base, and every filter base that is finer than a Cauchy filter base is a Cauchy filter base. If $\mathcal{U}$ and $\mathcal{V}$ are quasi-uniformities on X and $\mathcal{U} \subset \mathcal{V}$, then every $\mathcal{V}$-Cauchy filter base is a $\mathcal{U}$-Cauchy filter base. In particular, every $\mathcal{U}^*$-Cauchy filter base is a $\mathcal{U}$-Cauchy and a $\mathcal{U}^{-1}$-Cauchy filter base.

3.2 PROPOSITION. Let $(X,\mathcal{U})$ be a quasi-uniform space and let $\mathcal{B}$ be a filter base on X. Then $\mathcal{B}$ is a $\mathcal{U}^*$-Cauchy filter base if and only if for each $U \in \mathcal{U}$ there exists $B \in \mathcal{B}$ such that $B \times B \subset U$.

Proof. Suppose that for each $U \in \mathcal{U}$ there exists $B \in \mathcal{B}$ such that $B \times B \subset U$. Let $U \in \mathcal{U}$ and let $B \in \mathcal{B}$ such that $B \times B \subset U$. For each $x \in B$, $B \subset (U \cap U^{-1})(x)$ so that $\mathcal{B}$ is a $\mathcal{U}^*$-Cauchy filter base.

Conversely let $\mathcal{B}$ be a $\mathcal{U}^*$-Cauchy filter base and let $U \in \mathcal{U}$. Choose $V \in \mathcal{U}$ such that $V^2 \subset U$. There exists $B \in \mathcal{B}$ and $x \in X$ such that $B \subset (V \cap V^{-1})(x)$. Let $(a,b) \in B \times B$. Then $(a,x) \in V$ and $(x,b) \in V$; hence $(a,b) \in V^2$ so that $B \times B \subset U$. ∎

3.3 EXAMPLE. A locally symmetric quasi-uniform space $(X,\mathcal{U})$ and a $\mathcal{T}(\mathcal{U})$-convergent filter that is not a $\mathcal{U}^*$-Cauchy filter. Let $X = \{0\} \cup \{1/n \mid n \in \mathbb{N}\}$; for each $n \in \mathbb{N}$ let $U_n = \Delta \cup \{(0,1/i) \mid i > n\}$ and let $\mathcal{U}$ be the quasi-uniformity on X generated by $\langle U_n \rangle$. The $\mathcal{T}(\mathcal{U})$-neighborhood filter of 0 is not a $\mathcal{U}^*$-Cauchy filter.

3.4 PROPOSITION. Let $f : (X,\mathcal{U}) \to (Y,\mathcal{V})$ be a quasi-uniformly continuous function and let $\mathcal{B}$ be a $\mathcal{U}$-Cauchy filter base. Then $f(\mathcal{B})$ is a $\mathcal{V}$-Cauchy filter base.

Proof. Since $f(A \cap B) \subset f(A) \cap f(B)$, $f(\mathcal{B})$ is a filter base. Let $V \in \mathcal{V}$. There is an entourage $U \in \mathcal{U}$ such that $(f(x), f(y)) \in V$ whenever $(x,y) \in U$, and since $\mathcal{B}$ is a $\mathcal{U}$-Cauchy filter base there is $B \in \mathcal{B}$ and $x \in X$ such that $B \subset U(x)$. Then $f(B) \subset V(f(x))$ and $f(\mathcal{B})$ is a $\mathcal{V}$-Cauchy filter base. ∎

3.5 PROPOSITION. Let X be a set, let $\{(Y_i,\mathcal{V}_i) \mid i \in I\}$ be a family of uniform spaces, and for each $i \in I$ let $f_i : X \to Y_i$. Let $\mathcal{U}$ denote the coarsest uniformity on X for which each f_i is uniformly continuous. Then a filter base $\mathcal{B}$ on X is a $\mathcal{U}$-Cauchy filter base if and only if for each $i \in I$, $f_i(\mathcal{B})$ is a $\mathcal{V}_i$-Cauchy filter base.

COROLLARY. Let $(X,\mathcal{U})$ be a subspace of a quasi-uniform space $(Y,\mathcal{V})$. If a $\mathcal{V}^*$-Cauchy filter on Y induces a filter on X, this induced filter is a $\mathcal{U}^*$-Cauchy filter.

3.6 Example 3.3 shows that the induced filter of a Cauchy filter is not necessarily a Cauchy filter. Thus Proposition 3.5 does not hold for quasi-uniform spaces even when there is only one mapping involved. The following two propositions demonstrate that two important particular cases of Proposition 3.5 do generalize to quasi-uniform spaces.

PROPOSITION. Let $\{(X_i,\mathcal{V}_i) | i \in I\}$ be a family of quasi-uniform spaces and let $(X,\mathcal{U})$ denote the product quasi-uniform space. A filter base $\mathcal{B}$ on X is a $\mathcal{U}$-Cauchy filter base if and only if $\pi_i(\mathcal{B})$ is a $\mathcal{V}_i$-Cauchy filter base for each $i \in I$.

Proof. Let $\mathcal{B}$ be a filter base on X. It follows from Proposition 3.4 that if $\mathcal{B}$ is a $\mathcal{U}$-Cauchy filter base, then for each $i \in I$, $\pi_i(\mathcal{B})$ is a $\mathcal{V}_i$-Cauchy filter base. Conversely suppose that for each $i \in I$, $\pi_i(\mathcal{B})$ is a $\mathcal{V}_i$-Cauchy filter base. Let $U \in \mathcal{U}$. Then U contains an entourage of the form $\{(x,y) | \text{for each } j \in J, (\pi_j(x),\pi_j(y)) \in V_j\}$ where J is a finite subset of I and for each $j \in J$, $V_j \in \mathcal{V}_j$. For each $j \in J$ there exists $x_j \in X_j$ such that $V_j(x_j) \in \pi_j(\mathcal{B})$. Choose $x \in X$ such that for each $j \in J$, $\pi_j(x) = x_j$. Then $U(x) \in \mathcal{B}$. ∎

3.7 PROPOSITION. Let X be a set and let $(Y,\mathcal{V})$ be a quasi-uniform space. Let f be a map from X onto Y and let $\mathcal{U}$ denote the coarsest quasi-uniformity on X that makes f quasi-uniformly continuous. A filter base $\mathcal{B}$ on X is a $\mathcal{U}$-Cauchy filter base whenever $f(\mathcal{B})$ is a $\mathcal{V}$-Cauchy filter base.

Proof. Let $\mathcal{B}$ be a filter base on X and suppose that $f(\mathcal{B})$ is a $\mathcal{V}$-Cauchy filter base. Let $V \in \mathcal{V}$ and let $U = f_2^{-1}[V]$. There is a $y \in Y$ such that $V(y) \in f(\mathcal{B})$. Let $x \in f^{-1}(y)$. Then $f^{-1}(V(y)) = (f_2^{-1}[V])(x) = U(x)$. Thus $U(x) \in \mathcal{B}$ and $\mathcal{B}$ is a $\mathcal{U}$-Cauchy filter base. ∎

EXAMPLE. Let $X = Y_1 = Y_2 = \mathbb{R}$ and let $f_1 = f_2$ be the identity map on $\mathbb{R}$. Let $\mathcal{V}_1 = \{V | \text{for some } x \in \mathbb{R}, \Delta \cup (x,\infty) \times \mathbb{R} \subset V\}$ and let $\mathcal{V}_2 = \{V | \text{for some } x \in \mathbb{R}, \Delta \cup (-\infty,x) \times \mathbb{R} \subset V\}$. Let $\mathcal{U}$ denote the coarsest quasi-uniformity making both f_1 and f_2 quasi-uniformly continuous. Then $\mathcal{U}$ is the discrete uniformity on $\mathbb{R}$ and $\mathcal{B} = \{X\}$ is not a $\mathcal{U}$-Cauchy filter even though for $i = 1, 2$, $f_i(\mathcal{B}) = \mathcal{B}$ is a $\mathcal{V}_i$-Cauchy filter.

3.8 A quasi-uniform space is said to be *complete* provided that every Cauchy filter has a cluster point. If every Cauchy filter converges, the space is said to be *convergence complete*. A quasi-uniform space is complete provided every Cauchy ultrafilter converges, and a regular quasi-uniform space is complete provided every open Cauchy filter has a cluster point.

EXAMPLE. A point-symmetric quasi-uniform space that is complete but not convergence complete. Let $X = [0,1]$. For each $\epsilon > 0$ set $V_\epsilon = \Delta \cup \{0\} \times [0,\epsilon) \cup \{1\} \times (1 - \epsilon, 1] \cup (1/2 - \epsilon, 1/2) \times ((0,\epsilon) \cup (1 - \epsilon, 1))$. Note that for $\epsilon < 1/4$ if $(x,y), (y,z) \in V_\epsilon$ then $x = y$ or $y = z$. Let $\mathcal{U}$ denote the quasi-uniformity for which $\{V_\epsilon | \epsilon > 0\}$ is a (transitive) base. Then $\mathcal{T}(\mathcal{U}^{-1})$ is the discrete topology on X so that $(X,\mathcal{U})$ is point-symmetric. To see that $(X,\mathcal{U})$ is complete let $\mathcal{F}$ be a Cauchy ultrafilter. Suppose that $\mathcal{F}$ does not converge to 0 or 1. There exists $\epsilon > 0$ so that $[\epsilon, 1 - \epsilon] \in \mathcal{F}$. Let $x \in X$ such that $V_\epsilon(x) \in \mathcal{F}$. Then $V_\epsilon(x) = \{x\}$ and $\mathcal{F}$ converges to x.

Clearly $\mathcal{U}$ is not convergence complete, for $\{(0,\epsilon) \cup (1 - \epsilon, 1) | \epsilon > 0\}$ is a nonconvergent Cauchy filter base.

3.9 PROPOSITION. Let $(X,\mathcal{U})$ be a locally symmetric quasi-uniform space. Every cluster point of a $\mathcal{U}$-Cauchy filter $\mathcal{F}$ is a limit point of $\mathcal{F}$.

Proof. Let p be a cluster point of a $\mathcal{U}$-Cauchy filter $\mathcal{F}$, let $U \in \mathcal{U}$ and let V be a symmetric entourage such that $V^3(p) \subset U(p)$. Since $\mathcal{F}$ is a Cauchy filter, there exists $x \in X$ such that $V(x) \in \mathcal{F}$; and since p is a cluster point of $\mathcal{F}$, $V(x) \cap V(p) \neq \emptyset$. Thus $V(x) \subset U(p)$ so that $\mathcal{F}$ converges to p. ■

COROLLARY. Every locally symmetric complete quasi-uniform space is convergence complete.

3.10 PROPOSITION. Let $(X,\mathcal{U})$ be a (convergence) complete quasi-uniform space and let F be a closed subset of X. Then $(F,\mathcal{U}|F \times F)$ is a (convergence) complete space.

EXAMPLE. A complete subspace of a (locally symmetric) Hausdorff quasi-uniform space need not be closed. Let $(X,\mathcal{U})$ be the quasi-uniform space of Example 3.3. Then $\mathcal{U}$ induces the discrete quasi-uniformity on Y =

$\{1/n \mid n \in N\}$ and therefore Y is a complete subspace that is not closed. This example contrasts with Proposition 3.28.

3.11 PROPOSITION. Let $\{(X_i,\mathcal{V}_i) \mid i \in I\}$ be a family of quasi-uniform spaces and let $(X,\mathcal{U})$ denote the product quasi-uniform space. Then $(X,\mathcal{U})$ is (convergence) complete if and only if $(X_i,\mathcal{V}_i)$ is (convergence) complete for each $i \in I$.

Proof. In order to prove the parenthetical result, let $\mathcal{B}$ be a $\mathcal{U}$-Cauchy filter base. By Proposition 3.4, for each $i \in I$, $\pi_i(\mathcal{B})$ is a $\mathcal{V}_i$-Cauchy filter base and thus converges. It follows that $\mathcal{B}$ is convergent. Conversely suppose that $(X,\mathcal{U})$ is convergence complete, let $j \in I$ and let $\mathcal{B}_j$ be a $\mathcal{V}_j$-Cauchy filter base. For each $i \neq j$, let $\mathcal{B}_i$ be a $\mathcal{V}_i$-Cauchy filter on X_i. By Proposition 3.6, the product filter base $\Pi\mathcal{B}_i$ is a $\mathcal{U}$-Cauchy filter base and hence converges. Therefore $\mathcal{B}_j$ also converges.

A similar proof for completeness may be given using the characterization of completeness in terms of ultrafilters given in §3.8. ■

3.12 PROPOSITION. Let $f : (X,\mathcal{T}) \to (Y,\mathcal{T}')$ be a perfect map, let $\mathcal{U}$ denote the Pervin (resp. point-finite, locally finite, semi-continuous, fine transitive, fine) quasi-uniformity on $(X,\mathcal{T})$ and let $\mathcal{V}$ denote the corresponding quasi-uniformity on $(Y,\mathcal{T}')$. Then $(X,\mathcal{U})$ is complete whenever $(Y,\mathcal{V})$ is complete.

Proof. Suppose that $(Y,\mathcal{V})$ is complete and let $\mathcal{F}$ be a $\mathcal{U}$-Cauchy filter that has no cluster point. By Propositions 2.17 and 3.4, $f(\mathcal{F})$ is a $\mathcal{V}$-Cauchy filter base and has a cluster point, say p. Since $\mathcal{F}$ has no cluster point, there exists a closed set $F \in \mathcal{F}$ such that $F \cap f^{-1}(p) = \emptyset$. Then $f(F)$ is closed and $p \notin f(F)$ - a contradiction. ■

In light of Corollary 3.9, the previous proposition also holds for convergence completeness provided that the domain X is taken to be a regular space.

3.13 A quasi-uniform space $(X,\mathcal{U})$ is *precompact* provided that for each $U \in \mathcal{U}$ there exists a finite subset F of X such that $X = U(F)$. Precompactness is a productive property and any quasi-uniformity coarser than a precompact quasi-uniformity is precompact. The supremum of two precompact quasi-uniformities may fail to be precompact and a subspace of a precompact quasi-uniform space need not be precompact. Although precompactness was not defined in Chapter 1, the proof of Theorem 1.57 establishes the following proposition.

PROPOSITION. If $(X,\mathcal{U})$ is a precompact quasi-uniform space, every member of $Q(\mathcal{U})$ has a lower bound.

The converse of the preceding proposition fails since by Theorem 3.24, if X is countably compact but not compact, $\mathcal{FT}$ is not precompact.

3.14 PROPOSITION. Let $(X,\mathcal{U})$ be a quasi-uniform space.

(a) $(X,\mathcal{U})$ is precompact if and only if every ultrafilter on X is a $\mathcal{U}$-Cauchy filter.

(b) $(X,\mathcal{U})$ is totally bounded if and only if every ultrafilter on X is a $\mathcal{U}^*$-Cauchy filter.

Proof. (a) Let $(X,\mathcal{U})$ be a precompact quasi-uniform space, let $\mathcal{H}$ be an ultrafilter on X, and let $U \in \mathcal{U}$. There is a finite set F such that $U(F) = X$. Since $\mathcal{H}$ is an ultrafilter, there is an $x \in F$ such that $U(x) \in \mathcal{H}$ so that $\mathcal{H}$ is a $\mathcal{U}$-Cauchy filter. Conversely suppose that every ultrafilter on X is a $\mathcal{U}$-Cauchy filter. If $(X,\mathcal{U})$ is not precompact, there is $U \in \mathcal{U}$ such that for each finite subset F of X, $U(F) \neq X$. It follows that $\mathcal{B} = \{X - U(F) \mid F$ is a finite subset of $X\}$ is a filter base on X. Let $\mathcal{H}$ be an ultrafilter on X that contains $\mathcal{B}$. There exists $x \in X$ such that $U(x) \in \mathcal{H}$ and $X - U(x) \in \mathcal{H}$ - a contradiction.

A similar proof may be given for (b) using Proposition 3.2. ∎

COROLLARY. Let $(X,\mathcal{U})$ be a uniform space. Then $(X,\mathcal{U})$ is precompact if and only if it is totally bounded.

3.15 PROPOSITION. Let $(X,\mathcal{U})$ be a complete quasi-uniform space that contains a dense precompact subspace $(Y,\mathcal{V})$. Then every open filter on X has a cluster point.

Proof. Let $\mathcal{F}$ be an open filter on X. Then $\mathcal{F}|Y$ is contained in an ultrafilter $\mathcal{H}$ on Y, which by Proposition 3.14 is a $\mathcal{V}$-Cauchy filter. Therefore, $\mathcal{H}$ is a $\mathcal{U}$-Cauchy filter base and so $\mathcal{H}$ has a cluster point in X. Since $\mathcal{F} \subset \mathcal{H}$, $\mathcal{F}$ also has a cluster point. ∎

COROLLARY. If X is a regular complete quasi-uniform space that has a dense precompact subspace, X is compact.

3.16 It is dishearteningly easy to give examples of precompact quasi-uniform spaces that are not hereditarily precompact. On the other hand, we have already seen that the Pervin quasi-uniformity is totally bounded; hence every topological space admits a compatible hereditarily precompact quasi-uniformity. We now introduce a transitive quasi-uniformity, which like the Pervin quasi-uniformity is hereditarily precompact but which for infinite Hausdorff spaces is not totally bounded.

In the following lemma, for each $n \in \mathbb{N}$ let A_n denote $\cup_{i=1}^{n} a_i$.

LEMMA. Let $(X,\mathcal{T})$ be a topological space. For each $n \in \mathbb{N}$, for each collection $\{a_1, a_2, \ldots, a_n\}$ of point-finite open spectra in X, and for each $Y \subset X$, the collection $\{\cap(A_n)_y \mid y \in Y\}$ contains a finite subcover of Y.

Proof. For each $i \in \mathbb{N}$, set $a_i = \{A(i,j) \mid j \in \mathbb{Z}\}$ and suppose without loss of generality that for all $i \in \mathbb{N}$ and $j \geq 0$, $A(i,j) = X$. The proof is by induction on n. For $n = 1$ let $Y \subset X$ and let a_1 be a point-finite open spectrum in X. Set $m = \sup \{j \mid Y \cap (A(1,j) - A(1,j-1)) \neq \emptyset\}$ and let $y \in Y \cap (A(1,m) - A(1,m-1))$; then $Y \subset \cap(A_1)_y$.

Now let $Y \subset X$ and let $\{a_1, a_2, \ldots, a_n, a_{n+1}\}$ be a collection of point-finite open spectra in X. By the inductive hypothesis there exist $m \in \mathbb{N}$ and $y_i \in Y$ $(1 \leq i \leq m)$ such that $Y \subset \cup_{i=1}^{m} (\cap(A_n)_{y_i})$. For each $i \leq m$ there exists $j(i)$ such that $y_i \in A(n+1, j(i)) - A(n+1, j(i)-1)$. Set $M = \min \{j(i) \mid 1 \leq i \leq m\}$. Then $Y \cap A(n+1,M) \subset \cup_{i=1}^{m} (\cap(A_{n+1})_{y_i})$. Using the inductive hypothesis again, we see that for each k $(M < k \leq 0)$ such that $Y \cap (A(n+1,k) - A(n+1, k-1)) \neq \emptyset$ there exist m_k and $y(k,i)$ $(i = 1, 2, \ldots, m_k)$ such that $y(k,i) \in Y \cap (A(n+1,k) - A(n+1, k-1))$ and $Y \cap (A(n+1,k) - A(n+1, k-1)) \subset \cup_{i=1}^{m_k} \cap(A_n)_{y(k,i)}$. Hence for each k, $Y \cap (A(n+1,k) - A(n+1, k-1)) \subset \cup_{i=1}^{m_k} \cap(A_{n+1})_{y(k,i)}$. Since $Y = Y \cap A(n+1,M) \cup \cup_{k=M+1}^{0} Y \cap (A(n+1,k) - A(n+1, k-1))$, $\{\cap(A_{n+1})_{y_i} \mid 1 \leq i \leq m\} \cup \{\cap(A_{n+1})_{y(k,i)} \mid M < k \leq 0;\ 1 \leq i \leq m_k\}$ is the required finite subcover of $\{\cap(A_{n+1})_y \mid y \in Y\}$. ■

PROPOSITION. Let $(X,\mathcal{T})$ be a topological space and let $\mathcal{A}$ denote the collection of all point-finite open spectra in $(X,\mathcal{T})$. Then every subspace of $(X,\mathcal{U}_{\mathcal{A}})$ is precompact.

Proof. Let $U \in \mathcal{U}_{\mathcal{A}}$ and let $Y \subset X$. There exist $a_1, a_2, \ldots, a_n \in \mathcal{A}$ such that $\cap_{i=1}^{n} U_{a_i} \subset U$. By the preceding lemma there is a finite set $F \subset Y$ such that $Y \subset (\cap_{i=1}^{n} U_{a_i})(F) \subset U(F)$. ■

3.17 PROPOSITION. Let $(X,\mathcal{T})$ be a topological space. The following statements are equivalent.

(a) $(X,\mathcal{T})$ is countably compact.

(b) Every countable interior-preserving open cover of X has a finite subcover.

(c) $\mathcal{SC}$ is hereditarily precompact.

(d) $\mathcal{SC}$ is precompact.

(e) Every lower semi-continuous function has a lower bound.

Proof. It is evident that (a) ⇒ (b), and the implication (b) ⇒ (c) follows from Theorem 2.12 and the preceding proposition. Clearly (c) ⇒ (d) and the implication (d) ⇒ (e) follows from Proposition 3.13. It is well known that (e) ⇒ (a). ■

3.18 In order to establish the following proposition it is useful to note that if $(X,\mathcal{T})$ is a topological space then $\mathcal{PF}$ is precompact if and only if every point-finite open cover of X has a finite subcover.

PROPOSITION. Let $(X,\mathcal{T})$ be a regular T_1 topological space. Then $\mathcal{PF}$ is precompact if and only if $(X,\mathcal{T})$ is countably compact.

Proof. Suppose first that $(X,\mathcal{T})$ is not countably compact. Then there is a closed infinite discrete set $D = \{x_i\}_{i=1}^{\infty}$. Since $(X,\mathcal{T})$ is regular and T_1 there is a sequence $\langle V_i \rangle$ of pairwise disjoint closed sets such that for each $i \geq 1$, V_i is a neighborhood of x_i. Then $\{X - D\} \cup \{\text{int } V_i\}_{i=1}^{\infty}$ is a point-finite open cover of X that has no finite subcover. Hence $\mathcal{PF}$ is not precompact.

Now suppose that $(X,\mathcal{T})$ is countably compact and let $\mathcal{C}$ be a point-finite open cover of X. It suffices to show that $\mathcal{C}$ has a finite subcover. Let $\mathcal{G}$ be the collection of all subcollections $\mathcal{A}$ of $\mathcal{C}$ such that $\mathcal{C} - \mathcal{A}$ is a cover of X. Then G is partially ordered by inclusion, and by Zorn's lemma there is a maximal subcollection $M \in G$. As in the well-known proof of the Arens-Dugundji theorem, $C - M$ is a finite subcover of C. ∎

3.19 PROPOSITION. Let (X,T) be a T_1 space. The following statements are equivalent.

(a) LF is totally bounded.

(b) LF is precompact.

(c) Every locally finite open cover of X has a finite subcover.

Proof. We prove only that (c) ⇒ (a), since the remaining implications are evident. Suppose that C is an infinite locally finite open cover of X. Without loss of generality we assume that $C = \{C_n | n \in \mathbb{N}\}$. For each $n \in \mathbb{N}$ choose $x_n \in C_n$ and set $G_n = C_n - \{x_i | x_i \neq x_n\}$. Then $\{G_n | n \in \mathbb{N}\}$ is a locally finite open cover of X that has no finite subcover - a contradiction. Thus every locally finite open cover of X is finite and so $LF = P$, which is totally bounded. ∎

3.20 In a Tychonoff space, condition (c) of the previous proposition is equivalent to pseudocompactness, and in a normal Hausdorff space, pseudocompactness is equivalent to countable compactness. Thus in a normal Hausdorff space all the conditions of Propositions 3.17, 3.18 and 3.19 are equivalent. Even in the class of Tychonoff spaces the preceding propositions provide striking evidence that the quasi-uniformities discussed in §2.18 are ill behaved with respect to subspaces. We pause therefore to justify the assertions of the chart given below.

The quasi-uniformity indicated to the right restricts to	P	PF	LF	SC	FT	$FINE$
arbitrary subspaces	Yes	No	No	No	No	No
open subspaces	Yes	Yes	No	No	Yes	Yes
closed subspaces	Yes	No	No	No	Yes	Yes

It is evident that if G is an open subspace of a space X then $\mathcal{PF}_X|G \times G = \mathcal{PF}_G$, and the remaining affirmative statements of the chart have already been observed in §§2.18 and 2.19. It follows from Theorem 2.20 that neither $\mathcal{FT}$ nor $\mathcal{FINE}$ restricts to arbitrary subspaces, and as the closed unit interval has an open subspace that is not countably compact, we have from Propositions 3.17 and 3.19 that neither $\mathcal{SC}$ nor $\mathcal{LF}$ restricts to open subspaces. Further, as pseudocompactness is not a closed-hereditary property, it follows from Proposition 3.19 that $\mathcal{LF}$ does not restrict to closed subspaces. Thus the example given below concludes the verification of the assertions of the chart.

EXAMPLE. Let Q be the set of all rational numbers of the Michael Line $(\mathbb{R},\mathcal{T}(M))$. Then Q is a closed subspace of the Michael Line such that $\mathcal{PF}_{\mathbb{R}}|Q \times Q \neq \mathcal{PF}_Q$ and $\mathcal{SC}_{\mathbb{R}}|Q \times Q \neq \mathcal{SC}_Q$.

Suppose that $\mathcal{PF}_{\mathbb{R}}|Q \times Q = \mathcal{PF}_Q$; set $Q = \{x_i | i \in \mathbb{N}\}$ and for each $n \in \mathbb{N}$ set $G_n = Q - \{x_i | i < n\}$. Since $\{G_n | n \in \mathbb{N}\}$ is a point-finite open cover of Q, there is a point-finite open cover $\mathcal{C}$ of $\mathbb{R}$ such that for each $x_i \in Q$, $U_{\mathcal{C}}(x_i) \cap Q \subset G_i$. For each $i \in \mathbb{N}$ let H_i denote the Euclidean interior of $U_{\mathcal{C}}(x_i)$, let $\mathcal{C}' = \{H_i | i \in \mathbb{N}\} \cup \{\mathbb{R}\}$, and let $V = U_{\mathcal{C}'}$. Then $V \in \mathcal{PF}_{\mathbb{R}}$ and for each $q \in Q$, $V(q) \subset U_{\mathcal{C}}(q)$. Since $\mathcal{C}'$ is a countable point-finite open cover of $\mathbb{R}$, there is a countable subset $\{t_n | n \in \mathbb{N}\}$ of $\mathbb{R}$ such that $\{V(t_n) | n \in \mathbb{N}\} = \{V(x) | x \in \mathbb{R}\}$. For each $n \in \mathbb{N}$ set $A_n = \{x | V(x) = V(t_n)\}$. By the Baire Category Theorem there is an $m \in \mathbb{N}$ and a Euclidean open interval that is a subset of the Euclidean closure of A_m. It follows that there is an infinite subset S of $\mathbb{N}$ so that for each $j \in S$, $V(x_j) \cap A_m \neq \emptyset$. Thus $V(t_m) \cap Q \subset \cap_{j \in S} V(x_j) \cap Q \subset \cap\{G_n | n \in \mathbb{N}\}$ - a contradiction.

Now suppose $\mathcal{SC}_{\mathbb{R}}|Q \times Q = \mathcal{SC}_Q$. Since the family $\{G_n | n \in \mathbb{N}\}$ given above defines an open spectrum in Q, there is a collection $\{a_i | 1 \leq i \leq m\}$ of open spectra in $\mathbb{R}$ such that if $V = \cap_{i=1}^m V_{a_i}$ then for each $x_n \in Q$, $V(x_n) \cap Q \subset G_n$. For each $n \in \mathbb{N}$ let H_n denote the Euclidean interior of $V(x_n)$, let $\mathcal{C}' = \{H_n | n \in \mathbb{N}\} \cup \{\mathbb{R}\}$, and let $W = U_{\mathcal{C}'}$. Since, for each rational x_j, $x_j \in H_n$ only if $j > n$, $W(x_j)$ is a Euclidean neighborhood of x_j. There exists a countable subset $\{t_n | n \in \mathbb{N}\}$ of $\mathbb{R}$ such that $\{W(x) | x \in \mathbb{R}\} = \{W(t_n) | n \in \mathbb{N}\}$. Using the Baire Category Theorem we can establish, as in the previous argument, that for some $m \in \mathbb{N}$ there is an infinite subset S of $\mathbb{N}$ so that for each $j \in S$, $W(x_j) \cap A_m \neq \emptyset$; then $W(t_m) \cap Q \subset \cap\{G_n | n \in \mathbb{N}\}$ - a contradiction.

3.21 Since we have just observed that SC is particularly ill behaved with respect to subspaces, the following proposition is some consolation.

PROPOSITION. Let F be a closed G_δ-subspace of a topological space (X,T). Then $SC_X|F \times F = SC_F$.

Proof. It suffices to show that $SC_F \subset SC_X|F \times F$. Let $\{G_n | n \in \mathbb{Z}\}$ be an open spectrum in F and let $\langle A_n \rangle$ be a nested sequence of open sets such that $F = \cap_{n=1}^{\infty} A_n$. Let $\langle H_n \rangle$ be a nested sequence of open subsets of X such that $H_n \cap F = G_{-n}$. Define an open spectrum $\{K_n | n \in \mathbb{Z}\}$ in X as follows. For $n \geq 0$ set $K_n = G_n \cup X - F$ and for $n > 0$ set $K_{-n} = H_n \cap A_n \cap K_0$. For each $n \in \mathbb{Z}$, $K_n \cap F = G_n$. Thus $SC_F \subset SC_X|F \times F$. ■

3.22 The proposition given below is in marked contrast to Proposition 3.19.

PROPOSITION. Let (X,T) be a Hausdorff space. The following statements are equivalent.

(a) SC is totally bounded.

(b) PF is totally bounded.

(c) X is a finite set.

Proof. (a) ⇒ (b) : If SC is totally bounded, then by Theorem 1.57 every lower semi-continuous function is bounded. It follows that every point-finite open cover of X is finite (see the notes); by Proposition 2.8, $PF = P$. (b) ⇒ (c): If PF is totally bounded, $PF = P$ and every point-finite open cover of X is finite. Suppose that X is infinite. Then X has a countably infinite collection C of pairwise disjoint open sets, and $C \cup \{X\}$ is an infinite point-finite open cover of X. (c) ⇒ (a) : Obvious. ■

A remarkable consequence of Propositions 3.17 and 3.22 is that if X is an infinite countably compact Hausdorff space, then SC is hereditarily precompact but not totally bounded.

3.23 Proposition 3.17 suggests the possibility that a topological space is compact (Lindelöf) if and only if every interior-preserving open cover of X has a finite (countable) subcover. For the Lindelöf property, the matter is to the authors' knowledge unresolved. For compactness, a better result has been known for some time.

PROPOSITION. Let $(X,\mathcal{T})$ be a topological space. If every open cover of X that is well ordered by set inclusion has a finite subcover, then $(X,\mathcal{T})$ is compact.

Proof. Let $\mathcal{C}$ be an open cover of X and let $m = \min\{|\mathcal{C}'| \mid \mathcal{C}' \text{ is a subcover of } \mathcal{C}\}$. Suppose that m is infinite and let $\mathcal{C}' = \{G_\alpha \mid \alpha < m\}$ be a subcover of $\mathcal{C}$ having cardinality m. For each $\alpha < m$ let $H_\alpha = \cup_{\beta<\alpha} G_\beta$. Then $\{H_\alpha \mid \alpha < m\}$ is a well-ordered open cover of X that has no finite subcover - a contradiction. ■

3.24 THEOREM. Let $(X,\mathcal{T})$ be a topological space. The following statements are equivalent.

(a) Every admissible quasi-uniformity is complete.

(b) There is an admissible quasi-uniformity that is complete and precompact.

(c) Every admissible quasi-uniformity is complete and precompact.

(d) Every admissible quasi-uniformity is precompact.

(e) $\mathcal{FT}$ is precompact.

(f) $(X,\mathcal{T})$ is compact.

Proof. It is evident that (f) ⇒ (a) ⇒ (b) and (f) ⇒ (c) ⇒ (d) ⇒ (e). By Proposition 3.23, (e) ⇒ (f). We complete the proof by showing that (b) ⇒ (f). Let $\mathcal{U}$ be a quasi-uniformity compatible with $\mathcal{T}$ that is complete and precompact and let $\mathcal{F}$ be an ultrafilter on X. By Proposition 3.14, $\mathcal{F}$ is a $\mathcal{U}$-Cauchy filter; since $(X,\mathcal{U})$ is complete, $\mathcal{F}$ has a cluster point. ■

Theorem 3.24 remains true if complete is replaced by convergence complete in (a), (b) or (c). It follows from Proposition 3.17 and Theorem 3.24 that if X is a countably compact topological space that is not compact, then $\mathcal{SC} \subsetneq \mathcal{FT}$. If X is also a regular T_1 space, then by Proposition 3.18, $\mathcal{PF} \subsetneq \mathcal{FT}$.

3.25 The following table summarizes characterizations of precompactness and total boundedness of certain familiar quasi-uniformities.

	Precompactness	Total Boundedness
$\mathcal{P}$	Always	Always
$\mathcal{LF}$	Pseudocompactness (Tychonoff spaces)	Pseudocompactness (Tychonoff spaces)
$\mathcal{PF}$	Countable compactness (Regular T_1 spaces)	Finite ground set (Hausdorff spaces)
$\mathcal{SC}$	Countable compactness	Finite ground set (Hausdorff spaces)
$\mathcal{FT}$	Compactness	Finite ground set (Hausdorff spaces)

Except for the Pervin quasi-uniformity and the semi-continuous quasi-uniformity, little is known about topological characterizations of those topological spaces for which one of the quasi-uniformities listed on the previous page is (convergence) complete. Indeed it would be reasonable to pose as a problem that the table given on the previous page be extended to completeness and convergence completeness. Instead of pursuing this broad and vaguely posed problem, we pause to consider one fragment of it, which can be stated explicitly.

PROBLEM C. Does there exist a regular space for which no compatible quasi-uniformity is complete? (Note that the fine quasi-uniformity of a regular space is convergence complete whenever it is complete.) The corresponding problem for uniform spaces is solved by exhibiting a noncompact topological space, like ω_1 with the order topology, for which every compatible uniformity is precompact. This approach for quasi-uniform spaces is blocked by Theorem 3.24(d). The following observations may prove useful in answering the question. Every Hausdorff space can be embedded as a closed subspace of a minimal Hausdorff space and every minimal Hausdorff space is a closed subspace of any Hausdorff space in which it is embedded. Consequently, if every minimal Hausdorff space admits a (convergence) complete quasi-uniformity, then every Hausdorff space admits a (convergence) complete quasi-uniformity.

As we shall see in Chapter 6, it is difficult to construct a topological space for which $\mathcal{FT} \neq \mathcal{FINE}$ and so, as might be expected, it is also an unsolved problem whether there exists a regular space for which $\mathcal{FT}$ is not complete. We observe that an ultrafilter $\mathcal{F}$ on a space $(X,\mathcal{T})$ is an $\mathcal{FT}$-Cauchy ultrafilter if and only if every closure-preserving subcollection of $\mathcal{F}$ has a cluster point. A modification of the proof of Proposition 3.23 also provides some insight into the problem of determining whether the fine transitive quasi-uniformity of a regular space must be complete.

3.26 PROPOSITION. Let $(X,\mathcal{T})$ be a topological space. Then every closed $\mathcal{FT}$-Cauchy filter has a cluster point.

Proof. Suppose that $\mathcal{F}$ is a closed $\mathcal{FT}$-Cauchy filter that has no cluster point. Let $m = \min \{|\mathcal{E}| \mid \mathcal{E}$ is a closed subcollection of $\mathcal{F}$ and $\cap\mathcal{E} = \emptyset\}$. Let $\mathcal{E} = \{E_\alpha \mid \alpha < m\}$ be a closed subcollection of $\mathcal{F}$ such that $\cap\mathcal{E} = \emptyset$. For each $\alpha < m$ let $F_\alpha = \cap\{E_\beta \mid \beta < \alpha\}$. Since $\{X - F_\alpha \mid \alpha < m\}$ is an interior-preserving open cover, there exists $\alpha < m$ and a closed set $F \in \mathcal{F}$ such that $F \subset X - F_\alpha$. Then $\mathcal{E}' = \{F\} \cup \{E_\beta | \beta \leq \alpha\}$ is a closed subcollection of $\mathcal{F}$ consisting of fewer than m sets and $\cap\mathcal{E}' = \emptyset$ - a contradiction. ∎

COROLLARY. Let $(X,\mathcal{T})$ be a T_1 space such that for each interior-preserving open cover $\mathcal{C}$ of X there exists an interior-preserving open cover $\mathcal{R}$ such that $\{\overline{R} | R \in \mathcal{R}\}$ refines $\mathcal{C}$. Then $\mathcal{FT}$ is convergence complete.

Proof. Let $\mathcal{F}$ be an $\mathcal{FT}$-Cauchy filter. It follows from the hypothesis that $\mathcal{B} = \{\overline{F} | F \in \mathcal{F}\}$ is an $\mathcal{FT}$-Cauchy filter base. By the previous proposition $\mathcal{B}$ (hence $\mathcal{F}$) has a cluster point. Since the hypothesis implies that $(X,\mathcal{T})$ is regular, $\mathcal{FT}$ is convergence complete by Corollary 3.9. ∎

3.27 EXAMPLE. A T_1 space that admits no *convergence* complete quasi-uniformity.

For each nonnegative integer n let $X_n = \mathbb{R} \times \{n\}$ and let $X = \cup_{n=0}^{\infty} X_n$. Define $g : \mathbb{N} \times X \to \mathcal{P}(X)$ as follows. For each $x \in X$, $m \in \mathbb{N}$ and $p = (x,0)$, let $g(m,p) = \{p\} \cup \cup_{i=m}^{\infty} X_i - \{(x,i)\}$ and for each $p = (x,n) \in \mathbb{R} \times \mathbb{N}$ and $m \in \mathbb{N}$ let $g(m,p) = (x - 1/m,\ x + 1/m) \times \{n\}$. Let $\mathcal{T}$ be the topology for which $\{g(n,p) | n \in \mathbb{N},\ p \in X\}$ is a base.

Let $\mathcal{F} = \{F | \text{there is an } n \in \mathbb{N} \text{ so that } \cup_{i=n}^{\infty} X_i \subset F\}$ and let $\mathcal{U} = \mathcal{FINE}$. We assert that $\mathcal{F}$ is a $\mathcal{U}$-Cauchy filter. Suppose not; then there exists $U \in \mathcal{U}$ such that for all $p \in X$, $U(p) \notin \mathcal{F}$. Let $V \in \mathcal{U}$ such that $V^2 \subset U$. Then for each $x \in \mathbb{R}$, $\{n | U((x,0)) \cap X_n = V((x,0)) \cap X_n = X_n - \{(x,n)\}\}$ is infinite. For each $n \in \mathbb{N}$ let $A_n = \{x \in \mathbb{R} | U((x,0)) \cap X_n = V((x,0)) \cap X_n = X_n - \{(x,n)\}\}$. There is an $n \in \mathbb{N}$ so that A_n is uncountable and so there exists $z \in \mathbb{R}$ such that for each $\epsilon > 0$, $(z - \epsilon,\ z + \epsilon) \cap A_n - \{z\} \neq \emptyset$. Let $A = A_n - \{z\}$. Then $(z,n) \in \cap_{x \in A} V((x,0))$ and $X_n \cap V((z,n)) \subset X_n \cap [\cap_{x \in A} V^2((x,0))] \subset (\mathbb{R} - A) \times \{n\}$ - a contradiction. Thus $\mathcal{F}$ is a nonconvergent $\mathcal{U}$-Cauchy filter.

2. BICOMPLETENESS

3.28 In this section we present a theory of bicompleteness for quasi-uniform spaces, which extends the usual theory of completeness for uniform spaces, and which is in every respect analogous to this theory. As would be expected the keystone of the theory is the construction of the bicompletion of a quasi-uniform space. Our presentation of this construction closely parallels a standard construction of the completion of a uniform space by means of minimal Cauchy filters, and the reader who is familiar with this standard construction will understand that in our view the uniform completion is seen as if through binoculars. Such a reader may wish to omit many proofs. On the other hand the reader who is not familiar with the standard construction is assured that the construction presented herein is no more difficult than the standard construction and contains the usual completion of a uniform space as a special case.

A quasi-uniform space $(X,\mathcal{U})$ is said to be *bicomplete* provided that every $\mathcal{U}^*$-Cauchy filter has a $\mathcal{T}(\mathcal{U}^*)$-limit point. By Proposition 3.9, a quasi-uniform space $(X,\mathcal{U})$ is bicomplete if and only if $(X,\mathcal{U}^*)$ is complete. Thus $(X,\mathcal{U})$ is bicomplete if and only if $(X,\mathcal{U}^{-1})$ is bicomplete. The results on completeness established in Section 1 may therefore be regarded as results on bicompleteness.

If X is a set and $\mathcal{U}$ is the only quasi-uniformity on X under consideration, then for convenience we adopt the following notation. For each $x \in X$, $\eta^*(x)$ denotes the $\mathcal{T}(\mathcal{U}^*)$-neighborhood filter of x.

PROPOSITION. Let $(X,\mathcal{U})$ be a T_0 quasi-uniform space and let $(Y,\mathcal{V})$ be a bicomplete subspace of $(X,\mathcal{U})$. Then Y is a closed subspace of $(X,\mathcal{T}(\mathcal{U}^*))$.

Proof. Suppose that Y is not closed in $(X,\mathcal{T}(\mathcal{U}^*))$ and let $p \in \overline{Y} - Y$. Then $\eta^*(p)|Y$ is a $\mathcal{V}^*$-Cauchy filter and therefore $\eta^*(p)|Y$ converges to a point $q \in Y$. Since $(X,\mathcal{T}(\mathcal{U}^*))$ is a Hausdorff space, $p = q$ - a contradiction. ■

3.29 THEOREM. Let $(X,\mathcal{U})$ be a quasi-uniform space, let $(Y,\mathcal{V})$ be a bicomplete T_0 quasi-uniform space, let D be a dense subset of $(X,\mathcal{T}(\mathcal{U}^*))$, and let $f : (D,\mathcal{U}|D \times D) \to (Y,\mathcal{V})$ be a quasi-uniformly continuous function. Then there exists a unique continuous extension $g : (X,\mathcal{T}(\mathcal{U}^*)) \to (Y,\mathcal{T}(\mathcal{V}^*))$, and g is quasi-uniformly continuous.

Proof. We show first that f has a $\mathcal{T}(\mathcal{U}^*)$ - $\mathcal{T}(\mathcal{V}^*)$ continuous extension. Let $x \in X$. As is remarked in §3.1, $\eta^*(x)$ is a $\mathcal{U}^*$-Cauchy filter so that by Corollary 3.5, $\mathcal{F}_x = \{D \cap R | R \in \eta^*(x)\}$ is a Cauchy filter on D. Since f is quasi-uniformly continuous, it follows from Proposition 3.4 that $\{f(F) | F \in \mathcal{F}_x\}$ is a base for a $\mathcal{V}^*$-Cauchy filter $\mathcal{G}_x$. The space $(Y,\mathcal{V})$ is bicomplete and $(Y,\mathcal{T}(\mathcal{V}^*))$ is a Hausdorff space so that $\mathcal{G}_x$ has a unique limit, denoted by $g(x)$. It remains to be shown that $g : X \to Y$, defined in the manner described above, is a quasi-uniformly continuous extension of f.

To show that g extends f, let $x \in D$ and let G be a $\mathcal{T}(\mathcal{V}^*)$-neighborhood of $f(x)$. As $f^{-1}(G)$ is a $\mathcal{T}(\mathcal{U}^*)|D$-neighborhood of x, there exists a $\mathcal{T}(\mathcal{U}^*)$-neighborhood R of x such that $f(D \cap R) \subset G$. Thus $G \in \mathcal{G}_x$ and $\mathcal{G}_x$ converges to $f(x)$; that is $g(x) = f(x)$.

To show that g is quasi-uniformly continuous, let $V \in \mathcal{V}$. By Corollary 1.19 there exists $V_1 \in \mathcal{V}$ such that V_1 is closed in $\mathcal{T}(\mathcal{V} \times \mathcal{V}^{-1})$ and $V_1^3 \subset V$. Since f is quasi-uniformly continuous, there exists $U \in \mathcal{U}$ such that $(f(x), f(y)) \in V_1$ whenever $(x,y) \in U \cap D \times D$. Let $U_1 \in \mathcal{U}$ such that $U_1^3 \subset U$. The proof will be completed by showing that $(g(u), g(v)) \in V$ whenever $(u,v) \in U_1$.

Let $(u,v) \in U_1$, let $U_2 = U_1 \cap U_1^{-1}$, and let $V_2 = V_1 \cap V_1^{-1}$. We assert that if $u' \in D \cap U_2(u)$, then $g(u) \in V_2(f(u'))$. Let H be a $\mathcal{T}(\mathcal{V}^*)$-neighborhood of $g(u)$; then there exists a $\mathcal{T}(\mathcal{U}^*)$-neighborhood R of u such that $f(D \cap R) \subset H$. Let $w \in D \cap R \cap U_2(u)$, then $f(w) \in H \cap V_2(f(u'))$. Thus $H \cap V_2(f(u')) \neq \emptyset$, so that $g(u) \in \mathrm{cl}_{\mathcal{T}(\mathcal{V}^*)}(V_2(f(u'))) = V_2(f(u'))$ as was asserted.

Now let $u' \in D \cap U_2(u)$ and let $v' \in D \cap U_2(v)$; then $(u',v') \in U$ and $(f(u'), f(v')) \in V_1$. By the preceding assertion, $(g(u), f(u')) \in V_2 \subset V_1$ and $(f(v'), g(v)) \in V_2 \subset V_1$. Thus $(g(u), g(v)) \in V_1^3 \subset V$ as was to be shown.

Finally, since $\mathcal{T}(\mathcal{V}^*)$ is a Hausdorff topology, if h is a $\mathcal{T}(\mathcal{U}^*) - \mathcal{T}(\mathcal{V}^*)$ continuous function that agrees with g on the dense subspace D of X, then $g = h$. Thus g is the only $\mathcal{T}(\mathcal{U}^*) - \mathcal{T}(\mathcal{V}^*)$ continuous extension of f. ■

COROLLARY. Let $(X,\mathcal{U})$ and $(Y,\mathcal{V})$ be bicomplete T_0 quasi-uniform spaces and let D and E be dense subspaces of $(X,\mathcal{T}(\mathcal{U}^*))$ and $(Y,\mathcal{T}(\mathcal{V}^*))$ respectively. Let $f : (D,\mathcal{U}|D \times D) \to (E,\mathcal{U}|E \times E)$ be a quasi-unimorphism. Then f can be extended to a quasi-unimorphism between $(X,\mathcal{U})$ and $(Y,\mathcal{V})$.

Proof. By Theorem 3.29, f has a quasi-uniformly continuous extension $g_1 : X \to Y$, and f^{-1} has a quasi-uniformly continuous extension $g_2 : Y \to X$. Then $g_2 \circ g_1$ is a continuous extension of the identity map on D. Likewise $g_1 \circ g_2$ is a continuous extension of the identity map on E. As the identity maps on X and Y are the unique continuous extensions of the identity maps on D and E respectively, $g_2 \circ g_1$ is the identity map on X and $g_1 \circ g_2$ is the identity map on Y. It follows that g_1 is a quasi-unimorphism from $(X,\mathcal{U})$ onto $(Y,\mathcal{V})$. ■

3.30 Let $(X,\mathcal{U})$ be a T_0 quasi-uniform space and suppose that $(Y,\mathcal{V})$ is a T_0 quasi-uniform space such that $(Y,\mathcal{V}^*)$ is a complete Hausdorff uniform space that contains $(X,\mathcal{U}^*)$ as a dense subspace. If $(X,\mathcal{U})$ and $(Y,\mathcal{V})$ are both uniform spaces, then $(Y,\mathcal{V})$ must be the uniform completion of $(X,\mathcal{U})$. Thus any completion that assigns to each T_0 quasi-uniform space $(X,\mathcal{U})$ a T_0 quasi-uniform space $(Y,\mathcal{V})$ and that preserves symmetry (in the sense that $\mathcal{V}$ is a uniformity whenever $\mathcal{U}$ is a uniformity) must include the construction of the completion of a Hausdorff uniform space as a special case. In this section we present just such a construction. While it is comforting that our heretical construction recovers the canonical Hausdorff completion as a special case, it will become apparent in the next chapter that the importance of the construction is founded as much upon its preservation of transitivity as upon its preservation of symmetry. Indeed if $\cap\mathcal{U}$ is a linear order on X, then $\cap\mathcal{V}$ is a linear order on Y; and it is this remarkable attribute of the construction that establishes the importance of quasi-uniform spaces in the study of linearly ordered topological spaces and their subspaces.

A *bicompletion* of a quasi-uniform space $(X,\mathcal{U})$ is a bicomplete quasi-uniform space $(Y,\mathcal{V})$ that has a $\mathcal{T}(\mathcal{V}^*)$ dense subspace quasi-unimorphic to $(X,\mathcal{U})$. The key to our construction of a bicompletion of a quasi-uniform space is the use of minimal $\mathcal{U}^*$-Cauchy filters. A $\mathcal{U}^*$-Cauchy filter on a quasi-uniform space $(X,\mathcal{U})$ is said to be *minimal* provided that it contains no $\mathcal{U}^*$-Cauchy filter other than itself.

PROPOSITION. Let $\mathcal{F}$ be a $\mathcal{U}^*$-Cauchy filter on a quasi-uniform space $(X,\mathcal{U})$. There is exactly one minimal $\mathcal{U}^*$-Cauchy filter that is coarser than $\mathcal{F}$. Furthermore if $\mathcal{B}$ is any base for $\mathcal{F}$, then $\mathcal{B}_0 = \{U(B) \mid B \in \mathcal{B}$ and U is a symmetric member of $\mathcal{U}^*\}$ is a base for the minimal $\mathcal{U}^*$-Cauchy filter coarser than $\mathcal{F}$.

Proof. Let $\mathcal{B}$ be a base for $\mathcal{F}$ and let $U_1(B_1)$ and let $U_2(B_2)$ be members of $\mathcal{B}_0$. Let $U = U_1 \cap U_2$ and let $B \in \mathcal{B}$ such that $B \subset B_1 \cap B_2$; then $U(B) \subset U_1(B_1) \cap U_2(B_2)$. Thus $\mathcal{B}_0$ is a base for a filter $\mathcal{F}_0$. Given $V \in \mathcal{U}^*$, let U be a symmetric member of $\mathcal{U}^*$ such that $U^3 \subset V$. By Proposition 3.2 there exists $B \in \mathcal{B}$ such that $B \times B \subset U$. Then $U(B) \times U(B) \subset V$; hence $\mathcal{F}_0$ is a $\mathcal{U}^*$-Cauchy filter that is coarser than $\mathcal{F}$. Let $\mathcal{G}$ be a $\mathcal{U}^*$-Cauchy filter coarser than $\mathcal{F}$ and let $U(B) \in \mathcal{B}_0$. By Proposition 3.2 there exists $G \in \mathcal{G}$ such that $G \times G \subset U$. Since $G \in \mathcal{F}$, $G \cap B \neq \emptyset$. It follows that $G \subset U(B)$

and so $U(B) \in \mathcal{G}$. As any $\mathcal{U}^*$-Cauchy filter coarser than $\mathcal{F}$ contains $\mathcal{F}_0$, $\mathcal{F}_0$ is a minimal $\mathcal{U}^*$-Cauchy filter and is the only minimal Cauchy filter coarser than $\mathcal{F}$. ■

COROLLARY. Let $(X,\mathcal{U})$ be a quasi-uniform space. For each $x \in X$, $\eta^*(x)$ is a minimal $\mathcal{U}^*$-Cauchy filter.

3.31 THEOREM. Let $(X,\mathcal{U})$ be a quasi-uniform space. Every minimal $\mathcal{U}^*$-Cauchy filter has a base consisting of $\mathcal{T}(\mathcal{U}^*)$-open sets.

Proof. Let $V \in \mathcal{U}^*$. By Corollary 1.17 there is a symmetric entourage $U \in \mathcal{U}^*$ such that $U \subset V$ and such that for each $x \in X$, $U(x) \in \mathcal{T}(\mathcal{U}^*)$. For each subset A of X, U(A) is a $\mathcal{T}(\mathcal{U}^*)$-open subset of V(A); hence the result follows from Proposition 3.30. ■

3.32 PROPOSITION. Let $(X,\mathcal{U})$ be a quasi-uniform space and let D be a dense subset of $(X,\mathcal{T}(\mathcal{U}^*))$. If every Cauchy filter on $(D,\mathcal{U}^*|D \times D)$ converges in $(X,\mathcal{T}(\mathcal{U}^*))$, then $(X,\mathcal{U})$ is bicomplete.

Proof. It suffices to show that every minimal $\mathcal{U}^*$-Cauchy filter $\mathcal{F}$ on X converges. Since D is dense in $(X,\mathcal{T}(\mathcal{U}^*))$ and since, by Theorem 3.31 every member of $\mathcal{F}$ has a nonempty interior, $\mathcal{F}|D$ is a Cauchy filter on $(D,\mathcal{U}^*|D \times D)$ and so converges in $(X,\mathcal{T}(\mathcal{U}^*))$. As $\mathcal{F}$ is coarser than the filter on X determined by $\mathcal{F}|D$, it follows from Proposition 3.9 that $\mathcal{F}$ converges. ■

3.33 THEOREM. Every T_0 quasi-uniform space has a T_0 bicompletion.

Proof. Let $(X,\mathcal{U})$ be a quasi-uniform space and let $\tilde{X}$ be the set of all minimal $\mathcal{U}^*$-Cauchy filters on X. For each $U \in \mathcal{U}$, let $\tilde{U} = \{(\mathcal{F},\mathcal{G}) \mid$ there exist $F \in \mathcal{F}$ and $G \in \mathcal{G}$ such that $F \times G \subset U\}$. We show that $\{\tilde{U} \mid U \in \mathcal{U}\}$ is a base for a quasi-uniformity $\tilde{\mathcal{U}}$ on $\tilde{X}$. As every member of $\tilde{X}$ is a $\mathcal{U}^*$-Cauchy filter, for each $U \in \mathcal{U}$ and for each $\mathcal{F} \in \tilde{X}$, $(\mathcal{F},\mathcal{F}) \in \tilde{U}$. Let $V_1, V_2 \in \mathcal{U}$; then $V = V_1 \cap V_2 \in \mathcal{U}$ and $\tilde{V} = \tilde{V}_1 \cap \tilde{V}_2$. Given $U \in \mathcal{U}$ let $V \in \mathcal{U}$ such that $V^2 \subset U$. Let $(\mathcal{F},\mathcal{G}), (\mathcal{G},\mathcal{H}) \in \tilde{V}$; then there exist $F \in \mathcal{F}$, $G_1 \in \mathcal{G}$ such that $F \times G_1 \subset V$ and $G_2 \in \mathcal{G}$, $H \in \mathcal{H}$ such that $G_2 \times H \subset V$. Since $G_1 \cap G_2 \neq \emptyset$, $F \times H \subset V^2$; thus $(\tilde{V})^2 \subset \tilde{U}$.

To show that $(\tilde{X},\tilde{\mathcal{U}})$ is a T_0 quasi-uniform space, let $\mathcal{F}$ and $\mathcal{G}$ be minimal $\mathcal{U}^*$-Cauchy filters such that for each $U \in \mathcal{U}$, $(\mathcal{F},\mathcal{G}) \in \tilde{U}$ and $(\mathcal{G},\mathcal{F}) \in \tilde{U}$. Then $\mathcal{F} \cap \mathcal{G}$ is a $\mathcal{U}^*$-Cauchy filter. As $\mathcal{F}$ and $\mathcal{G}$ are minimal Cauchy filters and $\mathcal{F} \cap \mathcal{G}$ is coarser than both $\mathcal{F}$ and $\mathcal{G}$, $\mathcal{F} = \mathcal{F} \cap \mathcal{G} = \mathcal{G}$.

Let $i : X \to \tilde{X}$ be the function defined by $i(x) = \eta^*(x)$. It follows from Corollary 3.30 that $\eta^*(x) \in \tilde{X}$ for each $x \in X$. Thus i is well defined. Given $U \in \mathcal{U}$ let $V \in \mathcal{U}$ such that $V^3 \subset U$. Then $(x,y) \in U$ whenever $(\eta^*(x), \eta^*(y)) \in \tilde{U}$ and $(\eta^*(x), \eta^*(y)) \in \tilde{U}$ whenever $(x,y) \in V$. Thus i is a quasi-uniform embedding.

Since $(\tilde{\mathcal{U}})^* = (\mathcal{U}^*)^\sim$, we use the symbol $\tilde{\mathcal{U}}^*$ unambiguously hereafter. We assert that if $\mathcal{F} \in \tilde{X}$, then $i(\mathcal{F})$ converges to $\mathcal{F}$ in $(\tilde{X}, \mathcal{T}(\tilde{\mathcal{U}}^*))$. It follows at once from this assertion that $i(X)$ is dense in $(\tilde{X}, \mathcal{T}(\tilde{\mathcal{U}}^*))$. Given $\mathcal{F} \in \tilde{X}$ and $U_1 \in \mathcal{U}$, let $V_1 \in \mathcal{U}$ and $F \in \mathcal{F}$ such that $V_1^2 \subset U_1$ and $F \times F \subset V_1$. Let $U = U_1 \cap U_1^{-1}$ and $V = V_1 \cap V_1^{-1}$. For each $x \in F$, $F \times V(x) \subset U$; thus for each $x \in F$, $i(x) = \eta^*(x) \in \tilde{U}(\mathcal{F})$ and so $i(F) \subset \tilde{U}(\mathcal{F})$.

By Proposition 3.32 in order to show that $(\tilde{X}, \tilde{\mathcal{U}})$ is bicomplete it is sufficient to show that for every $\mathcal{U}^*$-Cauchy filter $\mathcal{F}$ on X, $i(\mathcal{F})$ converges in $(\tilde{X}, \mathcal{T}(\tilde{\mathcal{U}}^*))$. Let $\mathcal{F}$ be a $\mathcal{U}^*$-Cauchy filter. By Proposition 3.30 there is a minimal $\mathcal{U}^*$-Cauchy filter $\mathcal{G}$ coarser than $\mathcal{F}$. By the assertion of the previous paragraph, $i(\mathcal{G})$, converges to $\mathcal{G}$. As $i(\mathcal{F})$ is finer than $i(\mathcal{G})$, $i(\mathcal{F})$ converges as well. ∎

3.34 We call the map $i : X \to \tilde{X}$ defined in Theorem 3.33 the *canonical embedding* of $(X, \mathcal{U})$ into $(\tilde{X}, \tilde{\mathcal{U}})$. As a direct consequence of Corollary 3.29 we have the following uniqueness property for T_0 bicompletions.

THEOREM. If $(X, \mathcal{U})$ is a T_0 quasi-uniform space, any T_0 bicompletion of $(X, \mathcal{U})$ is quasi-unimorphic to $(\tilde{X}, \tilde{\mathcal{U}})$.

When $(X, \mathcal{U})$ is a T_0 quasi-uniform space, $(\tilde{X}, \tilde{\mathcal{U}})$ is said to be *the* bicompletion of $(X, \mathcal{U})$, and we may identify X with $i(X)$. When this identification is made, the minimal $\mathcal{U}^*$-Cauchy filters on X are the traces of the $\mathcal{T}(\tilde{\mathcal{U}}^*)$-neighborhood filters of the points of $\tilde{X}$.

3.35 THEOREM. Let $(X, \mathcal{U})$ be a Hausdorff uniform space. Then $(\tilde{X}, \tilde{\mathcal{U}})$ is a complete Hausdorff uniform space, and any complete Hausdorff uniform space that has a dense subspace unimorphic to $(X, \mathcal{U})$ is unimorphic to $(\tilde{X}, \tilde{\mathcal{U}})$.

3.36 PROPOSITION. Let $(X, \mathcal{U})$ be a T_0 quasi-uniform space. Then $(X, \mathcal{U})$ is totally bounded if and only if $(\tilde{X}, \tilde{\mathcal{U}}^*)$ is compact.

Proof. Suppose that $(\tilde{X},\tilde{U}^*)$ is compact. By Theorem 3.24, $(\tilde{X},\tilde{U}^*)$ is totally bounded. As total boundedness is a hereditary property, (X,U) is totally bounded.

The converse is an immediate consequence of Corollary 3.15. ■

3.37 The following proposition, which relies upon Corollary 1.20 and Theorem 3.35, establishes a one-to-one correspondence between the Hausdorff compactifications of a Tychonoff space (X,T) and the totally bounded uniformities on X that are compatible with T.

PROPOSITION. Let (X,T) be a Tychonoff space and let cX be a Hausdorff compactification of X. There is exactly one uniformity U compatible with T such that $T(\tilde{U})$ is the topology of cX.

Let (X,δ) be a T_0 quasi-proximity space. By Theorem 3.33, (X,U_δ) has a bicompletion $(\tilde{X},\tilde{U}_\delta)$. We let $\tilde{\delta}$ denote the quasi-proximity induced by $\tilde{U}_\delta$, the bicompletion of U_δ.

It is easy to prove that $(U^{-1})^\sim = (\tilde{U})^{-1}$ and that $(\tilde{U})^* = (U^*)^\sim$; indeed the second equality is observed in the proof of Theorem 3.33. These results together with equalities (a) through (d) of §1.40 are sufficient to establish that for each quasi-proximity δ, $(\tilde{\delta})^{-1} = (\delta^{-1})^\sim$ and $(\tilde{\delta})^* = (\delta^*)^\sim$. The reader is warned that $(\tilde{U})_\delta \neq \tilde{U}_\delta$, $(U_\omega)^\sim \neq (\tilde{U})_\omega$ and $\delta_{\tilde{U}} \neq (\delta_U)^\sim$.

3.38 THEOREM. Let (X,δ) be a T_0 quasi-proximity space.

(a) The canonical embedding $i : (X,U_\delta) \to (\tilde{X},\tilde{U}_\delta)$ is a qp-embedding of (X,δ) into $(\tilde{X},\tilde{\delta})$.

(b) $(\tilde{X},\tilde{\delta}^*)$ is compact.

(c) $A\ \delta\ B$ if and only if $\mathrm{cl}_{\tilde{\delta}^{-1}}(i(A)) \cap \mathrm{cl}_{\tilde{\delta}}(i(B)) \neq \emptyset$.

Proof.

(a) Proposition 1.51.

(b) Proposition 3.36.

(c) If $\mathrm{cl}_{\tilde{\delta}^{-1}}(i(A)) \cap \mathrm{cl}_{\tilde{\delta}}(i(B)) \neq \emptyset$, by Proposition 1.41, $i(A)\ \tilde{\delta}\ i(B)$ so that $A\ \delta\ B$. If $A\ \delta\ B$, then $i(A)\ \tilde{\delta}\ i(B)$ so that by Proposition 1.44 we have $\mathrm{cl}_{\tilde{\delta}^{-1}}(i(A)) \cap \mathrm{cl}_{\tilde{\delta}}(i(B)) \neq \emptyset$. ■

3. COMPLETIONS AND COMPACTIFICATIONS

3.39 Let $(X,\mathcal{U})$ be a T_1 quasi-uniform space. A *completion* of $(X,\mathcal{U})$ is a complete T_1 quasi-uniform space $(Y,\mathcal{V})$ that has a dense subspace quasi-unimorphic (relative to $\mathcal{U}$ and $\mathcal{V}$) to $(X,\mathcal{U})$. In the theory of uniform spaces it is common to insist that a completion of a uniform space be a T_1 space, since this restriction simplifies some of the proofs and makes it possible to strengthen the conclusion of several important theorems. In the study of quasi-uniform spaces it is necessary to insist that completions be T_1 spaces, for if we do not, every space $(X,\mathcal{U})$ admits the following trivial "compactification". Let $Y = X \cup \{\infty\}$ and for each $U \in \mathcal{U}$ let $U' = U \cup \{\infty\} \times Y$. Then $\{U' | U \in \mathcal{U}\}$ is a base for a complete precompact quasi-uniformity $\mathcal{V}$ on Y, and $(X,\mathcal{U})$ is a dense subspace of $(Y,\mathcal{V})$. Since we do insist that a completion must be a T_1 space, it is a consequence of Corollary 3.15 that a precompact quasi-uniformity compatible with a topological space that is not completely regular has no regular completion. This consequence of Corollary 3.15 serves as an early warning that the theory of completions for quasi-uniform spaces is not nearly so tidy as either the theory of completions for uniform spaces or the theory of bicompletions for quasi-uniform spaces.

3.40 In this section we present a completion, which is based on the construction of the Katětov extension of a Hausdorff space. In this construction we take certain nonconvergent filters as adjoined points and build open sets about an adjoined filter in terms of its members. To this end let $(X,\mathcal{U})$ be a quasi-uniform space, let $\mathbb{F} = \{\mathcal{F} | \mathcal{F}$ is a $\mathcal{U}$-Cauchy filter on X that has no cluster point$\}$, and let $\hat{X} = X \cup \mathbb{F}$. We need the collection Φ of all choice functions that pick a member of each filter in $\mathbb{F}$. Formally $\Phi = \{\phi | \phi : \mathbb{F} \to 2^X$ and $\phi(\mathcal{F}) \in \mathcal{F}\}$. For each $U \in \mathcal{U}$ and each $\phi \in \Phi$, set $S(U,\phi) = U \cup \Delta \cup \{(\mathcal{F},x) \in \mathbb{F} \times X | x \in U(\phi(\mathcal{F}))\}$ and let $\hat{\mathcal{U}}$ be the quasi-uniformity on $\hat{X}$ for which $\{S(U,\phi) | U \in \mathcal{U}, \phi \in \Phi\}$ is a base. Then $(\hat{X},\hat{\mathcal{U}})$ contains $(X,\mathcal{U})$ as a dense open subset.

3.41 LEMMA. Let $(X,\mathcal{U})$ be a T_1 quasi-uniform space. Then $(\hat{X},\hat{\mathcal{U}})$ is a T_1 space if and only if for each $\mathcal{F} \in \mathbb{F}$, $\operatorname{adh}_{\mathcal{T}(\mathcal{U}^{-1})} \mathcal{F} = \emptyset$.

Proof. Suppose that $(\hat{X},\hat{\mathcal{U}})$ is a T_1 space and let $\mathcal{F} \in \mathbb{F}$ and $p \in X$. Then there exist $U \in \mathcal{U}$ and $\phi \in \Phi$ such that $p \notin S(U,\phi)(\mathcal{F})$. Let $F = \phi(\mathcal{F})$; then $p \notin U(F)$ and $U^{-1}(p) \cap F = \emptyset$.

Now suppose that for each $F \in \mathbb{F}$, $\mathrm{adh}_{T(U^{-1})} F = \emptyset$ and let $p \in \hat{X}$. If $p \in X$, $\cap\hat{U}(p) = \cap U(p) = \{p\}$. If $p = F \in \mathbb{F}$, then $\cap\hat{U}(p) = \{p\} \cup X \cap (\cap\hat{U}(p))$. Suppose there exists $x \in X \cap (\cap\hat{U}(p))$. Since $\mathrm{adh}_{T(U^{-1})} F = \emptyset$, there exist $F \in F$ and $U \in U$ such that $U^{-1}(x) \cap F = \emptyset$. Let $\phi \in \Phi$ be such that $\phi(F) = F$. Then $x \notin S(U,\phi)(p)$ - a contradiction. ■

3.42 LEMMA. Let (X,U) be a quasi-uniform space and let M be a nonconvergent $\hat{U}$-Cauchy ultrafilter on $\hat{X}$. Then $X \in M$ and $M|X$ is a nonconvergent U-Cauchy ultrafilter on X.

Proof. Let $U_1 \in U$ and let $\phi_1 \in \Phi$. Since M is a $\hat{U}$-Cauchy filter, there exists $p_1 \in \hat{X}$ such that $S(U_1,\phi_1)(p_1) \in M$. Since M does not converge to p_1, there exist $U_2 \in U$, $\phi_2 \in \Phi$ and $p_2 \neq p_1$ such that $S(U_2,\phi_2)(p_2) \in M$. Then $S(U_1,\phi_1)(p_1) \cap S(U_2,\phi_2)(p_2) \subset X$ and $X \in M$. It follows that $M|X \subset M$ so that $M|X$ does not converge. To show that $M|X$ is a U-Cauchy filter, let $U \in U$ and let $V \in U$ such that $V^2 \subset U$. Choose $\phi \in \Phi$ such that for each $F \in \mathbb{F}$, $\phi(F) = V(x)$, where $V(x) \in F$. Since M is a $\hat{U}$-Cauchy filter, there exists $p \in \hat{X}$ such that $S(V,\phi)(p) \in M$. If $p \in X$, $S(V,\phi)(p) = V(p) \in M|X$; if $p = F \in \mathbb{F}$, then $S(V,\phi)(p) = \{F\} \cup V(\phi(F))$. There exists $x \in X$ so that $V(x) \in F$ and $\phi(F) = V(x)$; thus $U(x) \in M|X$. ■

3.43 THEOREM. Let (X,U) be a T_1 quasi-uniform space. The following statements are equivalent.

(a) (X,U) has a completion.

(b) For each U-Cauchy filter F, $\mathrm{adh}_{T(U^{-1})} F \subset \mathrm{adh}_{T(U)} F$.

(c) For each U-Cauchy filter F, if $\mathrm{adh}_{T(U^{-1})} F \neq \emptyset$, then $\mathrm{adh}_{T(U)} F \neq \emptyset$.

Proof. (a) ⇒ (b) : Suppose that $i : (X,U) \to (Y,V)$ is a dense embedding of (X,U) in a complete T_1 space (Y,V). Let F be a U-Cauchy filter and let $x \in \mathrm{adh}_{T(U^{-1})} F$. Then F is contained in an ultrafilter G that converges to x in $T(U^{-1})$. By Propositions 3.4 and 1.14, $i(G)$ is a V-Cauchy ultrafilter base that converges to $i(x)$ in $T(V^{-1})$. Since (Y,V) is complete, $i(G)$ converges in $T(V)$ to a point y. By Proposition 1.10, $y = i(x)$. Moreover $i^{-1}(i(G))$ is a filter base on X, coarser than G, that converges to x in $T(U)$; hence $x \in \mathrm{adh}_{T(U)} F$.

(c) ⇒ (a) : We now suppose that for each U-Cauchy filter F, $\mathrm{adh}_{T(U)} F \neq \emptyset$ whenever $\mathrm{adh}_{T(U^{-1})} F \neq \emptyset$, and we show that $(\hat{X},\hat{U})$ is a completion of $(X,\mathcal{U})$. By Lemma 3.41, $(\hat{X},\hat{\mathcal{U}})$ is a T_1 space. As noted in §3.40, $(X,\mathcal{U})$ is a dense open subspace of $(\hat{X},\hat{\mathcal{U}})$. To show that $(\hat{X},\hat{\mathcal{U}})$ is complete, suppose that there exists a nonconvergent $\hat{\mathcal{U}}$-Cauchy ultrafilter $\mathcal{M}$. By Lemma 3.42, $\mathcal{M}|X$ is a nonconvergent $\mathcal{U}$-Cauchy ultrafilter. Then $\mathcal{M}|X$, as a member of $\mathbb{F}$, is a $\mathcal{T}(\hat{\mathcal{U}})$-cluster point of $\mathcal{M}$ - a contradiction. ∎

3.44 COROLLARY. Let $(X,\mathcal{T})$ be a T_1 space and let $\mathcal{U}$ and $\mathcal{V}$ be quasi-uniformities that are compatible with $\mathcal{T}$. If $\mathcal{U} \subset \mathcal{V}$ and $(X,\mathcal{U})$ has a completion, $(X,\mathcal{V})$ also has a completion.

3.45 COROLLARY. Every point-symmetric T_1 quasi-uniform space has a completion.

3.46 The following proposition, which extends Theorem 3.24, shows that any topological space that admits a noncomplete quasi-uniformity also admits a quasi-uniformity that has no completion.

PROPOSITION. Let $(X,\mathcal{T})$ be a topological space for which every quasi-uniformity compatible with $\mathcal{T}$ has a completion. Then $(X,\mathcal{T})$ is compact.

Proof. Suppose that $\mathcal{F}$ is a filter on X that has no cluster point. By Proposition 2.4 the quasi-uniformity $\mathcal{U}$ generated by $\{T(X - F,F) \mid F \in \mathcal{F}$ and $X - F \in \mathcal{T}\}$ is compatible with $\mathcal{T}$. For each $U \in \mathcal{U}$ there is an $x \in X$ such that $U(x) = X$, and so $\mathcal{F}$ is a $\mathcal{U}$-Cauchy filter. To show that $(X,\mathcal{U})$ has no completion, by virtue of Theorem 3.43 it suffices to show that $\mathrm{adh}_{\mathcal{T}(\mathcal{U}^{-1})} \mathcal{F} \neq \emptyset$. We show indeed that $\mathrm{adh}_{\mathcal{T}(\mathcal{U}^{-1})} \mathcal{F} = X$. Let $x \in X$, $U \in \mathcal{U}$ and $F \in \mathcal{F}$. There exist closed sets $F_1, F_2, \ldots, F_n$ such that $\cap_{i=1}^n T(X - F_i,F_i) \subset U$. For each i, $T(F_i,X - F_i)(x) = X$ or F_i. Hence $\emptyset \neq F \cap (\cap_{i=1}^n F_i) \subset F \cap [\cap_{i=1}^n T(F_i,X - F_i)(x)] \subset F \cap U^{-1}(x)$. ∎

3.47 Let $(X,\mathcal{U})$ be a Tychonoff quasi-uniform space. A *compactification* of $(X,\mathcal{U})$ is a compact Hausdorff quasi-uniform space $(Y,\mathcal{V})$ that has a dense subspace quasi-unimorphic to $(X,\mathcal{U})$.

LEMMA. Let $(X,\mathcal{V})$ be a totally bounded Tychonoff quasi-uniform space, let cX be a Hausdorff compactification of X, and let $\mathcal{U}$ be the unique uniformity that is compatible with the topology of cX. Then $\mathcal{V}$ has an extension to a quasi-uniformity on cX that is compatible with $\mathcal{T}(\mathcal{U})$ if and only if $\mathcal{U}|X \times X \subset \mathcal{V}$.

Proof. Suppose that $\mathcal{U}|X \times X \subset \mathcal{V}$. Since $\mathcal{V}$ is totally bounded, by Corollary 1.57, $\{U_{(\epsilon,f)}|\epsilon > 0,\ f \in QB(\mathcal{V})\}$ is a subbase for $\mathcal{V}$. For each $f \in QB(\mathcal{V})$ we define a bounded lower semi-continuous extension f' of f to cX as follows: $f'(x) = \min\{\sup_{z\in X} f(z),\ \sup\{\inf_{z\in G} f(z)|G \in n_x\}\}$. Let $\mathcal{V}'$ be the quasi-uniformity on cX for which $\{U_{(\epsilon,f')}|\epsilon > 0,\ f \in QB(\mathcal{V})\}$ is a subbase. Then $\mathcal{V}'|X \times X = \mathcal{V}$ and $\mathcal{T}(\mathcal{V}') \subset \mathcal{T}(\mathcal{U})$ so that $\mathcal{U} \vee \mathcal{V}'$ is a quasi-uniformity on cX that is compatible with $\mathcal{T}(\mathcal{U})$. Moreover since $\mathcal{U}|X \times X \subset \mathcal{V}$ we have that $(\mathcal{U} \vee \mathcal{V}')|X \times X = \mathcal{V}$.

Now suppose that $\mathcal{W}$ is an extension of $\mathcal{V}$ to cX such that $\mathcal{T}(\mathcal{W}) = \mathcal{T}(\mathcal{U})$. By Proposition 1.45, $\mathcal{U} \subset \mathcal{W}$. Thus $\mathcal{U}|X \times X \subset \mathcal{W}|X \times X = \mathcal{V}$. ■

THEOREM. A totally bounded Tychonoff quasi-uniform space $(X,\mathcal{V})$ has a compactification if and only if $\mathcal{V}$ contains a uniformity compatible with $\mathcal{T}(\mathcal{V})$.

3.48 PROPOSITION. Let $(X,\mathcal{T})$ be a locally compact Hausdorff space and let $\mathcal{U}$ be a quasi-uniformity compatible with $\mathcal{T}$. Then $(X,\mathcal{U})$ has a compactification if and only if $(X,\mathcal{U})$ is locally symmetric.

Proof. If $(X,\mathcal{U})$ has a compactification, $(X,\mathcal{U})$ is locally symmetric by Proposition 2.27.

Now suppose that $(X,\mathcal{U})$ is locally symmetric. Let $X \cup \{\infty\}$ be the one-point compactification of X. For each $U \in \mathcal{U}$ and each compact subspace K of X define $W(U,K) = U \cup (\{\infty\} \times U(X - K)) \cup \{(\infty,\infty)\}$. Let $\mathcal{W}$ be the quasi-uniformity on $X \cup \{\infty\}$ for which $\{W(U,K)|U \in \mathcal{U},$ K is a compact subspace of X$\}$ is a base. Evidently $\mathcal{W}|X \times X = \mathcal{U}$ and $\mathcal{T}(\mathcal{W})$ is coarser than the compact Hausdorff topology of $X \cup \{\infty\}$. Therefore, we need only observe, using the local symmetry of $\mathcal{U}$, that $\mathcal{W}$ is a Hausdorff quasi-uniformity. ■

COROLLARY. Let $(X,\mathcal{T})$ be a locally compact Hausdorff space. A quasi-uniformity $\mathcal{W}$ compatible with $\mathcal{T}$ is locally symmetric if and only if $\mathcal{W}$ contains a uniformity compatible with $\mathcal{T}$.

Lemma 3.47 and Proposition 3.48 evidence the over-abundance of compactifications. Furthermore, it follows from a remark in §2.18 that if $(X,\mathcal{T})$ is a Tychonoff space and cX is any Hausdorff compactification of X, then $(cX,\mathcal{P})$ is a compactification of $(X,\mathcal{P})$. Moreover Proposition 3.48 establishes that a complete quasi-uniformity on a noncompact space can have a compactification; hence in the theory of quasi-uniformities the terminology "precompactness" is misleading. Obviously many questions concerning completions and compactifications remain unresolved. The following two questions are representative.

PROBLEM D. Theorem 3.43 characterizes those quasi-uniform spaces that have a completion. How can we characterize those quasi-uniform spaces that have a Hausdorff (completely regular) completion? Theorem 3.47 characterizes those *totally bounded* quasi-uniform spaces that have a compactification. Under what conditions does an arbitrary quasi-uniform space have a compactification?

Chapter 4

TOPOLOGICAL ORDERED SPACES

1. TOPOLOGICAL ORDERED SPACES

4.1 In this chapter we examine the symbiotic relationship existing between topologies and partial orders. We establish theories of completeness, compactness, and normality that include the usual uniform and topological theories in the special case that the partial order under consideration is equality. Although quasi-uniformities do not make their appearance until the third section of this chapter, their role is central to the study of the interdependence between topologies and orders. Indeed it is the theory of quasi-uniformities that enables us to consider ordered completions and compactifications of ordered spaces and to develop for generalized ordered spaces a theory of uniformities that reflects both the topological and order structures of these spaces.

A *topological ordered space* is a triple $(X,\mathcal{T},\leq)$ where X is a set, $\mathcal{T}$ is a topology on X, $\leq$ is a partial order on X, and $G(\leq) = \{(x,y)|x \leq y\}$ is a closed subset of $X \times X$. Some authors do not require that a topological ordered space have a closed order (see the notes). Since we make this requirement, it follows that the topology of a topological ordered space is a Hausdorff topology. When the topology and order are apparent or irrelevant, we write X in lieu of $(X,\mathcal{T},\leq)$. In particular, the set of real numbers equipped with the usual topology and order is denoted by $\mathbb{R}$.

If S is a subset of a topological ordered space $(X,\mathcal{T},\leq)$ and $\leq_S$ denotes the partial order induced on S, then $(S,\mathcal{T}|S,\leq_S)$ is a topological ordered space, called the *topological ordered subspace*. Let $\{(X_\alpha,\mathcal{T}_\alpha,\leq_\alpha)|\alpha \in A\}$ be a family of topological ordered spaces, let X denote ΠX_α, let T denote the product topology on X, and let $\le$ be (the product order) defined by $x \le y$ if and only if for each $\alpha \in A$, $x_\alpha \le y_\alpha$. Then $(X,T,\le)$ is a topological ordered space called the *product topological ordered space*.

Let $(X,T,\le)$ and $(X',T',\le')$ be topological ordered spaces. A function $f : X \to X'$ is said to be *increasing* (*decreasing*) provided $f(x) \le' f(y)$ whenever $x \le y (y \le x)$. A one-to-one continuous increasing function f mapping X onto X' is an *order homeomorphism* provided f^{-1} is also continuous and increasing. If f is an order homeomorphism of X onto a subspace of X', then f is said to be an *order embedding* of X into X'. We let $C(X,\uparrow) = \{f | f : X \to \mathbb{R}$, f is continuous and increasing$\}$ and $C^*(X,\uparrow) = \{f | f \in C(X,\uparrow)$ and f is bounded$\}$. It is easy to see that $C(X,\uparrow)$ is closed under the operations meet and join.

Let $(X,T,\le)$ be a topological ordered space. A subset Y of X is said to be *increasing* (*decreasing*) provided $y \in Y$ whenever $x \le y (y \le x)$ for some $x \in Y$. A subset Y of X is increasing if and only if $X - Y$ is decreasing. For each subset Y of X there is a smallest increasing set $i(Y)$ (decreasing set $d(Y)$) that contains Y.

A topological ordered space X is said to be a *normally ordered space* if, whenever A and B are disjoint closed subsets of X such that A is decreasing and B is increasing, there exist disjoint open sets G and H such that G is decreasing and contains A and H is increasing and contains B. We note that if the order of X is equality, then every subset of X is increasing and decreasing. Thus $(X,T,=)$ is normally ordered if and only if (X,T) is a normal T_1 space.

Let $(X,T,\le)$ be a topological ordered space. The family of all increasing (decreasing) open sets is a subtopology of T. For each subset A of X, there exists a smallest closed increasing (decreasing) subset $cl_i(A)(cl_d(A))$ that contains A. Moreover, cl_i and cl_d are Kuratowski closure operators on X.

LEMMA. A topological ordered space X is normally ordered if and only if it satisfies the following condition: If E is a closed decreasing subset of X and F is an open decreasing subset of X and $E \subset F$, then there exists a decreasing open subset G of X such that $E \subset G$ and $cl_d(G) \subset F$.

Proof. Suppose that the condition is satisfied. Let A be a closed decreasing subset of X that is disjoint from some closed increasing set B.

Then $X - B$ is an open decreasing set containing A so that, by the condition, there exists a decreasing open subset G of X such that $A \subset G$ and $cl_d(G) \subset X - B$. Since $X - cl_d(G)$ is increasing and contains B, we have shown that X is normally ordered.

Now suppose that X is normally ordered and let E be a closed and decreasing subset of an open and decreasing set F. Then $X - F$ is closed and increasing; thus there exists an open decreasing set G containing E and an open increasing set H containing $X - F$ such that $G \cap H = \emptyset$. Since $X - H$ is closed and decreasing, $E \subset G \subset cl_d(G) \subset X - H \subset F$ as was to be shown. ∎

4.2 We are now in a position to prove a generalization of Urysohn's Lemma. Our proof follows J. L. Kelley's presentation except that Lemma 4.1 is employed to permit the careful choice of certain open sets.

LEMMA. Let $(X,\mathcal{T},\leq)$ be a topological ordered space and let D be a dense subset of the positive real numbers. For each $t \in D$, let $F(t)$ be an open subset of X such that $\cup\{F(t) | t \in D\} = X$ and $cl_d(F(s)) \subset F(t)$ whenever $s < t$. Then the function f defined by $f(x) = \inf\{t | x \in F(t)\}$ is continuous and increasing.

Proof. The proof that f is continuous is well known. To show that f is increasing let $x \leq y$ and suppose that $f(y) < f(x)$. Then there exists $s, t \in D$ such that $f(y) < s < t < f(x)$ and $y \in F(s)$. Thus $x \in cl_d(F(s)) \subset F(t)$ - a contradiction. ∎

THEOREM. (Urysohn's Lemma) Let $(X,\mathcal{T},\leq)$ be a topological ordered space. Then X is normally ordered if and only if for any two disjoint closed sets A and B, where A is decreasing and B is increasing, there exists $f \in C^*(X,\uparrow)$ such that $f(A) = 0$ and $f(B) = 1$.

Proof. Suppose that X is normally ordered and let A and B be disjoint closed sets, where A is decreasing and B is increasing. Let $D = \{k \cdot 2^{-n} | k, n \in \mathbb{N}\}$. For $t \in D$ and $t > 1$ let $F(t) = X$, let $F(1) = X - B$, and let $F(0)$ be a decreasing open set containing A such that $cl_d(F(0)) \cap B = \emptyset$. For $t \in D$ and $0 < t < 1$ write t in the form $t = (2m + 1)\cdot 2^{-n}$ and use Lemma 4.1 to choose, inductively on n, $F(t)$ to be a decreasing open set containing $cl_d(F(2m \cdot 2^{-n}))$ such that $cl_d(F(t)) \subset F((2m + 2) \cdot 2^{-n})$. Define $f(x) = \inf\{t | x \in F(t)\}$. By the previous lemma, $f \in C^*(X,\uparrow)$. For each $t \in D$,

$A \subset F(t)$ so that $f(A) = 0$ and $f(B) = 1$ since $F(t) \subset X - B$ for $t \leq 1$ and $F(t) = X$ for $t > 1$.

Now suppose that X satisfies the condition and let A and B be disjoint closed sets where A is decreasing and B is increasing. By hypothesis there exists $f \in C^*(X,\uparrow)$ such that $f(A) = 0$ and $f(B) = 1$. Let $G = \{x \in X | f(x) < 1/2\}$ and $H = \{x \in X | f(x) > 1/2\}$. Then $G \cap H = \emptyset$, G is decreasing and contains A, and H is increasing and contains B. Therefore X is normally ordered. ∎

4.3 PROPOSITION. Let $(X,T,\leq)$ be a topological ordered space and let K be a compact subset of X. Then d(K) and i(K) are closed so that $d(K) = cl_d(K)$ and $i(K) = cl_i(K)$.

Proof. To show that d(K) is closed, let $p \in X - d(K)$. Since $G(\leq)$ is closed, for each $x \in K$ there exists an increasing neighborhood R_x of p and a decreasing neighborhood S_x of x such that $R_x \cap S_x = \emptyset$. Because K is compact, there exist $x_1, x_2, \ldots, x_n \in K$ such that $K \subset \cup_{i=1}^n S_{x_i}$. Let $R = \cap_{i=1}^n R_{x_i}$. Then X - R is decreasing and contains K so that $R \cap d(K) = \emptyset$. A similar proof establishes that i(K) is closed. ∎

4.4 PROPOSITION. Let $(X,T,\leq)$ be a compact topological ordered space, let F be a decreasing (increasing) subset of X, and let G be an open set containing F. Then there exists a decreasing (increasing) open set H such that $F \subset H \subset G$.

Proof. We establish this result in the case that F is decreasing; the proof in the case that F is increasing is similar. Set $H = X - i(X - G)$. By the preceding proposition H is open, and since $i(X - G)$ is increasing, H is decreasing. Clearly $H \subset G$. Suppose that F is not a subset of H. Then there exists $x \in F \cap i(X - G)$, so that for some $y \in X - G$, $y \leq x$. Since F is decreasing, $y \in F$ - a contradiction. ∎

4.5 THEOREM. Every compact topological ordered space is a normally ordered space.

Proof. We use the characterization of normally ordered spaces established in Lemma 4.1. Let $(X,T,\leq)$ be a compact topological ordered space. Let E be closed and decreasing, let F be open and decreasing and let $E \subset F$.

Since $(X,\mathcal{T})$ is a normal space, there exists an open set H such that $E \subset H \subset \bar{H} \subset F$. By the previous proposition there exists a decreasing open set G such that $E \subset G \subset H$. By Proposition 4.3, $d(\bar{G}) = cl_d(\bar{G})$. Thus $E \subset G \subset cl_d(G) \subset cl_d(\bar{G}) = d(\bar{G}) \subset d(\bar{H}) \subset F$. ∎

COROLLARY. Let $(X,\mathcal{T},\leq)$ be a compact topological ordered space and $\mathcal{S} = \{G \mid G \in \mathcal{T}$ and G is increasing or decreasing$\}$. Then $\mathcal{S}$ is a subbase for $\mathcal{T}$.

Proof. It is sufficient to show that $\mathcal{S}$ is a subbase for a Hausdorff topology on X. Let x and y be two points of X and without loss of generality assume that $y \leq x$ is false. Then $d(\{x\})$ and $i(\{y\})$ are disjoint; by Proposition 4.3, $d(\{x\})$ and $i(\{y\})$ are closed. Since $(X,\mathcal{T},\leq)$ is normally ordered, there exists a decreasing open set G containing $d(\{x\})$ and an increasing open set H containing $i(\{y\})$ such that $G \cap H = \emptyset$. ∎

2. UNIFORM ORDERED SPACES

4.6 The behavior of $\mathbb{R}$ is exemplary. It is natural to expect therefore that if X is a topological ordered space for which there exists a set A so that X is a topological ordered subspace of $\Pi\{\mathbb{R}_\alpha \mid \alpha \in A\}$, then X will inherit some measure of the well-behavior of $\mathbb{R}$ and its closed subintervals. Indeed, well-known results of general topology dictate that we should expect the class of all such spaces to consist of those topological ordered spaces that have order compactifications. This section meets this expectation, but we defer until Section 3 the study of quasi-uniformities and their use in meeting the equally natural expectation that the topological ordered spaces considered in this section are also the spaces for which there exists a theory of uniformities encompassing simultaneously both order and topology.

If A is a nonempty set and for each $\alpha \in A$, X_α is a closed and bounded interval of $\mathbb{R}$, then $\Pi\{X_\alpha \mid \alpha \in A\}$ with the product topology is called a *cube*. An *ordered cube* is a cube equipped with the product order.

Let $(X,\mathcal{T},\leq)$ be a topological ordered space and let F be a subset of $C(X,\uparrow)$. Then F is said to *separate* X provided

(a) $\mathcal{T}$ is the weak topology of F

and

(b) for $x, y \in X$, $x \leq y$ whenever $f(x) \leq f(y)$ for all $f \in F$.

EMBEDDING LEMMA. Let $(X,T,\le)$ be a topological ordered space and let F be a subcollection of $C(X,\uparrow)$. Then the evaluation map e that takes X into the product topological ordered space $\mathbb{R}^F$ is an order embedding if and only if F separates X.

Proof. Suppose that e is an order embedding. Then e is a topological embedding so that F satisfies (a). To show that F satisfies condition (b) let $x, y \in X$ such that $f(x) \le f(y)$ for all $f \in F$. Then $e(x) \le e(y)$ so that $x \le y$. Now suppose that F separates X and let $x, y \in X$. If $x \le y$, then $f(x) \le f(y)$ for all $f \in F$ so that $e(x) \le e(y)$. Conversely if $e(x) \le e(y)$, then $f(x) \le f(y)$ for all $f \in F$. Thus $x \le y$. ∎

4.7 The Embedding Lemma motivates the following definition. A topological ordered space $(X,T,\le)$ is a *uniform ordered space* provided $C^*(X,\uparrow)$ (equivalently $C(X,\uparrow)$) separates X. It is easy to show that a subspace of a uniform ordered space is a uniform ordered space. A wide and important class of uniform ordered spaces is identified by the following proposition.

PROPOSITION. Let $(X,T,\le)$ be a normally ordered space such that $\{G|G \in T$ and G is increasing or decreasing$\}$ is a subbase for T. Then X is a uniform ordered space.

Proof. Let $x \in X$ and let G be an open neighborhood of x. By hypothesis there exists an increasing open set A and a decreasing open set B such that $x \in A \cap B \subset G$. Then $X - A$ is closed and decreasing, and by Proposition 4.3, $i(x)$ is closed and increasing. Since X is normally ordered, there exists $f \in C^*(X,\uparrow)$ such that $f : X \to [0,1]$, $f(X - A) = 0$, and $f(i(x)) = 1$. Similarly there exists $g \in C^*(X,\uparrow)$ such that $g : X \to [0,1]$, $g(d(x)) = 0$ and $g(X - B) = 1$. Hence $x \in f^{-1}((1/2,1]) \cap g^{-1}([0,1/2)) \subset G$ so that T is the weak topology of $C^*(X,\uparrow)$. Finally, suppose that $x \le y$ is false; then $i(x) \cap d(y) = \emptyset$. As X is normally ordered, there exists $f \in C^*(X,\uparrow)$ such that $f : X \to [0,1]$, $f(d(y)) = 0$ and $f(i(x)) = 1$. Thus $f(x) = 1 > 0 = f(y)$ as is required. ∎

COROLLARY. Every subspace of a compact ordered space is a uniform ordered space.

4.8 EMBEDDING THEOREM. Let $(X,T,\le)$ be a topological ordered space. Then X is a uniform ordered space if and only if X is order embedded in an ordered cube.

Proof. Every ordered cube is a compact ordered space; hence each subspace of an ordered cube is a uniform ordered space by Corollary 4.7. Conversely, if X is a uniform ordered space, then by the Embedding Lemma, X is order embedded in $\mathbb{R}^{C^*(X,\uparrow)}$. ■

COROLLARY. The product topological ordered space of a family of uniform ordered spaces is a uniform ordered space.

4.9 Since a normal Hausdorff space is completely regular, it is reasonable to conjecture that every normally ordered space is a uniform ordered space. The following simple result is all that is required to produce a counterexample to this conjecture.

PROPOSITION. Let $(X,\mathcal{T},\le)$ be a topological ordered space such that $\le$ is a linear order. Then X is normally ordered.

Proof. Let A and B be disjoint closed subsets of X such that A is decreasing and B is increasing. If $A \cup B = X$, then both A and B are open. If there is an $x \in X - A \cup B$, then $A \subset (\leftarrow,x)$ and $B \subset (x,\rightarrow)$. ■

EXAMPLE. A normally ordered space that is not a uniform ordered space. Let X = [0,1] with the usual order and let all points of the form 1/n for $n \in \mathbb{N}$ be isolated. A neighborhood of 0 is any set containing 0 and all but finitely many members of $\{1/n \mid n \in \mathbb{N}\}$. All other points have their usual neighborhoods. It follows from the previous proposition that X is a normally ordered space and from Theorem 4.31 that X is not a uniform ordered space.

4.10 An *order compactification* of a topological ordered space X is a pair (f,Y) where Y is a compact ordered space and f is an order embedding of X as a dense subspace of Y. We sometimes identify X with $f(X) \subset Y$ and say simply that Y is an order compactification of X.

Let (f,Y) and (g,Z) be order compactifications of a uniform ordered space X. We say that (f,Y) *dominates* (g,Z) provided there exists a continuous increasing map $h : Y \to Z$ such that $h \circ f = g$. Two order compactifications are *equivalent* provided h can be taken to be an order homeomorphism. It is easy to show that (f,Y) and (g,Z) are equivalent if and only if (f,Y) dominates (g,Z) and (g,Z) dominates (f,Y). When we consider order compactifications it is convenient to rephrase the Embedding Lemma as follows.

PROPOSITION. Let $(X,\mathcal{T},\leq)$ be a uniform ordered space and let F be a subfamily of $C^*(X,\uparrow)$ that separates X. Let $e : X \to \mathbb{R}^F$ denote the evaluation map and let Y denote the closure of e(X) in $\mathbb{R}^F$. Then (e,Y) is an order compactification of X.

COROLLARY. Let X be a topological ordered space. Then X is a uniform ordered space if and only if X has an order compactification.

4.11 The following proposition is an immediate consequence of Corollaries 4.10 and 4.5.

PROPOSITION. If $(X,\mathcal{T},\leq)$ is a uniform ordered space, then $\{G|G \in \mathcal{T}$ and G is increasing or decreasing$\}$ is a subbase for $\mathcal{T}$.

4.12 Let $(X,\mathcal{T},\leq)$ be a uniform ordered space. The *Nachbin compactification* of X is the pair (e,nX) where $e : X \to \mathbb{R}^{C^*(X,\uparrow)}$ is the evaluation map and nX is the closure of e(X) in $\mathbb{R}^{C^*(X,\uparrow)}$. If $\leq$ is the discrete order, equality, then the Nachbin compactification is the well-known Stone-Čech compactification. Nonetheless the Nachbin compactification of X need not be the Stone-Čech compactification; for example if X = (0,1) with the usual topology and order, then nX = [0,1].

The most important characteristic of the Nachbin compactification is an extension property given by the following theorem. Our proof of this theorem follows the standard proof for the case of the discrete order.

THEOREM. Let X be a uniform ordered space, let Y be a compact ordered space, and let $f : X \to Y$ be a continuous increasing function. Then there is a continuous increasing function $F : nX \to Y$ such that $F \circ e = f$.

Proof. Since X and Y are uniform ordered spaces, the evaluation maps $e : X \to \mathbb{R}^{C^*(X,\uparrow)}$ and $e' : Y \to \mathbb{R}^{C^*(Y,\uparrow)}$ are order embeddings. Define $\phi : \mathbb{R}^{C^*(X,\uparrow)} \to \mathbb{R}^{C^*(Y,\uparrow)}$ as follows: for each $s \in \mathbb{R}^{C^*(X,\uparrow)}$, $(\phi(s))_g = s_{g\circ f}$. For each projection map π_g, $(\pi_g \circ \phi)(s) = \pi_{g\circ f}(s)$ so that ϕ is increasing and continuous. As $\phi(e(x)) = e'(f(x))$, ϕ takes e(X) into e'(Y).

Since $\phi(e(X))$ is dense in $\phi(nX)$ and since $e'(Y)$ is closed and contains $\phi(e(X))$, $\phi(nX) \subset e'(Y)$. Define $F : nX \to Y$ by $F = (e')^{-1} \circ (\phi|nX)$. Then $F : nX \to Y$ is continuous increasing function and $F \circ e = f$. ■

4.13 If we identify X with $e(X) \subset nX$, Theorem 4.12 becomes: Every continuous increasing function from X to a compact ordered space Y can be extended to nX. The following is an important consequence of Theorem 4.12.

COROLLARY. Let X be a uniform ordered space. Then the Nachbin compactification dominates every order compactification of X. Furthermore, any order compactification of X that satisfies the extension property given in Theorem 4.12 is equivalent to the Nachbin compactification.

3. QUASI-PROXIMITIES AND ORDER COMPACTIFICATIONS

4.14 Nachbin's discovery of the qp-invariant defined below was a primary motivation for him to initiate the study of quasi-uniform spaces, and his use of the qp-invariant remains a compelling argument that uniformities and quasi-uniformities should be studied together.

Let $(X,\mathcal{V})$ be a T_0 space. Then by Proposition 1.9(b), $\cap\mathcal{V}$ is a partial order, which by Corollary 1.19 is closed in $\mathcal{T}(\mathcal{V}^* \times \mathcal{V}^*)$. Consequently $(X,\mathcal{T}(\mathcal{V}^*), \cap\mathcal{V})$ is a topological ordered space. Let $(X,\mathcal{T},\leq)$ be a topological ordered space. We say that a quasi-uniformity $\mathcal{V}$ on X *determines* $(X,\mathcal{T},\leq)$ provided $\mathcal{T} = \mathcal{T}(\mathcal{V}^*)$ and $\cap\mathcal{V} = G(\leq)$. Similarly we say that a quasi-proximity δ on X *determines* $(X,\mathcal{T},\leq)$ provided $\mathcal{T} = \mathcal{T}(\delta^*)$ and $x \leq y$ if and only if $\{x\}\ \delta\ \{y\}$. It is easily seen that if $\mathcal{V}$ determines X then every member of $\mathcal{T}(\mathcal{V})$ is an increasing set and every member of $\mathcal{T}(\mathcal{V}^{-1})$ is a decreasing set.

PROPOSITION. Let $(X,\mathcal{T},\leq)$ be a topological ordered space and let $\mathcal{V}$ be a quasi-uniformity on X. Then the following are equivalent.

(a) $\mathcal{V}$ determines $(X,\mathcal{T},\leq)$.

(b) $\mathcal{V}_\omega$ determines $(X,\mathcal{T},\leq)$.

(c) $\delta_{\mathcal{V}}$ determines $(X,\mathcal{T},\leq)$.

Proof. (a) ⇔ (b) Since $(\mathcal{V}^{-1})_\omega = (\mathcal{V}_\omega)^{-1}$, we have that $\mathcal{T}((\mathcal{V}_\omega)^*) = \mathcal{T}(\mathcal{V}_\omega \vee (\mathcal{V}_\omega)^{-1}) = \mathcal{T}(\mathcal{V}_\omega) \vee \mathcal{T}((\mathcal{V}_\omega)^{-1}) = \mathcal{T}(\mathcal{V}_\omega) \vee \mathcal{T}((\mathcal{V}^{-1})_\omega) = \mathcal{T}(\mathcal{V}) \vee \mathcal{T}(\mathcal{V}^{-1}) = \mathcal{T}(\mathcal{V}^*)$. We now show that $\cap\mathcal{V}_\omega = \cap\mathcal{V}$. Since $\mathcal{V}_\omega \subset \mathcal{V}$, $\cap\mathcal{V} \subset \cap\mathcal{V}_\omega$. Suppose that

$(x,y) \notin \cap \mathcal{V}$. Then there is a $V \in \mathcal{V}$ such that $V \subset X \times X - \{(x,y)\}$. Thus $X \times X - \{(x,y)\} \in \mathcal{V}_\omega$ and $(x,y) \notin \cap \mathcal{V}_\omega$.

(b) $\Leftrightarrow$ (c) Since $\mathcal{T}(\mathcal{V}_\omega) = \mathcal{T}(\delta_\mathcal{V})$ and $\mathcal{T}((\mathcal{V}_\omega)^{-1}) = \mathcal{T}((\delta_\mathcal{V})^{-1})$, $\mathcal{T}((\mathcal{V}_\omega)^*) = \mathcal{T}((\delta_\mathcal{V})^*)$. Finally, $\cap \mathcal{V}_\omega = \{(x,y) | \{x\} \ \delta_\mathcal{V} \ \{y\}\}$. ■

In the previous proposition we established that $\mathcal{T}((\mathcal{V}_\omega)^*) = \mathcal{T}(\mathcal{V}^*)$ and since $(\mathcal{V}^*)_\omega$ is compatible with $\mathcal{T}(\mathcal{V}^*)$, it follows that $\mathcal{T}((\mathcal{V}_\omega)^*) = \mathcal{T}((\mathcal{V}^*)_\omega)$ and $\mathcal{T}((\delta_\mathcal{V})^*) = \mathcal{T}(\delta_{\mathcal{V}^*})$. We recall however that Example 1.59 shows that in general $(\mathcal{V}_\omega)^* \neq (\mathcal{V}^*)_\omega$ and $(\delta_\mathcal{V})^* \neq \delta_{\mathcal{V}^*}$.

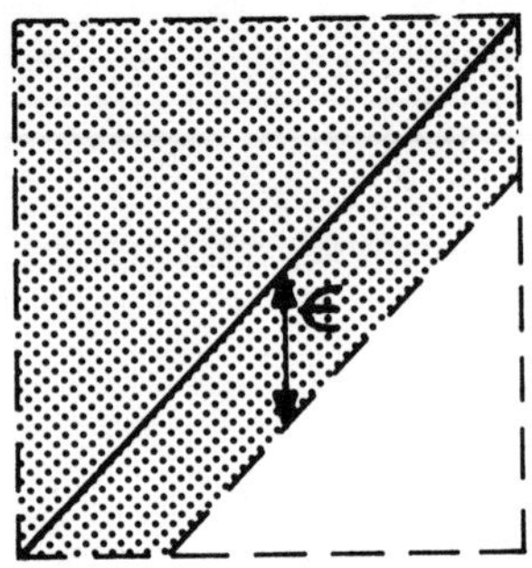

Figure 1. $Q_\varepsilon \cap (0,1) \times (0,1)$

4.15 EXAMPLES. It should be evident from Figure 1 that if $\leq$ denotes the usual order on $\mathbb{R}$, then $G(\leq) = \cap\{Q_\varepsilon | \varepsilon > 0\}$; and, as it has already been noted in §1.3 that $\mathcal{Q}^* = \mathcal{E}$, it follows that $\mathcal{Q}$ determines $\mathbb{R}$. It follows from the previous proposition that $\delta_\mathcal{Q}$ also determines $\mathbb{R}$. Note that $\mathcal{E}$ determines $\mathbb{R}$ with the usual topology and the discrete order and that $\mathcal{Z}$ determines $\mathbb{R}$ with the discrete topology and the discrete order. Let $\mathcal{P}$ denote the Pervin quasi-uniformity for the upper topology on $\mathbb{R}$. As noted in Example 2.2, $\mathcal{T}(\mathcal{P}^*)$ is the Sorgenfrey topology. Since $\cap \mathcal{P}^{-1}$ is the usual order on $\mathbb{R}$, $\mathcal{P}^{-1}$ determines the Sorgenfrey Line with its usual order.

4.16 PROPOSITION. Let δ and ρ be quasi-proximities that determine topological ordered spaces X and Y respectively and let $f : X \to Y$ be a qp-continuous mapping. Then f is a $\mathcal{T}(\delta^*) - \mathcal{T}(\rho^*)$ continuous increasing function.

Proof. If $x_1 \leq x_2$, then $\{x_1\} \ \delta \ \{x_2\}$. Since f is qp-continuous, $\{f(x_1)\} \ \rho \ \{f(x_2)\}$ so that $f(x_1) \leq f(x_2)$. The function f is $\mathcal{T}(\delta^*) - \mathcal{T}(\rho^*)$ continuous by Proposition 1.50. ■

COROLLARY. Let X be a topological ordered space that is determined by a quasi-proximity δ and by a quasi-uniformity $\mathcal{V}$. Then $Q(\delta) \subset C(X,\uparrow)$ and $Q(\mathcal{V}) \subset C(X,\uparrow)$.

Proof. The first result follows from the previous proposition. Since $Q(\mathcal{V}) \subset Q(\delta_{\mathcal{V}})$, the second result follows from the first result and Proposition 4.14. ∎

4.17 For each T_0 quasi-uniform space $(X,\mathcal{V})$ let $\lesssim$ denote the partial order $\cap\tilde{\mathcal{V}}$, defined on $\tilde{X}$, the bicompletion of $(X,\mathcal{V})$. For each T_0 quasi-proximity space (X,δ) let $\lesssim$ denote the relation on $\tilde{X}$ defined by $x \lesssim y$ if and only if $\{x\}\ \tilde{\delta}\ \{y\}$. If $\mathcal{V}$ is totally bounded and induces δ, then $\tilde{\mathcal{V}}$ induces $\tilde{\delta}$; thus the dual use of $\lesssim$ is justified.

If a quasi-uniformity $\mathcal{V}$ determines a topological ordered space $(X,\mathcal{T},\leq)$, then $\tilde{\mathcal{V}}$ determines the topological ordered space $(\tilde{X},\mathcal{T}(\tilde{\mathcal{V}}^*),\lesssim)$. Similarly, if a quasi-proximity δ determines a topological ordered space $(X,\mathcal{T},\leq)$, then $\tilde{\delta}$ determines the compact topological ordered space $(\tilde{X},\mathcal{T}(\tilde{\delta}^*),\lesssim)$. If $\mathcal{V}$ (respectively δ) determines a topological ordered space $(X',\mathcal{T}',\leq')$ and $(X,\mathcal{T})$ is a subspace of $(X',\mathcal{T}')$, then $\mathcal{V}|X \times X$ (respectively $\delta|X \times X$) determines $(X,\mathcal{T},\leq'|X \times X)$.

We are now prepared to examine the relationship between the order compactifications of a topological ordered space X and the quasi-proximities that determine X.

THEOREM. Let X be a topological ordered space determined by a quasi-proximity δ. Then $\tilde{\delta}$ determines an order compactification of X.

Proof. Since X is a topological ordered space, δ is a T_0 quasi-proximity. Thus by Theorem 3.38(b), $\tilde{\delta}$ determines a compact ordered space. The canonical embedding i is an order embedding by Theorem 3.38(a) and Proposition 4.16. ∎

4.18 The following is a fundamental result of the theory of uniform ordered spaces; in the particular case that the order under consideration is the discrete order, this result reduces to the classical result of L. Pontrjagin and A. Weil that a Hausdorff space admits a uniformity if and only if it is a Tychonoff space.

THEOREM. Let $(X,\mathcal{T},\leq)$ be a topological ordered space. Then X is a uniform ordered space if and only if there is a quasi-uniformity that determines $(X,\mathcal{T},\leq)$.

Proof. Let $\mathcal{V}$ be a quasi-uniformity that determines X; by Proposition 4.14 we have that $\delta_{\mathcal{V}}$ determines X. In view of Theorem 4.17, $\tilde{\delta}_{\mathcal{V}}$ determines an order compactification of X, and by Corollary 4.10, X is a uniform ordered space.

Now suppose that X is a uniform ordered space. Evidently $\{U_{(\epsilon,f)} \mid \epsilon > 0, f \in C^*(X,\uparrow)\}$ is a subbase for a quasi-uniformity $\mathcal{V}$ that determines X. ∎

4.19 Let $(X,\mathcal{T},\leq)$ be a uniform ordered space and let $\mathcal{U}$ be a uniformity compatible with $\mathcal{T}$. L. Nachbin raised the following question concerning the relationship between $(X,\mathcal{T},\leq)$ and $\mathcal{U}$: What are necessary and sufficient conditions in order that there exist a quasi-uniformity $\mathcal{V}$ determining X such that $\mathcal{V}^* = \mathcal{U}$? The status of this question is typical of many questions concerning uniform ordered spaces; a complete answer is known only in the case that the partial order is a linear order.

Let $\leq$ be a partial order on X. A subset A of X is said to be *convex* (with respect to $\leq$) provided that $z \in A$ whenever $x \leq z \leq y$ and $x, y \in A$. For each subset $A \subset X$ let conv $(A) = \{x \in X \mid \text{there exist } a, b \in A \text{ such that } a \leq x \leq b\}$; then conv (A) is a convex set containing A. We say that a quasi-uniformity $\mathcal{U}$ on X is *convex* (with respect to a partial order on X) provided that for each $U \in \mathcal{U}$ there is a $V \in \mathcal{V}$ such that $V \subset U$ and for each $x \in X$, $V(x)$ is convex.

PROPOSITION. Let $(X,\mathcal{V})$ be a T_0 quasi-uniform space. Then $\mathcal{V}^*$ is convex with respect to $\cap\mathcal{V}$.

Proof. Let $U \in \mathcal{V}^*$ and let $\leq$ denote the partial order $\cap\mathcal{V}$. There exists $V \in \mathcal{V}$ such that $V \cap V^{-1} \subset U$. Choose $V_1 \in \mathcal{V}$ such that $V_1^2 \subset V$ and let $W = V_1 \cap V_1^{-1}$. Let $M = \{(x,y) \mid y \in \text{conv } (W(x))\}$. Then $W \subset M$ so that $M \in \mathcal{V}^*$. To show that $M \subset U$ let $(x,y) \in M$. There exists $a, b \in W(x)$ such that $a \leq y \leq b$. Since $(x,a) \in W \subset V_1$ and $(a,y) \in \cap\mathcal{V} \subset V_1$, $(x,y) \in V$. Also $(b,x) \in W \subset V_1$ and $(y,b) \in \cap\mathcal{V} \subset V_1$ so that $(y,x) \in V$. Therefore $(x,y) \in V \cap V^{-1} \subset U$. ∎

4.20 THEOREM. Let $(X,\mathcal{T},\leq)$ be a uniform ordered space such that $\leq$ is a linear order and let $\mathcal{U}$ be a convex uniformity compatible with $\mathcal{T}$. There is exactly one quasi-uniformity $\mathcal{V}$ such that $\mathcal{V}$ determines X and $\mathcal{V}^* = \mathcal{U}$.

Proof. Let G denote the graph of $\leq$ and let $\mathcal{B}$ denote the filter base $\{G \circ U | U \in \mathcal{U}\}$. We show that $\mathcal{B}$ is a base for a quasi-uniformity $\mathcal{V}$. Let $U \in \mathcal{U}$ and let $W \in \mathcal{U}$ such that $W^2 \subset U$ and for each $x \in X$, $W(x)$ is convex. In order to show that $\mathcal{B}$ is a base for a quasi-uniformity it suffices to show that $G \circ W \circ G \circ W \subset G \circ U$. Let $(x,z) \in G \circ W \circ G \circ W$. There are s, y, and t in X such that $(x,s) \in G$, $(s,y) \in W$, $(y,t) \in G$ and $(t,z) \in W$. If $s \leq t$, then $(x,t) \in G$ and so $(x,z) \in G \circ U$. If $t < s$, then $(s,t) \in W$ so that $(s,z) \in W^2$ and $(x,z) \in G \circ U$.

To show that $\bigcap\mathcal{V} = G$ it suffices to prove that $\bigcap\mathcal{V} \subset G$. Suppose there exists $(x,z) \in \bigcap\mathcal{V} - G$. Let $U \in \mathcal{U}$, let W_1 be a symmetric entourage in $\mathcal{U}$ such that $W_1^2 \subset U$, and let W be an entourage such that $W \subset W_1$ and for each $x \in X$, $W(x)$ is convex. There is a $y \in X$ such that $(x,y) \in G$ and $(y,z) \in W$. Since $z < x \leq y$, $x \in W(y) \subset W_1(y)$. As $z \in W_1(y)$, $z \in U(x)$, thus $(x,z) \in \bigcap\mathcal{U}$ and so $x = z$ - a contradiction.

Because $\mathcal{V} \subset \mathcal{U}$, in order to show that $\mathcal{V}^* = \mathcal{U}$ it suffices to prove that $\mathcal{U} \subset \mathcal{V}^*$. Let $U \in \mathcal{U}$, let W_1 be a symmetric entourage in $\mathcal{U}$ such that $W_1^2 \subset U$, and let $W \in \mathcal{U}$ such that $W \subset W_1$ and for each $x \in X$, $W(x)$ is convex. Then $Q = (G \circ W) \cap (W \circ G^{-1}) \in \mathcal{V}^*$. Let $(x,y) \in Q$. There are s and t in X such that $(x,s) \in G$, $(s,y) \in W$, $(x,t) \in W$ and $(y,t) \in G$. If $x \leq y$, then since $W(x)$ is convex $(x,y) \in W \subset U$. If $y < x$, then since $W(s)$ is convex $(s,x) \in W \subset W_1$. Thus $(x,y) \in W_1^2 \subset U$.

Now suppose that $\mathcal{W}$ is a quasi-uniformity such that $\bigcap\mathcal{W} = G$ and $\mathcal{W}^* = \mathcal{U}$. For each $W \in \mathcal{W}$, $G \circ W \subset W^2$ so that $\mathcal{W} \subset \mathcal{V}$. Suppose that there is a $V \in \mathcal{V} - \mathcal{W}$. Since $V \in \mathcal{V}^* = \mathcal{W}^*$, there is a $W \in \mathcal{W}$ such that $W \cap W^{-1} \subset V$. Moreover there exists $(x,y) \in W - V$. Since $(x,y) \notin \bigcap\mathcal{V} = G$ and $\leq$ is a linear order, $(y,x) \in G = \bigcap\mathcal{W}$ so that in particular $(x,y) \in W^{-1}$. Then $(x,y) \in W \cap W^{-1} \subset V$ - a contradiction. ■

4.21 The previous theorem resolves the question of L. Nachbin, which we considered in §4.19, only for uniform ordered spaces whose partial order is a linear order; the next theorem identifies a class of uniform ordered spaces in which Nachbin's question is trivially answered.

THEOREM. A compact ordered space is determined by exactly one quasi-uniformity.

Proof. Let X be a compact ordered space. By Corollary 4.7, X is a uniform ordered space and so it follows from Theorem 4.18 that X is determined by a quasi-uniformity. By Theorem 1.20 a compact ordered space is determined by at most one quasi-uniformity. ∎

In light of Corollary 4.16, the previous theorem has an important corollary.

COROLLARY. Let X be a compact ordered space and let δ be the unique quasi-proximity that determines X. Then $Q(\delta) = C(X,\uparrow)$.

Proof. Let $\mathcal{V}$ denote the quasi-uniformity generated by $\{U_{(\epsilon,f)} | \epsilon > 0, f \in C(X,\uparrow)\}$. By the proof of Theorem 4.18, $\mathcal{V}$ determines X, and so $\delta = \delta_{\mathcal{V}}$. It is obvious that $C(X,\uparrow) \subset Q(\mathcal{V})$; hence by Corollary 4.16, $C(X,\uparrow) = Q(\mathcal{V})$. By Proposition 1.51, $Q(\mathcal{V}) \subset Q(\delta)$ and it follows from Corollary 4.16 that $C(X,\uparrow) = Q(\delta)$. ∎

PROBLEM E. Determine necessary and sufficient conditions so that an ordered space be determined by a unique quasi-proximity. By Theorem 4.24 an equivalent problem is to determine necessary and sufficient conditions so that an ordered space have a unique order compactification. The real line has a unique order compactification and, as noted in §4.30, so does the Sorgenfrey Line.

4.22 The following proposition characterizes the unique quasi-proximity guaranteed by Theorem 4.21.

PROPOSITION. Let X be a compact ordered space. Then the relation δ in $\mathcal{P}(X)$ defined by $A \,\delta\, B$ if and only if $cl_i(A) \cap cl_d(B) \neq \emptyset$ is the unique quasi-proximity that determines X. Furthermore $cl_i = cl_{\delta^{-1}}$ and $cl_d = cl_\delta$.

Proof. Let $\mathcal{V}$ be the quasi-uniformity for which $\{U_{(\epsilon,f)} | \epsilon > 0, f \in C^*(X,\uparrow)\}$ is a subbase. We show that $\delta = \delta_{\mathcal{V}}$. Let A and B be subsets of X such that $cl_i(A) \cap cl_d(B) = \emptyset$. By Theorem 4.2, there exists $f \in C^*(X,\uparrow)$ such that $f(cl_d(B)) = 0$ and $f(cl_i(A)) = 1$. Thus $A \not\delta_{\mathcal{V}} B$. Now suppose that A and B are subsets of X such that $A \not\delta_{\mathcal{V}} B$. Let $\bar{A}$ and $\bar{B}$ denote the $\mathcal{T}(\mathcal{V}^*)$-closures of A and B. By Proposition 1.41, $\bar{A} \not\delta_{\mathcal{V}} \bar{B}$ so that there exists an entourage $V \in \mathcal{V}$ such that $V(\bar{A}) \cap V^{-1}(\bar{B}) = \emptyset$. By Proposition 4.3, $cl_i(\bar{A}) = i(\bar{A})$ so that $cl_i(A) \subset V(\bar{A})$. Similarly $cl_d(B) \subset V^{-1}(\bar{B})$. Thus $cl_i(A) \cap cl_d(B) = \emptyset$.

Since the arguments needed to obtain that $cl_i = cl_{\delta^{-1}}$ and that $cl_d = cl_\delta$ are similar, we conclude by establishing that $cl_i = cl_{\delta^{-1}}$. Let $x \in X$ and let $A \subset X$. Then $x \in cl_{\delta^{-1}}(A)$ if and only if $cl_d\{x\} \cap cl_i(A) \neq \emptyset$. By Proposition 4.3, $cl_d\{x\} = d(\{x\})$ and $cl_i(\bar{A}) = i(\bar{A})$. Thus $x \in cl_i(A)$ if and only if $x \in cl_{\delta^{-1}}(A)$. ■

4.23 Let $(f_\alpha, \alpha X)$ be an order compactification of a uniform ordered space X and let ρ_α be the quasi-proximity that determines αX. Then the relation δ_α in $\mathcal{P}(X)$ defined by $A \;\delta_\alpha\; B$ if and only if $f_\alpha(A) \;\rho_\alpha\; f_\alpha(B)$ is a quasi-proximity that determines X. This quasi-proximity δ_α is called *the quasi-proximity associated* with $(f_\alpha, \alpha X)$. Clearly $f_\alpha : (X, \delta_\alpha) \to (\alpha X, \rho_\alpha)$ is a qp-embedding.

PROPOSITION. Let X be a uniform ordered space and let δ be a quasi-proximity that determines X. Then δ is the quasi-proximity associated with the order compactification of X determined by $\tilde{\delta}$.

Proof. By Theorem 4.17, $\tilde{\delta}$ determines an order compactification, Y of X. In view of Theorem 4.21, $\tilde{\delta}$ is the only quasi-proximity that determines Y. Thus δ is the quasi-proximity associated with Y. ■

4.24 THEOREM. Let X be a uniform ordered space and let $(f_\alpha, \alpha X)$ and $(f_{\alpha'}, \alpha' X)$ be order compactifications of X. Then $(f_\alpha, \alpha X)$ dominates $(f_{\alpha'}, \alpha' X)$ if and only if $\delta_\alpha \subset \delta_{\alpha'}$. In particular $(f_\alpha, \alpha X)$ and $(f_{\alpha'}, \alpha' X)$ are equivalent if and only if $\delta_\alpha = \delta_{\alpha'}$.

Proof. Suppose that $\delta_\alpha \subset \delta_{\alpha'}$. If $A \;\delta_\alpha\; B$, then $A \;\delta_{\alpha'}\; B$ and $f_{\alpha'}(A) \;\tilde{\delta}_{\alpha'}\; f_{\alpha'}(B)$. Thus $f_{\alpha'} : (X, \delta_\alpha) \to (\alpha' X, \tilde{\delta}_{\alpha'})$ is a qp-continuous function, and since f_α is a qp-embedding, $f_{\alpha'} \circ f_\alpha^{-1} : f_\alpha(X) \to \alpha' X$ is qp-continuous. It follows from Theorem 3.29 and Corollary 1.55 that there is a (unique) qp-continuous function $F : \alpha X \to \alpha' X$ such that $F \circ f_\alpha = f_{\alpha'}$. By Proposition 4.16, F is an increasing function; consequently $(f_\alpha, \alpha X)$ dominates $(f_{\alpha'}, \alpha' X)$.

Now suppose that $(f_\alpha, \alpha X)$ dominates $(f_{\alpha'}, \alpha' X)$. Let $F : \alpha X \to \alpha' X$ be a continuous increasing function such that $F \circ f_\alpha = f_{\alpha'}$, and suppose that $A \;\not\delta_{\alpha'}\; B$. By Proposition 4.22 and Theorem 4.2, there exists a function $g \in C(\alpha' X, \uparrow)$ such that $(g \circ f_{\alpha'})(A) = 1$ and $(g \circ f_{\alpha'})(B) = 0$. Then $g \circ F$ is a member of $C(\alpha X, \uparrow)$ for which $(g \circ F)(f_\alpha(A)) = 1$ and $(g \circ F)(f_\alpha(B)) = 0$. Thus $A \;\not\delta_\alpha\; B$ so that $\delta_\alpha \subset \delta_{\alpha'}$. ■

COROLLARY. Let X be a uniform ordered space, let (f,Y) be an order compactification of X, and let δ be the quasi-proximity associated with (f,Y). Then (f,Y) is equivalent to $(i,\tilde{X})$, where i is the canonical embedding of (X,δ) into $(\tilde{X},\tilde{\delta})$.

4.25 PROPOSITION. Let X be a uniform ordered space and let $K = \{\alpha X | \alpha \in A\}$ be a nonempty family of order compactifications of X. Then there is an order compactification $\alpha_0 X$ of X such that:

(a) $\alpha_0 X$ dominates every member of K

and

(b) $\alpha_0 X$ is dominated by any order compactification of X that dominates every member of K.

Proof. For each $\alpha X \in K$ let δ_α be the quasi-proximity associated with αX and let $\delta_{\alpha_0} = \sup \{\delta_\alpha | \alpha \in A\}$. By Proposition 1.38 and Corollary 1.39, δ_{α_0} determines X. Let $\alpha_0 X$ be the order compactification determined by $\tilde{\delta}_{\alpha_0}$. Since for each $\alpha \in A$, $\delta_{\alpha_0} \subset \delta_\alpha$, condition (a) is satisfied.

Now suppose that $\alpha' X$ is an order compactification of X that dominates every member of K. For each $\alpha \in A$, $\delta_{\alpha'} \subset \delta_\alpha$ so that $\delta_{\alpha'} \subset \delta_{\alpha_0}$. ■

4.26 Because the results of this chapter make possible a short proof of Lemma 1.57 we have postponed the proof of this lemma until the present chapter and have agreed to establish its truth only in the case that (X,δ) is a T_0 quasi-proximity space. As the use of this lemma is now imminent we pause for a moment to establish this fundamental result from Chapter 1.

LEMMA. Let (X,δ) be a T_0 quasi-proximity space and let A and B be subsets of X. If $A \not\delta B$, then there exists $f : X \to [0,1]$ such that $f \in QB(\delta)$, $f(A) = 1$ and $f(B) = 0$.

Proof. Suppose that $A \not\delta B$. By Proposition 3.38(c), $cl_{\tilde{\delta}^{-1}}(i(A)) \cap cl_{\tilde{\delta}}(i(B)) = \emptyset$. Thus, in light of Proposition 4.22, $cl_i(i(A))$ and $cl_d(i(B))$ are disjoint subsets of the compact ordered space $\tilde{X}$ determined by $\tilde{\delta}$. As $\tilde{X}$ is normally ordered it follows from Theorem 4.2 that there is a $g \in C^*(\tilde{X},\uparrow)$ such that $g : \tilde{X} \to [0,1]$, $g(i(B)) = 0$, and $g(i(A)) = 1$. By Corollary 4.21, $g \in QB(\tilde{\delta})$ so that $g \circ i$ is the required mapping. ■

THEOREM. Let X be a uniform ordered space and let δ be the quasi-proximity given by $A \not\delta B$ if and only if there is an $f \in C^*(X,\uparrow)$ such that $f : X \to [0,1]$, $f(B) = 0$ and $f(A) = 1$. Then δ is the quasi-proximity associated with the Nachbin compactification of X and is the finest quasi-proximity that determines X.

Proof. By the previous lemma δ is the finest quasi-proximity that determines X. The theorem follows from Corollary 4.13 and Theorem 4.24. ∎

4.27 An order compactification (f,Y) of a uniform ordered space X is said to be *strict* provided that the partial order of (f,Y) is a subset of any closed partial order on Y for which f is an order embedding. Strict order compactifications are more natural than nonstrict order compactifications. For example, if X is a uniform ordered space whose partial order is equality, then an order compactification Y of X is a strict order compactification if and only if the partial order on Y is also equality. We now describe the quasi-proximities associated with strict order compactifications.

PROPOSITION. Let X be a uniform ordered space, let (f,Y) be an order compactification of X, and let δ be the quasi-proximity associated with (f,Y). Then (f,Y) is a strict order compactification of X if and only if $\rho = \delta$ whenever ρ is a quasi-proximity that determines X such that $\rho \subset \delta$ and $\rho^* = \delta^*$.

Proof. Suppose that (f,Y) is a strict order compactification of X and suppose that ρ is a quasi-proximity that determines X such that $\rho \subset \delta$ and $\rho^* = \delta^*$. By Proposition 4.23 there is an order compactification (g,Z) of X whose associated quasi-proximity is ρ. Since $\rho^* = \delta^*$, we assume without loss of generality that as topological spaces Y = Z and that f(x) = g(x) for each $x \in X$. We show that $\tilde{\delta}$ and $\tilde{\rho}$ define the same partial order. Suppose $\{p\}\ \tilde{\rho}\ \{q\}$. Then since $\rho \subset \delta$, $(p,q) \in \bigcap\{\tilde{U} | U \in \mathcal{U}_\rho\} \subset \bigcap\{\tilde{U} | U \in \mathcal{U}_\delta\}$ and so $\{p\}\ \tilde{\delta}\ \{q\}$. As (f,Y) is a strict order compactification of X, $\{p\}\ \tilde{\rho}\ \{q\}$ whenever $\{p\}\ \tilde{\delta}\ \{q\}$. Since $\tilde{\rho}$ and $\tilde{\delta}$ determine the same partial order, it follows from Theorem 4.21 that $\rho = \delta$.

Now suppose that δ satisfies the condition above and let $\leq'$ be a closed order on Y for which f is an order embedding. By Theorem 4.21 there is a quasi-proximity ρ' that determines $(Y,\mathcal{T}_Y,\leq')$. Set $\rho = \sup\{\delta,\rho'|X\}$. It follows from Corollary 1.39 that ρ is a quasi-proximity that determines

X such that $\rho \subset \delta$. Since $\delta^* = (\rho'|X)^*$, $\delta^* \subset \rho$ and so $\delta^* = \rho^*$. Therefore by hypothesis $\rho = \delta$ so that $\tilde{\delta} \subset (\rho'|X)^{\sim} = \rho'$. Hence $\leq'$ is larger than the order of (f,Y). ■

The first corollary below follows from Theorem 4.26; the second follows from Theorem 4.20.

COROLLARY. Let X be a uniform ordered space. The Nachbin compactification of X is a strict order compactification.

COROLLARY. Let $(X,\mathcal{T},\leq)$ be a uniform ordered space such that $\leq$ is a linear order. Every order compactification of X is a strict order compactification.

4. LINEAR ORDERS

4.28 In this section we consider those uniform ordered spaces whose partial order is a linear order. We show that these spaces are the generalized ordered spaces (GO spaces). It follows from this result and Theorem 4.20 that a space $(X,\mathcal{T})$ is a GO space if and only if there is a linear order $\leq$ on X such that $(X,\mathcal{T})$ admits a uniformity $\mathcal{U}$ that is convex with respect to $\leq$. Thus for GO spaces the appropriate theory of uniform spaces is the theory of convex uniform spaces; and, as might be expected, some insight into both order convex uniform spaces and generalized ordered spaces is provided by the interplay that exists between these two classes of spaces.

A *linearly ordered space* (abbreviated LOTS) is a triple $(X,\lambda,\leq)$ in which $\leq$ is a linear order on X and λ is the usual open-interval topology of the order $\leq$. A *generalized ordered space* (abbreviated GO space) is a triple $(X,\mathcal{T},\leq)$ in which $\leq$ is a linear order on X and $\mathcal{T}$ is a topology on X such that the open-interval topology of $\leq$ is coarser than $\mathcal{T}$ and for each $x \in X$ the $\mathcal{T}$-neighborhood filter of x has a base consisting of (possibly degenerate) intervals of X. Every GO space is a topological ordered space; however Example 4.9 shows that a topological ordered space whose order is a linear order need not be a GO space.

Let $X = (X,\mathcal{T},\leq)$ be a GO space and let λ denote the open-interval topology of $\leq$. Define a subset $X^+ = (X,\mathcal{T},\leq)^+$ of $X \times \{-1,0,1\}$ by $X^+ = (X \times \{0\}) \cup \{(x,-1) \mid [x,\rightarrow) \in \mathcal{T} - \lambda\} \cup \{(x,1) \mid (\leftarrow,x] \in \mathcal{T} - \lambda\}$. The order of X^+ is the lexicographic order and its topology is the open-interval topology of the lexicographic order. Then X^+ is a LOTS in which X is dense and order embedded. Note that if the topology of X is the open-interval topology, then X^+ is order homeomorphic to X. Further, in a sense made precise in the following proposition, whose proof we omit, X^+ is the smallest LOTS in which X is dense and order embedded. (See the notes.)

PROPOSITION. Let X be a GO space and let f be an order embedding of X as a dense subspace of a LOTS Y. Let $\phi : X \to X^+$ be defined by $\phi(x) = (x,0)$. Then there is an order homeomorphism $F : X^+ \to Y$ such that $F \circ \phi = f$.

4.29 Before relating GO spaces to uniform ordered spaces, we recall one more result from the theory of GO spaces.

THEOREM. Let $(X,\mathcal{T})$ be a topological space and let $\leq$ be a linear order on X. Then the following statements are equivalent.

(a) $(X,\mathcal{T},\leq)$ is a GO space.

(b) $(X,\mathcal{T},\leq)$ is topologically and order embedded as a (closed) subspace of a LOTS.

(c) $(X,\mathcal{T},\leq)$ is topologically and order embedded as a dense subspace of a compact LOTS.

4.30 The bicompletion of a quasi-uniform space provides an important tie between GO spaces and those uniform ordered spaces that we are presently considering, and the explicit description of the bicompletion given in Theorem 3.33 enables us to establish the following important result without imposing an order upon the bicompletion *deus ex machina*.

THEOREM. Let $(X,\mathcal{T},\leq)$ be a uniform ordered space in which $\leq$ is a linear order and let $\mathcal{V}$ be a quasi-uniformity that determines X. Then $\lesssim\, = \bigcap\tilde{\mathcal{V}}$ is a linear order on $\tilde{X}$.

Proof. The proof is by contradiction. By Theorem 3.33, $\tilde{X} = \{\mathcal{F} \mid \mathcal{F}$ is a minimal $\mathcal{V}^*$-Cauchy filter on X$\}$ and for each $V \in \mathcal{V}$, $\tilde{V} = \{(\mathcal{F},\mathcal{G}) \in \tilde{X} \times \tilde{X} \mid$ there is an $F \in \mathcal{F}$ and a $G \in \mathcal{G}$ with $F \times G \subset V\}$. Since $\tilde{X}$ is a T_0 space, $\lesssim$ is a partial order. Suppose that $\lesssim$ is not a linear order on $\tilde{X}$. Then there are $\mathcal{F}$, $\mathcal{G} \in \tilde{X}$ and $V \in \mathcal{V}$ such that $(\mathcal{F},\mathcal{G}) \notin \tilde{V}$ and $(\mathcal{G},\mathcal{F}) \notin \tilde{V}$. Let $W \in \mathcal{V}$ such that $W^3 \subset V$. There exists an $F \in \mathcal{F}$ and a $G \in \mathcal{G}$ such that $F \times F \subset W$ and $G \times G \subset W$. Since $(\mathcal{F},\mathcal{G}) \notin \tilde{W}$, there exists $(x,y) \in F \times G - W$ and since

$\leq$ is a linear order on X and $G(\leq) = \cap V$, $(y,x) \in W$. Since $(y,x) \in W$, $G \times F \subset W^{-1}(y) \times W(x) \subset W^3 \subset V$. It follows that $(G,F) \in \tilde{V}$ - a contradiction. ■

EXAMPLE. Let P be the Pervin quasi-uniformity for the upper topology on $\mathbb{R}$ and let $V = P^{-1}$. Recall from Example 4.15 that V determines the Sorgenfrey Line with its usual order. Then $\tilde{V}$ determines the compact LOTS $\{0\} \times (0,1] \cup \{1\} \times [0,1)$ with the lexicographic order. This order compactification is the Nachbin compactification of the Sorgenfrey Line; indeed it is the only order compactification of the Sorgenfrey Line.

4.31 A topological ordered space X is *locally convex* provided that for each $x \in X$ and for each neighborhood G of x there exists a convex neighborhood H of x such that $H \subset G$.

THEOREM. Let $(X,T,\leq)$ be a topological ordered space in which $\leq$ is a linear order. The following statements are equivalent.

(a) $(X,T,\leq)$ is a GO space.

(b) $(X,T,\leq)$ is locally convex.

(c) $(X,T,\leq)$ is a uniform ordered space.

Proof. (a) ⇒ (b) Evident.

(b) ⇒ (c) Since $G(\leq)$ is closed and $\leq$ is a linear order, the open interval-topology is coarser than T. Thus it is an immediate consequence of Propositions 4.9 and 4.7 that (b) ⇒ (c) .

(c) ⇒ (a) Let δ be a quasi-proximity that determines X. By Theorem 4.17, $(\tilde{X},T(\tilde{\delta}^*),\lesssim)$ is an order compactification of $(X,T,\leq)$ and by Theorem 4.30, $\lesssim$ is a linear order on $\tilde{X}$. Since the open interval topology is a Hausdorff topology coarser than $T(\delta^*)$ and $T(\delta^*)$ is compact, it follows that $(\tilde{X},T(\tilde{\delta}^*),\lesssim)$ is a compact LOTS. By Theorem 4.29, $(X,T,\leq)$ is a GO space. ■

4.32 THEOREM. Let $(X,T,\leq)$ be a GO space and let U be a convex uniformity compatible with T. There is exactly one linear order on the completion $(\tilde{X},\tilde{U})$ of (X,U) extending $\leq$ such that $\tilde{U}$ is convex.

Proof. Since U is convex, by Theorem 4.20 there is a quasi-uniformity V on X such that V determines X and $V^* = U$. By Theorems 4.30 and 4.31,

$(\tilde{X},\mathcal{T}(\tilde{\mathcal{U}}),\lesssim)$ is a uniform ordered space and $\lesssim$ is a linear order on $\tilde{X}$. Thus $\lesssim$ is a linear order on $\tilde{X}$ extending $\leq$ and, by Proposition 4.19, $\tilde{\mathcal{U}}$ is convex. Let $\lesssim'$ be another such order. By Theorem 4.20 there is a quasi-uniformity $\mathcal{W}$ on $\tilde{X}$ such that $\mathcal{W}^* = \tilde{\mathcal{U}}$ and $\cap\mathcal{W}$ is $\lesssim'$. It also follows from Theorem 4.20 that $\tilde{\mathcal{W}}|X \times X = \mathcal{V}$. By Theorem 3.34, $\tilde{\mathcal{V}} = \tilde{\mathcal{W}}$ and so $G(\lesssim') = \cap\tilde{\mathcal{W}} = \cap\tilde{\mathcal{V}} = G(\lesssim)$. ∎

4.33 The proof of the following theorem makes use of the concept of almost 2-full normality, which is defined and discussed in §5.21.

THEOREM. The fine uniformity of a GO space is convex.

Proof. Let $(X,\mathcal{T},\leq)$ be a GO space. Since every GO space is almost 2-fully normal, the fine uniformity of X consists of all neighborhoods of the diagonal. Let U be a neighborhood of Δ and for each $x \in X$ let W_x be a convex open set about x such that $W_x \times W_x \subset U$. Then $W = \cup\{W_x \times W_x | x \in X\}$ is a member of the fine uniformity, $W \subset U$ and for each $x \in X$, $W(x)$ is convex. ∎

4.34 COROLLARY. Every metrizable GO space admits a convex uniformity with a countable base.

Proof. Let $(X,\mathcal{T},\leq)$ be a GO space and $\langle U_n\rangle$ be a nested base for a uniformity compatible with $\mathcal{T}$. By the proof of the previous theorem we may define a sequence $\langle V_n\rangle$ of symmetric neighborhoods of Δ such that for each $n \in \mathbb{N}$, $V_n \subset U_n$, $V_{n+1}^2 \subset V_n$ and for each $x \in X$, $V_n(x)$ is convex. Then $\{V_n | n \in \mathbb{N}\}$ is a base for the required uniformity. ∎

4.35 COROLLARY. Every metrizable GO space can be embedded as a dense subspace of a completely metrizable GO space.

Proof. Let $(X,\mathcal{T},\leq)$ be a metrizable GO space. By the previous corollary, $(X,\mathcal{T})$ admits a convex uniformity $\mathcal{U}$ with a countable base. By the proof of Theorem 4.20, there is a quasi-uniformity $\mathcal{V}$ that has a countable base such that $\mathcal{V}$ determines X and $\mathcal{V}^* = \mathcal{U}$. Consequently, $(\tilde{X},\mathcal{T}(\tilde{\mathcal{V}})^*,\lesssim)$ is a completely metrizable GO space in which $(X,\mathcal{T},\leq)$ is embedded as a dense subspace. ∎

4.36 THEOREM. Let $(X,\mathcal{T},\leq)$ be a GO space. The following statements are equivalent.

(a) X is a pseudocompact.

(b) C*(X) is a convex uniformity.

(c) There is a closed linear order on βX that extends $\leq$.

(d) nX and βX are topologically equivalent compactifications.

Proof. (a) ⇒ (b) Since X is pseudocompact, C*(X) is the fine uniformity, which is convex.

(b) ⇒ (c) This implication is an immediate consequence of Theorem 4.32.

(c) ⇒ (d) This implication follows from Corollary 4.13.

(d) ⇒ (a) Suppose that X is not pseudocompact. Then X contains a C*-embedded copy A of $\mathbb{N}$. Since $C^*(A) = C^*(X)|A \times A$, C*(A) is a convex uniformity. By Theorems 4.31 and 4.32, βA, and hence $\beta\mathbb{N}$, is a GO space - a contradiction. ■

4.37 The theory of convex uniformities for GO spaces provides problems that arise naturally from the theory and so require no insight to formulate. For example if $(X,\mathcal{T},\leq)$ is a GO space and $(X,\mathcal{T})$ admits a uniformity with a certain uniform property, it is of evident concern whether one or all the convex uniformities for $(X,\mathcal{T},\leq)$ also possess the given uniform property. The following problem is of equal interest but in general of greater difficulty. Suppose we assume that all the uniformities on a topological space X that are convex with respect to at least one of the linear orders on X that renders X a GO space have a certain uniform property. What topological properties are consequences of our assumption? We do not consider all problems such as those suggested above; we are content to point out some related questions.

4.38 Let $(X,\mathcal{T})$ be a completely regular space and let $\mathcal{U}$ be the fine uniformity. If card (X) is a nonmeasurable cardinal, then the realcompactification υX of X is homeomorphic to the completion of $(X,\mathcal{U})$. Thus it follows from Theorems 4.32 and 4.33 that if X is a topological space of nonmeasurable cardinality then X is a GO space if and only if υX is a GO space. This result and the result of Theorem 4.36 that a GO space is pseudocompact if and only if C*(X) is a convex uniformity motivate the following question.

PROBLEM F. Let $(X,\mathcal{T},\leq)$ be a GO space. Is C(X) necessarily convex with respect to $\leq$? If not, what are necessary and sufficient conditions that C(X) be convex?

It might seem reasonable to conjecture that if $(X,\mathcal{T},\leq)$ is a LOTS determined by a quasi-uniformity $\mathcal{V}$, then the GO space $\tilde{X}$ determined by $\tilde{\mathcal{V}}$ would also be a LOTS. L. Gillman and M. Henriksen have given an example of a LOTS X of nonmeasurable cardinality for which υX is not a LOTS (see the notes). Thus this conjecture is false and so it is natural to ask the following question.

PROBLEM G. Let $(X,\mathcal{T},\leq)$ be a GO space determined by a quasi-uniformity $\mathcal{V}$. Under what conditions on $\mathcal{V}$ is $(\tilde{X},\mathcal{T}(\tilde{\mathcal{V}}^*),\leq)$ necessarily a LOTS?

In section 4.37 we pointed out that to each problem in the study of uniform spaces and completely regular topological spaces, there corresponds a problem in the study of convex uniform spaces and uniform ordered spaces. We end this chapter with two problems that illustrate this correspondence.

PROBLEM H. For which topological spaces $(X,\mathcal{T})$ is it true that for each uniformity $\mathcal{U}$ compatible with $\mathcal{T}$ there is a linear order $\leq$ on X such that $(X,\mathcal{T},\leq)$ is a GO space and $\mathcal{U}$ is convex with respect to $\leq$? For which GO spaces $(X,\mathcal{T},\leq)$ is every uniformity compatible with $\mathcal{T}$ convex? We conjecture that for a GO space $(X,\mathcal{T},\leq)$ every uniformity compatible with $\mathcal{T}$ is convex if and only if $(X,\mathcal{T})$ is locally compact, pseudocompact and its one-point compactification is orderable. The conjecture obtains if $(X,\mathcal{T},\leq)$ is LOTS (see the notes).

PROBLEM I. Suppose that $(X,\mathcal{T},\leq)$ is a uniform ordered space. If every convex uniformity compatible with $\mathcal{T}$ is totally bounded, is $(X,\mathcal{T})$ pseudocompact? By Theorem 4.33, if $(X,\mathcal{T},\leq)$ is a GO space for which every convex uniformity compatible with $\mathcal{T}$ is totally bounded, then $(X,\mathcal{T})$ is pseudocompact.

Chapter 5

COVERING PROPERTIES OF QUASI-UNIFORM SPACES

1. ORTHOCOMPACT AND PREORTHOCOMPACT SPACES

5.1 If $(X,\mathcal{T})$ is a compact metrizable space, then every metric that induces $\mathcal{T}$ has the following property. For each open cover $\mathcal{C}$ of X there is a positive real number $\lambda(\mathcal{C})$, called a Lebesgue number of $\mathcal{C}$, such that if $A \subset X$ and diam $(A) \le \lambda(\mathcal{C})$, then there exists $C \in \mathcal{C}$ such that $A \subset C$. In general, if $(X,\mathcal{T})$ is a metrizable space that is not compact, an open cover $\mathcal{C}$ may have a Lebesgue number for one metric and fail to have a Lebesgue number for some other metric. What is required of a metric in order that every open cover have a Lebesgue number is that for each open cover $\mathcal{C}$ there is an entourage U of the induced metric uniformity such that $\{U(x) \mid x \in X\}$ refines $\mathcal{C}$. This property of metric uniformities is important for quasi-uniformities as well.

A cover $\mathcal{C}$ of a subset A of a quasi-uniform space $(X,\mathcal{U})$ is a *quasi-uniform cover* of A provided that there is a $U \in \mathcal{U}$ such that for each $x \in A$ there is a $C \in \mathcal{C}$ such that $U(x) \subset C$. A quasi-uniformity $\mathcal{U}$ on a set X is a *Lebesgue quasi-uniformity* provided that every $\mathcal{T}(\mathcal{U})$-open cover is a quasi-uniform cover. An open cover of a topological space $(X,\mathcal{T})$ is a *quasi-normal cover* provided it is a quasi-uniform cover of $(X,\mathcal{U})$ for some quasi-uniformity $\mathcal{U}$ compatible with $\mathcal{T}$. Thus an open cover $\mathcal{C}$ of X is a quasi-normal cover provided there is a normal neighbornet U of X such that $\{U(x) \mid x \in X\}$ refines $\mathcal{C}$.

The characterization of uniform spaces, originally given in terms of entourages by Andre Weil and L. Pontrjagin, can also be formulated by axioms that determine which covers of a set are "of the same size." In this approach, a uniform space consists of a set X together with a collection $\mathcal{A}$ of covers of X, and if $(X,\mathcal{U})$ is the corresponding uniform space as given by Weil and Pontrjagin, then $\mathcal{A}$ is the collection of all uniform covers as defined above. Thus our interest in quasi-uniform covers is well motivated even though, in contrast to the uniform case, two quasi-uniformities may have the same quasi-uniform covers.

PROPOSITION. Let K be a compact subspace of a quasi-uniform space $(X,\mathcal{U})$ and let $\mathcal{C}$ be a collection of open subsets of X such that $K \subset \cup\mathcal{C}$. Then there is an entourage V in $\mathcal{U}$ such that for each $x \in K$ there is a $C \in \mathcal{C}$ such that $V(x) \subset C$.

COROLLARY. Let $(X,\mathcal{T})$ be a compact topological space. Then every quasi-uniformity compatible with $\mathcal{T}$ is a Lebesgue quasi-uniformity.

5.2 In Proposition 2.27, we established that a regular space is compact if and only if every admissible quasi-uniformity is locally symmetric. This result and the previous corollary suggest that if every open cover of a quasi-uniform space is a quasi-uniform cover, then the quasi-uniform space should possess some degree of symmetry. The following three propositions pursue this line of investigation.

PROPOSITION. Every Lebesgue quasi-uniformity is equinormal.

Proof. Let $(X,\mathcal{U})$ be a quasi-uniform space in which every open cover is a quasi-uniform cover, let F be a closed subset of $(X,\mathcal{T}(\mathcal{U}))$, and let $G \in \mathcal{T}(\mathcal{U})$ such that $F \subset G$. Since $\{X - F, G\}$ is a quasi-uniform cover, there is an entourage $U \in \mathcal{U}$ such that $U(F) \subset G$. ■

The previous theorem and Proposition 2.26 yield the following corollary.

COROLLARY. Let $(X,\mathcal{U})$ be a Lebesgue quasi-uniform space. If $(X,\mathcal{U})$ is an R_0 space, it is point-symmetric; if it is a regular space, it is locally symmetric.

5.3 It follows from Corollary 5.4 that if $(X,\mathcal{U})$ is a completely regular Lebesgue quasi-uniform space then $\mathcal{U}$ contains the fine uniformity. If $(X,\mathcal{T})$

is a topological space that is not paracompact, $\mathcal{LF}$ is not a Lebesgue quasi-uniformity; nonetheless, $\mathcal{LF}$ contains the fine uniformity of any completely regular space.

PROPOSITION. Let $(X,\mathcal{T})$ be a completely regular space and let $\mathcal{U}$ be the fine uniformity. Then $\mathcal{U} \subset \mathcal{LF}$.

Proof. Let $U \in \mathcal{U}$ and let V be a symmetric member of $\mathcal{U}$ such that $V^2 \subset U$. It follows from Lemma 1.5 that there is a pseudo-metric d for X such that $\mathcal{T}(d) \subset \mathcal{T}$ and $\{(x,y) \mid d(x,y) < 1\} \subset V$. Since every pseudo-metrizable space is paracompact, there is a locally finite open cover $\mathcal{C}$ of X such that $\{st(x,\mathcal{C}) \mid x \in X\}$ refines $\{V(x) \mid x \in X\}$. Then $U_{\mathcal{C}} \in \mathcal{LF}$ and $U_{\mathcal{C}} \subset U$. ■

5.4 PROPOSITION. A Lebesgue quasi-uniformity contains every neighborhood of the diagonal.

Proof. Let X be a set and let $\mathcal{U}$ be a Lebesgue quasi-uniformity on X. Let W be a neighborhood of Δ and for each $x \in X$ let $G_x \in \mathcal{T}(\mathcal{U})$ such that $x \in G_x$ and $G_x \times G_x \subset W$. Let $U \in \mathcal{U}$ such that $\{U(x) \mid x \in X\}$ refines $\mathcal{C} = \{G_x \mid x \in X\}$. For each $x \in X$, $U(x) \subset st(x,\mathcal{C}) \subset W(x)$ so that $U \subset W$. ■

If W is a symmetric neighbornet, W^2 is a neighborhood of the diagonal so that we have the following corollary.

COROLLARY. A Lebesgue quasi-uniformity contains the square of every symmetric neighbornet.

5.5 The role of Lebesgue uniformities in the study of uniform spaces can be summarized succinctly. It follows from Proposition 5.4 that only the fine uniformity of a completely regular space $(X,\mathcal{T})$ can be a Lebesgue uniformity. Moreover the fine uniformity of X is a Lebesgue uniformity if and only if X is a paracompact. The situation in quasi-uniform spaces cannot be stated so nicely. As Corollary 5.1 implies, a topological space may admit many Lebesgue quasi-uniformities, and no familiar topological property characterizes those topological spaces for which the fine quasi-uniformity is a Lebesgue quasi-uniformity. Nonetheless, there is an elegant topological characterization of the class of topological spaces for which the fine transitive quasi-uniformity is a Lebesgue quasi-uniformity, and it has taken considerable effort to establish that this class is strictly smaller than the class of topological spaces for which the fine quasi-uniformity is a Lebesgue quasi-uniformity. A topological space is *orthocompact* provided that every open cover has an interior-preserving open refinement. It is easy to show that a topological space is orthocompact if and only if $\mathcal{FT}$ is a Lebesgue quasi-uniformity.

It is often easier to establish that a topological space is orthocompact than to establish the weaker property that it admits a Lebesgue quasi-uniformity. In this regard, it is useful to observe that if each nonisolated point of a space $(X,\mathcal{T})$ has a neighborhood in which it is the only nonisolated point, then $(X,\mathcal{T})$ is orthocompact. Many theorems concerning paracompact spaces have natural counterparts that involve orthocompact spaces. For example, we obtain quasi-uniform analogues of both Stone's theorem and Tamano's theorem.

5.6 A collection $\mathcal{C}$ of subsets of a set X is *monotone* provided that any two members of $\mathcal{C}$ are comparable under set inclusion. A collection $\mathcal{C}$ of subsets of X is of *subinfinite rank* provided that if $\mathcal{A}$ is an infinite subcollection of $\mathcal{C}$ and $\cap\mathcal{A} \neq \emptyset$, then there are two members of $\mathcal{A}$ that are comparable under set inclusion.

Let $\mathcal{C} = \{C_\alpha | \alpha \in A\}$ be a cover of a set X. A cover $\mathcal{R} = \{R_\alpha | \alpha \in A\}$ of X is said to be *precise refinement* of $\mathcal{C}$ provided that for each $\alpha \in A$, $R_\alpha \subset C_\alpha$.

THEOREM. Let $(X,\mathcal{T})$ be a topological space. The following statements are equivalent.

(a) $(X,\mathcal{T})$ is orthocompact.

(b) $\mathcal{FT}$ is a Lebesgue quasi-uniformity.

(c) Every open cover of X has a precise interior-preserving open refinement.

(d) If $\mathcal{C}$ is an open cover of X, there is an open refinement $\mathcal{R}$ of $\mathcal{C}$ such that for each $x \in X$, either $\cap\mathcal{R}_x \in \mathcal{T}$ or $\mathcal{R}_x$ is the finite union of monotone collections.

(e) If $\mathcal{C}$ is an open cover of X, there is an open refinement $\mathcal{R}$ of $\mathcal{C}$ such that for each $x \in X$, either $\cap\mathcal{R}_x \in \mathcal{T}$ or $\mathcal{R}_x$ is of subinfinite rank.

Proof. (a) $\Rightarrow$ (b) : Let $\mathcal{C}$ be an open cover of X and let $\mathcal{R}$ be an interior-preserving open refinement of $\mathcal{C}$. Then $U_{\mathcal{R}}$ is the required entourage.

(b) ⇒ (c) : Let $\mathcal{C} = \{C_\alpha | \alpha \in A\}$ be an open cover of X. There is a transitive entourage $U \in \mathcal{FT}$ such that $\{U(x) | x \in X\}$ refines $\mathcal{C}$. Define $f: X \to A$ by choosing for each x in X an α in A such that $U(x) \subset C_\alpha$. For each $\alpha \in A$ let $R_\alpha = \cup\{U(x) | f(x) = \alpha\}$. Then $\mathcal{R} = \{R_\alpha | \alpha \in A\}$ is an open refinement of $\mathcal{C}$. For each $p \in X$, $U(p) \subset \cap\mathcal{R}_p$. Hence $\mathcal{R}$ is an interior-preserving open refinement of $\mathcal{C}$ such that for each $\alpha \in A$, $R_\alpha \subset C_\alpha$.

(c) ⇒ (d) ⇒ (e): Evident.

(e) ⇒ (a) : Let $\mathcal{C}$ be an open cover of X. There is an open refinement $\mathcal{R}$ of $\mathcal{C}$ such that for each $x \in X$, either $\cap\mathcal{R}_x \in \mathcal{T}$ or $\mathcal{R}_x$ is of subinfinite rank. Well order $\mathcal{R}$ and for each $x \in X$ let R_x be the least member of $\mathcal{R}_x$. The family $\mathcal{R}_0 = \{R_x | x \in X\}$ is a subcover of $\mathcal{R}$ with the property that each $R \in \mathcal{R}_0$ contains a point of X that belongs to no predecessor of R. Suppose that $\mathcal{R}_0$ is not interior preserving. Then there is an $x \in X$ such that $\mathcal{R}_1 = \{R \in \mathcal{R}_0 | x \in R\}$ is of subinfinite rank and $\cap\mathcal{R}_1 \notin \mathcal{T}$. Let R_1 denote the least member of $\mathcal{R}_1$ and let $\mathcal{R}_2 = \{R \in \mathcal{R}_1 | R_1 \text{ is not a subset of } R\}$. Then $\cap\mathcal{R}_1 = R_1 \cap (\cap\mathcal{R}_2)$. Thus $\cap\mathcal{R}_2 \notin \mathcal{T}$ and $\mathcal{R}_2$ is infinite. Inductively let R_n denote the least member of $\mathcal{R}_n$ and set $\mathcal{R}_{n+1} = \{R \in \mathcal{R}_n | R_n \text{ is not a subset of } R\}$; then $\cap\mathcal{R}_n = R_n \cap (\cap\mathcal{R}_{n+1})$ and since $\cap\mathcal{R}_{n+1} \notin \mathcal{T}$, $\mathcal{R}_{n+1}$ is infinite. In this way we obtain an infinite subset $\{R_n | n \in N\}$ of $\mathcal{R}_0$ no two members of which are comparable - a contradiction. Thus $\mathcal{R}_0$ is an interior-preserving open refinement of $\mathcal{C}$. ∎

We have a few words of caution with regard to condition (c) of the previous theorem. If $\mathcal{C}$ is an open cover of X indexed by X itself, then condition (c) provides an interior-preserving open refinement $\{R_x | x \in X\}$. For a given x there is no guarantee that $R_x \neq \emptyset$ and even if $R_x \neq \emptyset$, there is no guarantee that $x \in R_x$.

5.7 PROPOSITION. Every Lebesgue quasi-uniformity is convergence complete.

Proof. Let X be a set, let $\mathcal{U}$ be a Lebesgue quasi-uniformity on X, and let $\mathcal{F}$ be a Cauchy filter on $(X,\mathcal{U})$ that does not converge. Then for each $x \in X$ there is a $G_x \in \eta_x - \mathcal{F}$ and there is a $U \in \mathcal{U}$ such that $\{U(x) | x \in X\}$ refines $\{G_x | x \in X\}$. Since $\mathcal{F}$ is a Cauchy filter, for some $x \in X$, $U(x) \in \mathcal{F}$ - a contradiction. ∎

COROLLARY. If $(X,\mathcal{T})$ is orthocompact, then $\mathcal{FT}$ is convergence complete.

5.8 PROPOSITION. Let $(X,\mathcal{T})$ be a topological space and let $\mathcal{T}'$ be a hereditarily orthocompact Hausdorff topology on X that is coarser than $\mathcal{T}$. Then the fine transitive quasi-uniformity on $(X,\mathcal{T})$ is convergence complete.

Proof. Let $\mathcal{F}$ be an $\mathcal{FT}$-Cauchy filter on $(X,\mathcal{T})$ and let $\mathcal{U}$ denote the fine transitive quasi-uniformity on $(X,\mathcal{T}')$. Then $\mathcal{F}$ is a $\mathcal{U}$-Cauchy filter and so converges to a point p in $(X,\mathcal{T}')$. To show that $\mathcal{F}$ converges to p in $(X,\mathcal{T})$, let $H \in \mathcal{T}$ such that $p \in H$. Since $(X,\mathcal{T}')$ is a Hausdorff space, for each $x \neq p$ there exists $G_x \in \mathcal{T}'$ such that $x \in G_x$ and $p \notin \mathrm{cl}_{\mathcal{T}'}(G_x)$. Let $\mathcal{C} = \{G_x \mid x \neq p\}$ and let $\mathcal{R}$ be an interior-preserving open refinement of $\mathcal{C}$. Then $\mathcal{R} \cup \{H\}$ is a cover of X that is open and interior preserving with respect to $\mathcal{T}$. Thus $H \in \mathcal{F}$ and $\mathcal{F}$ converges to p in $(X,\mathcal{T})$. ∎

Since every Hausdorff space can be embedded as a closed subspace of a minimal Hausdorff space, it is easy to construct a Hausdorff space $(X,\mathcal{T})$ with the property that if $\mathcal{T}'$ is a Hausdorff topology on X coarser than $\mathcal{T}$, then $(X,\mathcal{T}')$ is not orthocompact.

5.9 A topological space $(X,\mathcal{T})$ is *σ-orthocompact* provided that if $\mathcal{C}$ is an open cover of X there is an open refinement $\mathcal{R} = \cup_{n=1}^{\infty} \mathcal{R}_n$ of $\mathcal{C}$ such that for each $n \in \mathbb{N}$, $\mathcal{R}_n$ is interior preserving.

PROPOSITION. Every countably metacompact σ-orthocompact space is orthocompact.

Proof. Let $(X,\mathcal{T})$ be a σ-orthocompact countably metacompact space and let $\mathcal{C}$ be an open cover of X. For each $n \in \mathbb{N}$ let $\mathcal{V}_n$ be an interior-preserving open collection so that $\mathcal{V} = \cup_{n=1}^{\infty} \mathcal{V}_n$ is a refinement of $\mathcal{C}$. For each $n \in \mathbb{N}$ let $G_n = \cup\mathcal{V}_n$. Then $\{G_n \mid n \in \mathbb{N}\}$ is a countable open cover of X. Since $(X,\mathcal{T})$ is countably metacompact, there is an open point-finite refinement $\mathcal{W} = \{W_n \mid n \in \mathbb{N}\}$ of $\{G_n \mid n \in \mathbb{N}\}$ such that for each $n \in \mathbb{N}$, $W_n \subset G_n$. For each n let $\mathcal{R}_n = \{V \cap W_n \mid V \in \mathcal{V}_n\}$ and let $\mathcal{R} = \cup_{n=1}^{\infty} \mathcal{R}_n$. Evidently $\mathcal{R}$ is an open refinement of $\mathcal{C}$. Let $x \in X$. Since $(\mathcal{R}_n)_x = \emptyset$ for all but finitely many n, $\cap\mathcal{R}_x \in \mathcal{T}$. ∎

PROBLEM I. Is a countably orthocompact σ-orthocompact space necessarily orthocompact? An affirmative answer even in normal Hausdorff spaces would be pleasant and unexpected. We note, however, that it follows from Propositions 5.9 and 5.13 that every separable countably orthocompact σ-orthocompact T_1 space is orthocompact.

5.10 Although orthocompactness is a natural generalization of paracompactness, it does not provide a topological characterization of those spaces in which every open cover is a quasi-normal cover. A characterization of paracompactness, due to E. Michael, suggests a likely candidate for a covering property that characterizes such spaces. If $\mathcal{C}$ and $\mathcal{R}$ are collections of subsets of a topological space X, we say that $\mathcal{R}$ is *cushioned in* $\mathcal{C}$ if one can assign to each $R \in \mathcal{R}$ a $C_R \in \mathcal{C}$ such that for every subcollection $\mathcal{R}'$ of $\mathcal{R}$, $\overline{\cup\{R \mid R \in \mathcal{R}'\}} \subset \cup\{C_R \mid R \in \mathcal{R}'\}$. A refinement of $\mathcal{C}$ that is cushioned in $\mathcal{C}$ is called a *cushioned refinement* of $\mathcal{C}$. Michael showed that a space X is paracompact provided every open cover of X has a cushioned refinement or equivalently provided for each open cover $\mathcal{C}$ of X, there is a neighbornet V of X such that V^{-1} is a neighbornet of X and $\{V(x) \mid x \in X\}$ refines $\mathcal{C}$. The relationship between closure-preserving collections and interior-preserving collections given in Proposition 2.5 suggests a concept that is complementary to the concept of a cushioned refinement. We say that $\mathcal{R}$ is *cocushioned in* $\mathcal{C}$ if one can assign to each $R \in \mathcal{R}$ a $C_R \in \mathcal{C}$ such that for every subcollection $\mathcal{R}'$ of $\mathcal{R}$, $\cap\{R \mid R \in \mathcal{R}'\} \subset \operatorname{int}(\cap\{C_R \mid R \in \mathcal{R}'\})$. A refinement of $\mathcal{C}$ that is cocushioned in $\mathcal{C}$ is called a *cocushioned refinement* of $\mathcal{C}$. A topological space $(X,\mathcal{T})$ is *preorthocompact* provided that every open cover of X has an open cocushioned refinement. We first consider the relationship between neighbornets and cocushioned refinements.

PROPOSITION. Let V be a neighbornet of a topological space X and let $\mathcal{R}$ be a collection of subsets of X. Then $\mathcal{R}$ is cocushioned in $\{V(R) \mid R \in \mathcal{R}\}$.

Proof. Let $\mathcal{R}'$ be a subcollection of $\mathcal{R}$ and let $x \in \cap\{R \mid R \in \mathcal{R}'\}$. Then $V(x) \subset \cap\{V(R) \mid R \in \mathcal{R}'\}$ so that $x \in \operatorname{int}(\cap\{V(R) \mid R \in \mathcal{R}'\})$. ∎

5.11 PROPOSITION. Let $\mathcal{R}$ and $\mathcal{C}$ be collections of subsets of a space X and suppose that $\mathcal{R}$ is cocushioned in $\mathcal{C}$. Then there exists a neighbornet U of X such that for each $x \in \cup\mathcal{R}$, there is a $C \in \mathcal{C}$ such that $U^2(x) \subset C$.

Proof. For each $R \in \mathcal{R}$ choose a $C_R \in \mathcal{C}$ such that for each subcollection $\mathcal{R}'$ of $\mathcal{R}$, $\cap\{R \mid R \in \mathcal{R}'\} \subset \operatorname{int}(\cap\{C_R \mid R \in \mathcal{R}'\})$. For each $x \in \cup\mathcal{R}$ choose $R_x \in \mathcal{R}$ such that $x \in R_x$. Let $U = \{(x,y) \mid y \in R_x \text{ and } y \in C_R \text{ whenever } x \in R\}$. Let $x \in \cup\mathcal{R}$ and let $z \in U^2(x)$. There exists $y \in X$ such that $z \in U(y)$ and $y \in U(x)$. Since $y \in U(x)$, $y \in R_x$ and so $z \in C_{R_x}$. ∎

COROLLARY. A topological space X is preorthocompact if and only if for each open cover $\mathcal{C}$ of X there is a neighbornet V of X such that $\{V^2(x) | x \in X\}$ refines $\mathcal{C}$.

If $\mathcal{C}$ is an open cover of a preorthocompact space X, then for each $n \in \mathbb{N}$ there is a neighbornet V of X such that $\{V^n(x) | x \in X\}$ refines $\mathcal{C}$. If a topological space X admits a Lebesgue quasi-uniformity, X is preorthocompact. It is an open question whether the converse of this statement holds, and it is not until Example 7.25 that we have the machinery to construct a preorthocompact space that is not orthocompact.

5.12 PROPOSITION. Every closed subspace of a preorthocompact space is preorthocompact, and if every open subspace of a topological space X is preorthocompact then X is hereditarily preorthocompact.

The obvious analogues of Proposition 5.12 hold for both σ-orthocompact and orthocompact spaces. Moreover, every σ-orthocompact perfect space is hereditarily orthocompact.

2. OPEN-FINITE COVERS

5.13 An open cover $\mathcal{C}$ of a space is *open finite* provided that no nonempty open set is a subset of infinitely many members of $\mathcal{C}$. Our interest in open-finite covers stems from the lemma given below, which enables us to construct many examples of spaces that are not preorthocompact and to establish that a product of metrizable spaces is normal if and only if it is preorthocompact.

LEMMA. Every open-finite cover of a preorthocompact space has a point-finite open refinement.

Proof. Let X be a preorthocompact space and let $\mathcal{C}$ be an open-finite cover of X. By Corollary 5.11 there is a neighbornet V of X such that $\{V^2(x) | x \in X\}$ refines $\mathcal{C}$. For each $C \in \mathcal{C}$ let $G(C) = \text{int}\{x | V(x) \subset C\}$ and let $\mathcal{G} = \{G(C) | C \in \mathcal{C}\}$. Let $p \in X$ and let $C \in \mathcal{C}$ such that $V^2(p) \subset C$. Then $p \in G(C)$ so that $\mathcal{G}$ is a cocushioned open refinement of $\mathcal{C}$. Since $\mathcal{C}$ is open finite, $\mathcal{G}$ is point finite. ■

PROPOSITION. Every countably orthocompact separable T_1 space is countably metacompact.

Proof. Let X be a countably orthocompact separable T_1 space, let $\{d_n | n \in \mathbb{N}\}$ be a countable dense subset of X, and let $\{G_n | n \in \mathbb{N}\}$ be a countable open cover of X. For each $n \in \mathbb{N}$ set $R_n = G_n - \{d_k | k \leq n$ and $d_k \in \cup_{i<n} G_i\}$. Then $\{R_n | n \in \mathbb{N}\}$ is an open-finite refinement of $\{G_n | n \in \mathbb{N}\}$, and so by the proof of the preceding lemma, $\{G_n | n \in \mathbb{N}\}$ has a point-finite open refinement. ■

5.14 R. H. Bing's Example G suggests a technique that may be used to construct from any infinite T_1 space X, a T_1 space $\mathcal{G}(X)$ that is preorthocompact if and only if X is metacompact. If X is T_2, T_3, normal, transitive, perfect, or metacompact, then $\mathcal{G}(X)$ has the respective property. Thus for the space F of Bing's Example H, $\mathcal{G}(F)$ is a perfectly normal space that is not preorthocompact. We construct the space $\mathcal{G}(X)$ from X as follows. Let $\mathcal{T}$ be a T_1 topology on an infinite set X and let $D = \{\Gamma | \Gamma$ is a finite subset of $\mathcal{T}\}$. For each $G \in \mathcal{T}$ set $G' = G \cup \{\Gamma \in D | G \in \Gamma\}$ and set $G'' = (X - \bar{G}) \cup \{\Gamma \in D | G \notin \Gamma\}$. The underlying set for $\mathcal{G}(X)$ is $X \cup D$ and the topology for $\mathcal{G}(X)$ has a subbase $\{G' | G \in \mathcal{T}\} \cup \{G'' | G \in \mathcal{T}\} \cup \{\{\Gamma\} | \Gamma \in D\}$. As a subspace of $\mathcal{G}(X)$, D is dense and open.

We establish that if $\mathcal{G}(X)$ is preorthocompact then X is metacompact. Let $\mathcal{C}$ be an open cover of X and let $\mathcal{C}' = \{C' | C \in \mathcal{C}\} \cup \{\{\Gamma\} | \Gamma \in D\}$. We show that $\mathcal{C}'$ is an open-finite cover of $\mathcal{G}(X)$. Let G be a nonempty open subset of $\mathcal{G}(X)$ and let $\mathcal{H} = \{C \in \mathcal{C} | G \subset C'\}$. Since D is dense in $\mathcal{G}(X)$, there is a $\Gamma \in G \cap D$ and so $\mathcal{H} \subset \Gamma$. Thus $\mathcal{C}'$ is an open-finite cover, and by the preceding lemma $\mathcal{C}'$ has a point-finite open refinement $\mathcal{R}$. Then $\mathcal{R}|X$ is a point-finite open refinement of $\mathcal{C}$.

5.15 A space X is *semi-metacompact* provided that every open cover of X has an open-finite refinement. A space X is *nearly metacompact* provided that if $\mathcal{C}$ is an open cover of X, there is a dense subset D of X and an open refinement $\mathcal{R}$ of $\mathcal{C}$ such that $\mathcal{R}_x$ is finite for each $x \in D$. Evidently, every nearly metacompact space is semi-metacompact.

THEOREM. Let $(X,\mathcal{T})$ be a topological space. Then the following statements are equivalent.

(a) $(X,\mathcal{T})$ is metacompact.

(b) $(X,\mathcal{T})$ is preorthocompact and nearly metacompact.

(c) $(X,\mathcal{T})$ is preorthocompact and semi-metacompact.

Proof. The implications (a) $\Rightarrow$ (b) and (b) $\Rightarrow$ (c) are evident. (c) $\Rightarrow$ (a) Let $(X,\mathcal{T})$ be a preorthocompact semi-metacompact space. Let $\mathcal{V}$ be an open cover of X and let $\mathcal{C}$ be an open-finite refinement of $\mathcal{V}$. By Lemma 5.13 there is a point-finite open refinement of $\mathcal{C}$. ■

5.16 EXAMPLE. A Moore space X that is not nearly metacompact whose topology has an open-finite base.

Let $D = \{k/2^n | 0 < k < 2^n, n \in \mathbb{N}\}$ and for each $n \in \mathbb{N}$ set $D_n = \{k/2^i \in D | i \le n\}$. Since D is countable, there is a decreasing sequence $\langle U_n \rangle$ of open subsets of $\mathbb{R}$ such that for each $n \in \mathbb{N}$ the measure of $U_n = 2^{-n}$ and $D \subset U_n$. Let M be a Euclidean dense subset of $(0,1) - \cap_{n=1}^{\infty} U_n$ of cardinality $\aleph_1$ such that $x - y \notin M$ whenever x and $y \in M$. Index M as $\{x_\alpha | \alpha < \omega_1\}$ and set $X = M \cup M \times M$. For each $\alpha < \omega_1$ and $n \in \mathbb{N}$ set $T(n,x_\alpha) = \{x_\alpha\} \cup \{(y,z) \in M \times M | z \le 1/n \text{ and } |x_\alpha - y| \le z\}$, set $L_\alpha = \{(y,z) \in M \times M | y = x_\alpha\}$, and set $G(n,x_\alpha) = T(n,x_\alpha) - \cup\{L_\beta | \beta < \alpha\}$.

For each $\gamma < \omega_1$, let the ordinals that precede γ be simply ordered as $\langle \gamma_n \rangle$. Let $i \in \mathbb{N}$ and let $\alpha < \omega_1$. For each $\gamma > \alpha$, where $\alpha = \gamma_n$, and $L_\gamma \cap G(i,x_\alpha) \ne \emptyset$ and for each $p \in \{x_\gamma\} \times (D_n \cap (|x_\gamma - x_\alpha|, 1/i))$, choose an open interval I (depending on i, α, γ and p) with the following properties: I is centered at p, I has end points in $\{x_\gamma\} \times D$, and $I \subset \{x_\gamma\} \times U_n$. Let $H(i,x_\alpha)$ denote the points of $G(i,x_\alpha)$ that are not members of any interval chosen above. For each $x = (p,q) \in M \times M$ and $i \in \mathbb{N}$ let k be that nonnegative integer such that $k/2^i < q < (k+1)/2^i$ and set $H(i,x) = \{p\} \times (k/2^i, (k+1)/2^i) \cap M$. For each $n \in \mathbb{N}$ set $\mathcal{H}_n = \{H(n,x) | x \in X\}$. Then $\mathcal{H} = \cup_{n=1}^{\infty} \mathcal{H}_n$ is a point-countable base for a topology on X and $\langle \mathcal{H}_n \rangle$ is a development for X.

To see that $\mathcal{H}$ is an open-finite base let W be a nonempty open set and suppose that W is a subset of infinitely many members of $\mathcal{H}$. Let $z = (x_\gamma, y) \in W \cap M \times M$. Then $A = \{\alpha | \text{there exists } i_\alpha \in \mathbb{N} \text{ such that } W \subset H(i_\alpha, x_\alpha) \text{ and } x_\alpha \in M\}$ is infinite and since $z \in W$, for each $\alpha \in A$, $\gamma > \alpha$. There is an $\epsilon > 0$ such that $\{x_\gamma\} \times M \cap (y - \epsilon, y + \epsilon) \subset W$ and there exists $n \in \mathbb{N}$ such that $D_n \cap (y - \epsilon, y + \epsilon) \ne \emptyset$. Choose $d \in D_n \cap (y - \epsilon, y + \epsilon)$ and set $p = (x_\gamma, d)$. Choose $\alpha \in A$ such that for some $k \ge n$, $\alpha = \gamma_k$. Let I denote the interval depending on i_α, α, γ, and p that is deleted from $G(i_\alpha, x_\alpha)$ to form

$H(i_\alpha,x_\alpha)$. Then $I \cap H(i_\alpha,x_\alpha) = \emptyset$ whereas $I \cap W \neq \emptyset$ since M is a Euclidean dense subset of $(0,1) - \cap_{n=1}^{\infty} U_n$. As $W \subset H(i_\alpha,x_\alpha)$, we have reached a contradiction.

To show that X is not nearly metacompact let $C = \{H(i_\alpha,x_\alpha) | \alpha \in A\}$ and let $C' = \{G(i_\alpha,x_\alpha) | \alpha \in A\}$. We show that for each $p \in X$, C_p is finite if and only if C'_p is finite. Let $p = (x_\gamma,y)$ and suppose that for some $\alpha \in A$, $p \in G(i_\alpha,x_\alpha) - H(i_\alpha,x_\alpha)$. There is an $n \in \mathbb{N}$ such that $\alpha = \gamma_n$. Since y belongs to only finitely many U_n's, $\{\alpha | p \in G(i_\alpha,x_\alpha) - H(i_\alpha,x_\alpha)\}$ is finite.

We conclude the proof that X is not nearly metacompact by establishing that for each cover C by sets of the form $G(i,x)$, $\{x \in X | C_x \text{ is finite}\}$ is not dense in X. Let $C = \{G(i_\alpha,x_\alpha) | \alpha \in A\}$ be a cover of M. There is an $n \in \mathbb{N}$ such that $E = \{x_\alpha | i_\alpha \leq n\}$ is uncountable. There is $x_\gamma \in E$ and a sequence $<x_{\alpha_n}>$ in E such that for each $n \in \mathbb{N}$, $\alpha_n < \gamma$ and $<x_{\alpha_n}>$ converges to x_γ in $(\mathbb{R},T(E))$. (See the notes.) There exists a natural number $m \in \mathbb{N}$ and an open interval $S \subset \{x_\gamma\} \times [0,1/n)$ such that for each $j > m$, $S \subset G(i_{\alpha_j},x_{\alpha_j})$. Thus C_x is infinite for each $x \in S$.

5.17 EXAMPLE. *The Niemytzki plane* is an orthocompact Moore space that is not semi-metacompact. Let X be the union of the upper half plane and the x-axis. The upper half plane has the Euclidean subspace topology T'. For each $x \in \mathbb{R}$ and $n \in \mathbb{N}$ set $G(n,x) = \{(x,0)\} \cup G_n$ where G_n is the open disk of radius $1/n$ in the upper half plane, tangent to the x-axis at $(x,0)$. Then $\{G(n,x) | n \in \mathbb{N}, x \in \mathbb{R}\} \cup T'$ is a base for the topology T of the Niemytzki plane. Since X is a separable Moore space that is not metrizable, X is not metacompact. To show that X is not semi-metacompact, in light of Proposition 5.9 and Theorem 5.15 it suffices to show that the topology of X has a σ-interior-preserving base. For each $n \in \mathbb{N}$ let C_{n+1} be a point-finite open cover of the upper half plane such that for each $x \in \mathbb{R}$, $st(G(n + 1,x),C_{n+1}) \subset G(n,x)$. Since T' has a countable base, in order to show that T has a σ-interior-preserving base it suffices to show that for each $n \in \mathbb{N}$, the family $G_n = \{\{(x,0)\} \cup st(G(n + 1,x), C_{n+1}) | x \in \mathbb{R}\}$ is interior preserving. Let $n \in \mathbb{N}$ and let $x = (p,q) \in X$. If $q = 0$, it is evident that $\cap(G_n)_x \in T$. If $q \neq 0$, then $\cap(C_{n+1})_x \subset \cap(G_n)_x$ and since C_{n+1} is point finite $\cap(C_{n+1})_x \in T$.

5.18 Let (X,T) be a topological space and let B be a base for T. Then B is a *uniform base* provided that if $U \in T$ and $x \in U$ then there are at most finitely many members of $\mathcal{B}$ containing x that are not subsets of U. A space $(X,\mathcal{T})$ is metacompact and developable if and only if $\mathcal{T}$ has a uniform base. Clearly if $\mathcal{B}$ is a uniform base, then $\mathcal{B}$ is an open-finite base. Example 5.16 shows that a Moore space with an open-finite base need not have a uniform base. It is evident that the property of having an open-finite base, like many properties involving bases, is open hereditary. What is more significant is that unlike most properties involving bases, the property of having an open-finite base is also productive. In this respect the property is better behaved than the associated property involving covers, semi-metacompactness. Indeed, the Sorgenfrey Line provides a simple example of a paracompact space whose square is not semi-metacompact.

In the proof of the following proposition we make use of A. H. Stone's classic result that $\mathbb{N}^{\omega_1}$ is not normal.

PROPOSITION. Let X be a preorthocompact space that is the product of nonempty T_1 spaces. Then all but countably many factors are countably compact.

Proof. Suppose that $X = \Pi\{X_i \mid i \in I\}$ where for uncountably many i, X_i is not countably compact. Then X contains a closed copy of $\mathbb{N}^{\omega_1}$. As $\mathbb{N}^{\omega_1}$ is separable but not normal, it is not metacompact. Since $\mathbb{N}^{\omega_1}$ has an open-finite base, it follows from Theorem 5.15 that $\mathbb{N}^{\omega_1}$ is not preorthocompact - a contradiction to Proposition 5.12. ■

COROLLARY. Let X be a product of metrizable spaces. Then X is normal if and only if it is preorthocompact.

Proof. It is a well-known result of A. H. Stone that a product of metrizable spaces is normal if and only if all but countably many factors are compact. ■

5.19 The following theorem parallels N. Noble's result that a T_1 space $(X,\mathcal{T})$ is compact provided there exists an uncountable cardinal λ not less than the weight of X such that X^λ is normal.

LEMMA. Let X be a topological space and let $\{0,1\}$ have the discrete topology. Suppose that λ is a cardinal number such that $X \times \{0,1\}^\lambda$ is preorthocompact; then each open cover of X having cardinality not greater than λ has a point-finite open refinement.

Proof. Let $\mathcal{C} = \{C_\alpha | \alpha < \lambda\}$ be an open cover of X. For each $\alpha < \lambda$ and $i = 0, 1$ set $D(\alpha,i) = C_\alpha \times \{y \in \{0,1\}^\lambda | y(\alpha) = i\}$. Let $\mathcal{D} = \{D(\alpha,i) | \alpha < \lambda, i = 0,1\}$. Then $\mathcal{D}$ is an open-finite cover of $X \times \{0,1\}^\lambda$. Since $X \times \{0,1\}^\lambda$ is preorthocompact, it follows from Lemma 5.13 that $\mathcal{D}$ has a point-finite open refinement $\mathcal{R}$. Let $y \in \{0,1\}^\lambda$. Then $\{\pi_1(R \cap (X \times \{y\})) | R \in \mathcal{R}\}$ is a point-finite open refinement of $\mathcal{C}$. ■

THEOREM. Let X be a T_1 space and let λ be a cardinal number. Suppose that X^λ is preorthocompact.

(a) If λ is infinite, then X is countably metacompact.

(b) If λ is uncountable, then X is countably compact.

(c) If λ is uncountable and the Lindelöf degree of X does not exceed λ, then X is compact.

Proof. (a) Without loss of generality we assume that X has two points, which we denote by 0 and 1. Then $X \times \{0,1\}^{\omega_0}$ is a closed subspace of X^λ and hence $X \times \{0,1\}^{\omega_0}$ is preorthocompact. By the previous lemma X is countably metacompact.

(b) This assertion is an immediate consequence of Proposition 5.18.

(c) We have from (b) that X is countably compact so that by the well-known theorem of R. Arens and J. Dugundji it suffices to show that X is metacompact. As in the proof of assertion (a) we have that $X \times \{0,1\}^\lambda$ is preorthocompact so that by the previous lemma each open cover of X having cardinality not greater than λ has a point-finite open refinement. Since the Lindelöf degree of X does not exceed λ, X is metacompact. ■

The method of proof of the following proposition is that of assertion (a) of the previous theorem.

5.20 PROPOSITION. For each $n \in \mathbb{N}$, let X_n be a nondegenerate T_1 space. If $\Pi\{X_n | n \in \mathbb{N}\}$ is preorthocompact, then for each $n \in \mathbb{N}$, X_n is countably metacompact.

5.21 In Corollary 5.18 we caught a glimpse of the curious parallel between preorthocompactness and normality. Therefore, although the construction of 5.14 establishes that a perfectly normal space need not be preorthocompact, we are well motivated to search for some form of normality weaker than full normality that implies preorthocompactness.

A topological space X is *almost 2-fully normal* provided that if $\mathcal{C}$ is an open cover of X then there is an open refinement $\mathcal{R}$ of $\mathcal{C}$ such that if x_1 and $x_2 \in X$ and $st(x_1,\mathcal{R}) \cap st(x_2,\mathcal{R}) \neq \emptyset$ then there is a set $C \in \mathcal{C}$ such that $\{x_1,x_2\} \subset C$. A topological space X is *2-fully normal* provided that if $\mathcal{C}$ is an open cover of X then there is an open refinement $\mathcal{R}$ of $\mathcal{C}$ such that if R_1 and $R_2 \in \mathcal{R}$ and $R_1 \cap R_2 \neq \emptyset$ then there is a set $C \in \mathcal{C}$ such that $R_1 \cup R_2 \subset C$. Every almost 2-fully normal space is collectionwise normal, and a Hausdorff space X is almost 2-fully normal if and only if the collection of all neighborhoods of the diagonal of X is a base for the fine uniformity on X. (See the notes.)

THEOREM. Every almost 2-fully normal preorthocompact space is 2-fully normal.

Proof. Let $(X,\mathcal{T})$ be an almost 2-fully normal preorthocompact space and let $\mathcal{C}$ be an open cover of X. By Corollary 5.11 there is a neighbornet R of X such that $\{R^2(x) | x \in X\}$ refines $\mathcal{C}$. Since $(X,\mathcal{T})$ is almost 2-fully normal, the open cover $\{int[R(x)] | x \in X\}$ has a refinement $\mathcal{G}$ with the property that if x_1 and $x_2 \in X$ and $st(x_1,\mathcal{G}) \cap st(x_2,\mathcal{G}) \neq \emptyset$ then there is an $x \in X$ such that $\{x_1,x_2\} \subset int[R(x)]$. For each $x \in X$ let $G(x) \in \mathcal{G}$ such that $x \in G(x)$ and set $H(x) = G(x) \cap int[R(x)]$. We assert that $\mathcal{H} = \{H(x) | x \in X\}$ is an open refinement of $\mathcal{C}$ that satisfies the requirements imposed by the definition of 2-full normality. To establish this assertion let x and y be members of X such that $H(x) \cap H(y) \neq \emptyset$. Since $G(x) \cap G(y) \neq \emptyset$, there is some $z \in X$ such that $\{x,y\} \subset int[R(z)]$. Let $C \in \mathcal{C}$ such that $R^2(z) \subset C$. Then $H(x) \cup H(y) \subset R(x) \cup R(y) \subset R^2(z) \subset C$. ∎

It is an unanswered question whether the converse of Theorem 5.21 holds. Indeed we have no example of a nonorthocompact 2-fully normal space.

5.22 A. H. Stone's proof that $\mathbb{N}^{\omega_1}$ is not normal, which we have mentioned previously, relies on establishing that the following two subsets F_0 and F_1 of $\mathbb{N}^{\omega_1}$ are disjoint closed subsets of $\mathbb{N}^{\omega_1}$ that cannot be separated in $\mathbb{N}^{\omega_1}$: $F_0 = \{x \in \mathbb{N}^{\omega_1} | \text{for any } n \neq 0,\ x_\alpha = n \text{ for at most one } \alpha < \omega_1\}$

$F_1 = \{x \in \mathbb{N}^{\omega_1} | \text{for any } n \neq 1,\ x_\alpha = n \text{ for at most one } \alpha < \omega_1\}$.

As H. H. Corson has observed, Stone's space F_0 (equivalently F_1) is of interest in its own right since it provides a counterexample to a conjecture of J. Kelley that an almost 2-fully normal space that admits a complete uniformity must be paracompact.

EXAMPLE. F_0 is an almost 2-fully normal space that is not preorthocompact. Since F_0 is collectionwise normal but not paracompact, in light of Theorem 5.15 in order to show that F_0 is not preorthocompact, it suffices to show that F_0 has an open-finite base. For each $x \in F_0$ and for each finite subset A of ω_1, set $T(x,A) = \{y \in F_0 | \text{for each } \alpha \in A,\ x_\alpha = y_\alpha\}$ and let $\mathcal{B} = \{T(x,A) | x \in F_0 \text{ and } A \text{ is a finite subset of } \omega_1\}$. Then $\mathcal{B}$ is a base for the topology of F_0. We show that $\mathcal{B}$ is an open-finite base. Let $T(x,A)$ and $T(y,B)$ be members of $\mathcal{B}$ and suppose that $T(x,A) \subset T(y,B)$. To show that $B \subset A$, suppose that $\beta \in B - A$. Define $p \in F_0$ such that $p_\alpha = x_\alpha$ for each $\alpha \in A$, $p_\beta \neq y_\beta$, and $p_\alpha = 0$ for each $\alpha \notin A \cup \{\beta\}$. Then $p \in T(x,A) - T(y,B)$ - a contradiction. Thus $B \subset A$. For each $\beta \in B$, $x_\beta = y_\beta$ and A has only finitely many subsets; hence only finitely many members of $\mathcal{B}$ contain $T(x,A)$. ■

5.23 Besides the metacompact spaces, there is another important class of spaces that is easily shown to be contained in the class of orthocompact spaces.

THEOREM. Every GO space is orthocompact.

Proof. Let $(X,\mathcal{T},\leq)$ be a GO space and let $\mathcal{C}$ be a cover of X by possibly degenerate open intervals. Let Y be a subset of X that is maximal with respect to the property that for each $y \in Y$, $st(y,\mathcal{C}) \cap Y = \{y\}$. For each $y \in Y$ set $\mathcal{L}(y) = \{(\leftarrow,y) \cap C | C \in \mathcal{C}_y\}$, set $\mathcal{R}(y) = \{(y,\rightarrow) \cap C | C \in \mathcal{C}_y\}$, choose $C_y \in \mathcal{C}_y$, and set $\mathcal{A}(y) = \{C_y\}$. Let $\mathcal{G} = \bigcup\{\mathcal{L}(y) \cup \mathcal{A}(y) \cup \mathcal{R}(y) | y \in Y\}$. It follows from the maximality of Y that $\mathcal{G}$ is an open refinement of $\mathcal{C}$. For each $x \in X$ there is at most one $y \in Y$ such that $y \leq x$ and $x \in st(y,\mathcal{C})$ and there is at most one $y \in Y$ such that $y \geq x$ and $x \in st(y,\mathcal{C})$. Therefore for each $x \in X$, $\mathcal{G}_x$ is the union of at most four monotone subfamilies and so, by Theorem 5.6, X is orthocompact. ■

COROLLARY. Every GO-space is hereditarily orthocompact.

Every collectionwise normal metacompact space is paracompact, and if there are no measurable cardinals then every normal metacompact space is realcompact. The linearly ordered space ω_1 shows that neither of these results can be generalized by writing "orthocompact" in place of "metacompact."

5.24 We close this section by considering the behavior of preorthocompactness with respect to mappings. Since preorthocompactness is related to metacompactness and paracompactness, it might appear reasonable to conjecture that preorthocompactness is preserved by perfect maps.

EXAMPLE. An orthocompact space X and a perfect mapping f taking X onto a space Y that is not preorthocompact. Let $X_0 = \omega_1 \times [0,1] \times \{0\}$, $X_1 = \omega_1 \times [0,1] \times \{1\}$, and $X = X_0 \cup X_1$. For $\alpha \leq \beta < \omega_1$, where α is a nonlimit ordinal, for $0 \leq x \leq 1$ and $\epsilon > 0$ define $B(\alpha,\beta,x,\epsilon) = [\alpha,\beta] \times (x - \epsilon, x + \epsilon) \times \{0,1\} - [\alpha,\beta] \times \{x\} \times \{0\}$. The family $\{\{p\} | p \in X_0\} \cup \{B(\alpha,\beta,x,\epsilon) | \alpha < \beta < \omega_1$, α a nonlimit ordinal, $0 \leq x \leq 1, \epsilon > 0\}$ is a base for the topology of X.

We first establish that X is orthocompact. Let $\mathcal{C}$ be an open cover of X. For each $x \in [0,1]$ define $g_x : \omega_1 \to \omega_1$ so that $g_x(\beta)$ is a nonlimit ordinal $\alpha \leq \beta$ where for some $\epsilon > 0$ and $C \in \mathcal{C}$, $B(\alpha,\beta,x,\epsilon) \subset C$. By the Pressing Down Lemma there is an $\alpha_x < \omega_1$ such that $\{\beta | g_x(\beta) = \alpha_x\}$ is uncountable. Thus there is an uncountable subset A_x of (α_x,ω_1) and an $\epsilon_x > 0$ such that for each $\beta \in A_x$ there is a $C \in \mathcal{C}$ so that $B(\alpha_x,\beta,x,\epsilon_x) \subset C$. Note that the collection $A(x) = \{B(\alpha_x,\beta,x,\epsilon_x) | \beta \in A_x\}$ is well ordered by set inclusion and so is interior preserving. Let F be a finite subset of [0,1] such that $[0,1] = \bigcup\{(x - \epsilon_x, x + \epsilon_x) | x \in F\}$ and let $\beta_0 = \max\{\alpha_x | x \in F\}$. Then $Z = [0,\beta_0] \times [0,1] \times \{0,1\}$ is an open Lindelöf subspace of X, and there is an interior-preserving open cover $\mathcal{R}$ of Z such that for each $R \in \mathcal{R}$ there is a $C \in \mathcal{C}$ such that $R \subset C$. It follows that $\mathcal{R} \cup (\bigcup\{A(x) | x \in F\}) \cup \{\{p\} | p \in X_0\}$ is an interior-preserving open refinement of $\mathcal{C}$.

Let $\mathcal{D} = \{\{p\} | p \in X_0\} \cup \{\{\beta\} \times [0,1] \times \{1\} | \beta < \omega_1\}$ and let f be the projection of X onto $\mathcal{D}$. The space Y is the decomposition space $\mathcal{D}$ of X, and we regard Y as $X_0 \cup \omega_1$. Since $\mathcal{D}$ is an upper semi-continuous decomposition of X, f is a perfect map.

To see that Y is not orthocompact let $\{x_\alpha | \alpha < \omega_1\}$ be a subset of (0,1) such that $x_\alpha \neq x_\beta$ whenever $\alpha \neq \beta$. For each $\beta < \omega_1$ let $G_\beta = [0,\beta] \cup [X_0 \cap B(0,\beta,x_\beta,1)]$ and let $\mathcal{G} = \{G_\beta | \beta < \omega_1\}$. Suppose that Y is orthocompact. Then there is an interior-preserving open cover $\mathcal{H} = \{H_\beta | \beta < \omega_1\}$ of Y such that for each $\beta < \omega_1$, $H_\beta \subset G_\beta$. It is well known that the open cover $\{[0,\beta] | \beta < \omega_1\}$ of ω_1 has no point-finite open refinement. Thus there exists $\delta \in \omega_1$ so that $\{\beta | \delta \in H_\beta\}$ is infinite. Hence $\bigcap\{H \in \mathcal{H} | \delta \in H\}$ is not open - a contradiction.

Finally, we show that Y is not preorthocompact. It suffices to show that for each neighbornet U of Y there is a transitive neighbornet contained in U^6. Let U be a neighbornet of Y. By Theorem 6.29, there is a transitive neighbornet V of ω_1 such that $V \subset (U \cap (\omega_1 \times \omega_1))^2$. Set $W = V \cup \Delta_{X_0} \cup (U^2 \cap \omega_1 \times X_0)$. Then $W^\infty = W^3 \subset U^6$.

5.25 The following simple example demonstrates that preorthocompactness is also ill behaved with respect to compact open maps.

EXAMPLE. An orthocompact Moore space X and a compact open mapping onto a Hausdorff space Y that is not preorthocompact. Let X be the subspace of the Niemytzki plane (Example 5.17) that is the union of the x-axis and the set to which (x,y) belongs provided that x is an irrational number and $y = 1/n$ for some $n \in \mathbb{N}$. The ground set of the space Y is the real line. Let $\mathcal{T}(E)$ denote the Euclidean topology for Y and let $\mathcal{B}$ be an open-finite base for $\mathcal{T}(E)$. Let J denote the set of all irrational numbers and enumerate the rational numbers as $\{q_n | n \in \mathbb{N}\}$. The topology $\mathcal{T}$ of Y is the topology for which $\{B \cap J | B \in \mathcal{B}\} \cup \{\{q_n\} \cup (q_n - 1/i, q_n + 1/i) \cap J | n \in \mathbb{N} \text{ and } i \geq n\}$ is an open-finite base. Although $(Y,\mathcal{T})$ is not preorthocompact, it is interesting to note that $\mathcal{T}$ has a countable base so that $(Y,\mathcal{T})$ is σ-orthocompact. Clearly the projection map π_1 is a compact open map from X onto Y.

5.26 PROPOSITION. The open finite-to-one image of a (pre)orthocompact space is (pre)orthocompact.

Proof. Let f be an open finite-to-one map from a space X onto a space Y. First suppose that X is orthocompact. Let $\mathcal{C}$ be an open cover of Y and let $\mathcal{R}$ be an interior-preserving open refinement of $\{f^{-1}(C) | C \in \mathcal{C}\}$. To show that $\{f(R) | R \in \mathcal{R}\}$ is an interior-preserving open refinement of $\mathcal{C}$, let $y \in Y$. For each $x \in f^{-1}(y)$, $\cap\mathcal{R}_x$ is open and $y \in \cap\{f(\cap\mathcal{R}_x) | x \in f^{-1}(y)\} \subset \cap\{f(R) | R \in \mathcal{R} \text{ and } y \in f(R)\}$.

Now suppose that X is preorthocompact and let $\mathcal{C}$ be an open cover of Y. There is a neighbornet R of X such that for each $x \in X$ there exists a $C_x \in \mathcal{C}$ such that $R^2(x) \subset f^{-1}(C_x)$. Set $W = \{(y,z) | z \in f(R(p)) \text{ whenever } f(p) = y\}$. Evidently, for each $y \in Y$, $W(y) \in \eta_y$. Let $y \in Y$, let $v \in W^2(y)$ and let $x \in f^{-1}(y)$. There is $u \in W(y)$ such that $v \in W(u)$ and there exists $z \in f^{-1}(u) \cap R(x)$. Similarly, there is $t \in f^{-1}(v) \cap R(z)$. Then $t \in R^2(x) \subset f^{-1}(C_x)$. We have that $v = f(t) \in C_x$ so that $W^2(y) \subset C_x \in \mathcal{C}$. ■

5.27 A topological space X is θ-*refinable* provided that if $\mathcal{C}$ is an open cover of X there is a sequence $<\mathcal{G}_n>$ of open refinements of $\mathcal{C}$ so that for each $x \in X$ there is an $n \in \mathbb{N}$ for which $(\mathcal{G}_n)_x$ is finite. Example 5.17 shows that an orthocompact θ-refinable space need not be semi-metacompact. Although orthocompactness is not preserved by perfect maps, we now establish that the closed continuous image Y of an orthocompact space X is orthocompact provided either X or Y is θ-refinable.

LEMMA. Let X be an orthocompact space and let Y be the image of X under a closed continuous map f. Let $\mathcal{F}$ be a discrete collection of closed subsets of Y and for each $F \in \mathcal{F}$ let G_F be an open subset of Y containing F. Then there is an interior-preserving collection $\{H_F | F \in \mathcal{F}\}$ of open sets such that $F \subset H_F \subset G_F$ for each $F \in \mathcal{F}$.

Proof. Set $\mathcal{K} = \{f^{-1}(F) | F \in \mathcal{F}\}$ and for each $F \in \mathcal{F}$ set $G_F' = f^{-1}(G_F) - \cup(\mathcal{K} - \{f^{-1}(F)\})$. Since $\mathcal{K}$ is a discrete collection of closed sets and $f^{-1}(F) \subset f^{-1}(G_F)$ for each $F \in \mathcal{F}$, $\mathcal{G} = \{X - \cup\{f^{-1}(F) | F \in \mathcal{F}\}\} \cup \{G_F' | F \in \mathcal{F}\}$ is an open cover of X and so $\mathcal{G}$ has an interior-preserving open refinement $\mathcal{R}$. Note that $st(f^{-1}(F),\mathcal{R}) \subset G_F'$ for each $F \in \mathcal{F}$. For each $F \in \mathcal{F}$ set $H_F = Y - f(X - st(f^{-1}(F),\mathcal{R})$. Then $F \subset H_F \subset G_F$ for each $F \in \mathcal{F}$. Since $\{st(f^{-1}(F),\mathcal{R}) | F \in \mathcal{F}\}$ is an interior-preserving collection of open sets and f is closed and continuous, it follows from Proposition 2.5 that $\{H_F | F \in \mathcal{F}\}$ is an interior-preserving collection of open sets. ∎

It is convenient to make use of the following definition in proving Theorem 5.27. Let $\mathcal{G}$ be a collection of subsets of a set X. A collection $\mathcal{R}$ of subsets of X is a *partial refinement* of $\mathcal{G}$ provided that if $R \in \mathcal{R}$ there is a $G \in \mathcal{G}$ such that $R \subset G$.

THEOREM. Let X be an orthocompact space and let Y be the image of X under a closed continuous map f. If Y is θ-refinable, then Y is orthocompact.

Proof. Let $\mathcal{C}$ be an open cover Y and let $<\mathcal{G}_n>$ be a sequence of open refinements of $\mathcal{C}$ such that for each $y \in Y$ there is an $n \in \mathbb{N}$ for which $(\mathcal{G}_n)_y$ is finite.

Let $\mathcal{G}$ be a member of the sequence $<\mathcal{G}_n>$. We show first that there is an interior-preserving open partial refinement $\mathcal{R}$ of $\mathcal{G}$ such that for each $y \in Y$, if $\mathcal{G}_y$ is finite and nonempty then $y \in \cup\mathcal{R}$. For each $n \in \mathbb{N}$ set $S_n = \{y \in Y | \text{card } (\mathcal{G}_y) < n\}$ and note that S_n is closed. We use induction to show that for each $n \in \mathbb{N}$ there is an interior-preserving open partial refinement

$\mathcal{R}_n$ of $\mathcal{G}$ such that $S_n \subset \cup\mathcal{R}_n$. Since $S_1 = \emptyset$, we may choose $\mathcal{R}_1$ to be the empty collection. Let $n > 1$ and suppose that $\mathcal{G}$ has an interior-preserving open partial refinement $\mathcal{R}_{n-1}$ such that $S_{n-1} \subset \cup\mathcal{R}_{n-1}$. Let $\mathcal{W} = \{\cap\mathcal{G}' | \mathcal{G}' \in \mathcal{G}$ and card $(\mathcal{G}') = n - 1\}$. Note that if $y \in Y - S_{n-1}$ then $y \in \cup\mathcal{W}$. Consequently $\mathcal{W}' = \mathcal{W} \cup \{\cup\mathcal{R}_{n-1}\}$ is an open cover of Y. For each $W \in \mathcal{W}$ let $F_W = Y - \cup(\mathcal{W}' - \{W\})$. Then $\{F_W | W \in \mathcal{W}\}$ is a discrete collection of closed sets and $F_W \subset W$ for each $W \in \mathcal{W}$. By the preceding lemma there exists an interior-preserving family $\mathcal{H} = \{H_W | W \in \mathcal{W}\}$ of open sets such that $F_W \subset H_W \subset W$ for each $W \in \mathcal{W}$. Set $\mathcal{R}_n = \mathcal{R}_{n-1} \cup \mathcal{H}$. It remains to show that $S_n \subset \cup\mathcal{R}_n$. Let $y \in S_n - \cup\mathcal{R}_{n-1}$. Then card $(\mathcal{G}_y) = n - 1$ and $W = \cap\mathcal{G}_y$ is the only member of $\mathcal{W}$ to which y belongs. Since $y \notin \cup\mathcal{R}_{n-1}$, $y \in F_W \subset H_W$, and so $S_n \subset \cup\mathcal{R}_n$.

For each $m \in \mathbb{N}$, we have constructed a sequence $\langle\mathcal{R}_{m,n}\rangle$ of interior-preserving partial refinements of $\mathcal{G}_m$ such that if card $((\mathcal{G}_m)_y) < n$, then $y \in \cup\mathcal{R}_{m,n}$. Since Y is θ-refinable, $\cup\{\mathcal{R}_{m,n} | m, n \in \mathbb{N}\}$ is a σ-interior-preserving open cover of Y that refines $\mathcal{C}$. Every θ-refinable space is countably metacompact so that Y is orthocompact by Proposition 5.9. ∎

Since the closed continuous image of a θ-refinable space is θ-refinable, we have the following corollary.

COROLLARY. The closed continuous image of an orthocompact θ-refinable space is orthocompact.

3. QUASI-UNIFORM COVERS

5.28 Let $\mathcal{C}$ be a collection of subsets of a set X. Then $\mathcal{C}^F$ denotes the collection of all finite unions of members of $\mathcal{C}$. A cover $\mathcal{C}$ is *directed* (under set inclusion) provided that the union of any two members of $\mathcal{C}$ is a subset of some member of $\mathcal{C}$. In particular if $\mathcal{C}$ is a cover, then $\mathcal{C}^F$ is a directed cover and if $\mathcal{C}$ is a directed cover, then $\mathcal{C}^F$ refines $\mathcal{C}$.

If $(X, \mathcal{C}(X))$ is a Lebesgue uniformity, then, as is easily verified, for each open cover $\mathcal{C}$ of X there is a natural number n and an open refinement $\mathcal{R}$ of $\mathcal{C}$ so that card $(\mathcal{R}_x) \le n$ for every $x \in X$. The following theorem establishes that $(X, \mathcal{C}(X))$ is a Lindelöf space if and only if every directed open cover of X is a uniform cover. Thus if X is Hilbert space, every directed open cover of $(X, \mathcal{C}(X))$ is a uniform cover, even though $\mathcal{C}(X)$ is not a Lebesgue uniformity. (See the notes.)

THEOREM. For a completely regular topological space $(X,\mathcal{T})$ the following statements are equivalent.

(a) $(X,\mathcal{T})$ is a Lindelöf space.

(b) Every directed open cover of $(X,C(X))$ is a uniform cover.

(c) Every directed open cover of $(X,\mathcal{SC})$ is a quasi-uniform cover.

Proof. (a) ⇒ (b) Suppose that $(X,\mathcal{T})$ is a Lindelöf space. In order to establish condition (b) it suffices to show that every countable directed open cover of X by cozero sets is a uniform cover. Let $\mathcal{C} = \{C_n | n \in \mathbb{N}\}$ be such a cover and for each $n \in \mathbb{N}$ let $f_n : X \to [0,1]$ be a continuous function such that $C_n = X - f_n^{-1}(0)$. Without loss of generality we assume that $C_n \subset C_{n+1}$ for each $n \in \mathbb{N}$. Define $g = 1/(\Sigma_{i=1}^{\infty} f_i/2^i)$. Then g is a continuous function so that $V = \{(x,y) | \, |g(x) - g(y)| \leq 1\} \in C(X)$. To show that $\mathcal{C}$ is a uniform cover, suppose there is a $p \in X$ such that for each $i \in \mathbb{N}$, $V(p) - C_i \neq \emptyset$. For each $i \in \mathbb{N}$ there is an $x_i \in V(p)$ such that $f_j(x_i) = 0$ for $1 \leq j \leq i$. Then for each $n \in \mathbb{N}$, $g(x_n) > 2^n$ and $g(p) \geq g(x_n) - 1$ - a contradiction.

(b) ⇒ (c) This implication holds because $C(X) \subset \mathcal{SC}$.

(c) ⇒ (a) This implication follows from Corollary 2.15. ■

5.29 PROPOSITION. Let X be a topological space.

(a) X is compact if and only if every directed open cover of $(X,\mathcal{P})$ is a quasi-uniform cover.

(b) X is paracompact if and only if every directed open cover of $(X,\mathcal{LF})$ is a quasi-uniform cover.

(c) X is metacompact if and only if every directed open cover of $(X,\mathcal{PF})$ is a quasi-uniform cover.

5.30 The omission of the quasi-uniformity $\mathcal{FT}$ in the preceding proposition suggests a generalization of orthocompactness. A topological space X is *weakly orthocompact* provided that every directed open cover of $(X,\mathcal{FT})$ is a quasi-uniform cover. Thus a space X is weakly orthocompact provided that each directed open cover of X has an interior-preserving open refinement. A motivation for defining weak orthocompactness is provided by the following theorem, which shows that, in contrast to orthocompactness, weak orthocompactness is well behaved under perfect maps.

THEOREM. Let $f : X \to Y$ be a perfect map of X onto Y. Then X is weakly orthocompact if and only if Y is weakly orthocompact.

Proof. We prove only that if X is weakly orthocompact then so is Y, since the converse of this implication is part of a more general result stated below. Suppose that X is weakly orthocompact and let $\mathcal{C}$ be a directed open cover of Y. Since $\mathcal{H} = \{f^{-1}(C) | C \in \mathcal{C}\}$ is a directed open cover of X, there is an interior-preserving open refinement $\mathcal{R}$ of $\mathcal{H}$. Then $\mathcal{R}^F$ is a directed interior-preserving open refinement of $\mathcal{H}$. Set $\mathcal{A} = \{Y - f(X - R) | R \in \mathcal{R}^F\}$. Since $\mathcal{R}^F$ is interior preserving, by Proposition 2.5 we have that $\{X - R | R \in \mathcal{R}^F\}$ is closure preserving so that $\{f(X - R) | R \in \mathcal{R}^F\}$ is closure preserving and $\mathcal{A}$ is interior preserving. Let $\mathcal{B} = \{A | A \in \mathcal{A}$ and there is a $C \in \mathcal{C}$ such that $A \subset C\}$. In order to establish that $\mathcal{B}$ is an open interior-preserving refinement of $\mathcal{C}$, we need only to show that $\mathcal{B}$ is a cover of Y.

Let $y \in Y$. Since $f^{-1}(y)$ is compact, there exists $\{R_i | 1 \le i \le n\} \subset \mathcal{R}$ and $\{C_i | 1 \le i \le n\} \subset \mathcal{C}$ such that $f^{-1}(y) \subset \cup_{i=1}^n R_i$ and for each i with $1 \le i \le n$, $R_i \subset f^{-1}(C_i)$. Set $R = \cup_{i=1}^n R_i$ and $C = \cup_{i=1}^n C_i$. Then $y \in Y - f(X - R)$. Moreover $f(R) \subset C$ and there is a $C' \in \mathcal{C}$ such that $C \subset C'$. Since $Y - f(X - R) \subset f(R) \subset C'$, $y \in \cup\mathcal{B}$. ■

EXAMPLE. Let Y be the range of the perfect map of Example 5.24. Then Y is a weakly orthocompact space that is not preorthocompact.

5.31 THEOREM. Let $f : (X,\mathcal{U}) \to (Y,\mathcal{V})$ be a quasi-uniformly continuous function. If f is a perfect map and every directed open cover of $(Y,\mathcal{V})$ is a quasi-uniform cover, then every directed open cover of $(X,\mathcal{U})$ is a quasi-uniform cover.

Proof. Let $\mathcal{C}$ be a directed open cover of X. Since f is a perfect map and $\mathcal{C}$ is a directed open cover, $\mathcal{C}' = \{Y - f(X - C) | C \in \mathcal{C}\}$ is a directed open cover of Y and there is a $V \in \mathcal{V}$ such that $\{V(y) | y \in Y\}$ refines $\mathcal{C}'$. There is a $U \in \mathcal{U}$ such that $\{(f(x), f(y)) | (x,y) \in U\} \subset V$. Let $x \in X$. There is a $C \in \mathcal{C}$ so that $f(U(x)) \subset V(f(x)) \subset Y - f(X - C)$. Hence $U(x) \subset C$ and $\mathcal{C}$ is a quasi-uniform cover of $(X,\mathcal{U})$. ■

COROLLARY. Let X be a compact Hausdorff space and let Y be a topological space.

(a) If Y is compact, so is $X \times Y$.

(b) If Y is paracompact, so is $X \times Y$.

(c) If Y is metacompact, so is $X \times Y$.

(d) If Y is weakly orthocompact, so is $X \times Y$.

EXAMPLE. Let X be the Sorgenfrey line. Since X is paracompact, every open cover of $(X, \mathcal{LF})$ is a quasi-uniform cover, whereas some directed open cover of $(X \times X, \mathcal{LF} \times \mathcal{LF})$ is not a quasi-uniform cover since the Sorgenfrey plane is not paracompact. (Implicit in this argument is the use of the remark of §2.18 that $f(X) \times f(Y) \subset f(X \times Y)$.)

5.32 Proposition 5.7 suggests that there should be some completeness condition that is a necessary and sufficient condition that every directed open cover of a quasi-uniform space be a quasi-uniform cover.

Let $(X, \mathcal{U})$ be a quasi-uniform space and let $\mathcal{F}$ be a filter on X. Then $\mathcal{F}$ is a *weakly Cauchy* filter provided that for each $U \in \mathcal{U}$ there is a $p \in X$ such that for each $F \in \mathcal{F}$, $F \cap U(p) \neq \emptyset$.

PROPOSITION. Every directed open cover of a quasi-uniform space $(X, \mathcal{U})$ is a quasi-uniform cover if and only if every weakly Cauchy filter has a cluster point.

Proof. Suppose that every weakly Cauchy filter has a cluster point and let $\mathcal{C}$ be a directed open cover of X. Without loss of generality we assume that $X \notin \mathcal{C}$. Then $\{X - C | C \in \mathcal{C}\}$ is a filter base of closed sets that has no cluster point. Consequently, there is a $U \in \mathcal{U}$ such that for each $x \in X$ there is a $C_x \in \mathcal{C}$ such that $U(x) \cap (X - C_x) = \emptyset$. Evidently $\{U(x) | x \in X\}$ refines $\mathcal{C}$.

Now suppose that every directed open cover of $(X, \mathcal{U})$ is a quasi-uniform cover and let $\mathcal{F}$ be a weakly Cauchy filter that has no cluster point. Then $\{X - \overline{F} | F \in \mathcal{F}\}$ is a directed open cover of X. Let $U \in \mathcal{U}$ such that $\{U(x) | x \in X\}$ refines $\{X - \overline{F} | F \in \mathcal{F}\}$. There is a $p \in X$ such that for each $F \in \mathcal{F}$, $U(p) \cap F \neq \emptyset$ and there is $F_p \in \mathcal{F}$ such that $U(p) \subset X - \overline{F}_p$ - a contradiction. ∎

COROLLARY. If every directed open cover of a quasi-uniform space $(X, \mathcal{U})$ is a quasi-uniform cover, then $(X, \mathcal{U})$ is complete.

5.33 We illustrate the use of the completeness condition of §5.32 to establish an elegant characterization of those locally compact quasi-uniform spaces in which every directed open cover is a quasi-uniform cover.

A quasi-uniform space $(X,\mathcal{U})$ is *uniformly locally compact* provided there is a $V \in \mathcal{U}$ such that for each $x \in X$, $\overline{V(x)}$ is compact.

THEOREM. A locally compact quasi-uniform space is uniformly locally compact if and only if every directed open cover is a quasi-uniform cover.

Proof. Let $(X,\mathcal{U})$ be a locally compact quasi-uniform space for which every directed open cover is a quasi-uniform cover. For each $x \in X$ let $N(x)$ be an open neighborhood of x with compact closure. Let $\mathcal{C} = \{N(x) | x \in X\}$. Then there is an entourage $U \in \mathcal{U}$ such that $\{U(x) | x \in X\}$ refines $\mathcal{C}^F$. Let $x \in X$. There is a finite subset $\{y_i | 1 \leq i \leq n\}$ of X such that $U(x) \subset \cup_{i=1}^n N(y_i)$. Since $\overline{N(y_i)}$ is compact for each i with $1 \leq i \leq n$, $\overline{U(x)}$ is compact.

Now suppose that $(X,\mathcal{U})$ is a uniformly locally compact quasi-uniform space. Let $\mathcal{F}$ be a weakly Cauchy filter on X. By Proposition 5.32, it suffices to show that $\mathcal{F}$ has a cluster point. Let $V \in \mathcal{U}$ such that for each $x \in X$, $\overline{V(x)}$ is compact. There is a $p \in X$ such that for each $F \in \mathcal{F}$, $V(p) \cap F \neq \emptyset$. Let $\mathcal{F}' = \{\overline{V(p)} \cap F | F \in \mathcal{F}\}$. Then $\mathcal{F}'$ is a filter on $\overline{V(p)}$ and, since $\overline{V(p)}$ is compact, $\mathcal{F}'$ has a cluster point. It follows that $\mathcal{F}$ has a cluster point. ■

5.34 EXAMPLE. A locally compact Moore space X, which admits a complete uniformity, such that there is a directed open cover of X that is not a quasi-normal cover. Let the set of all rational numbers be enumerated as $\{q_n | n \in \mathbb{N}\}$; set $B = \{(q_n, 1/n) | n \in \mathbb{N}\} \cup \{(q_n + 1/k, 1/n) | n, k \in \mathbb{N}\}$. Let $A = \{(x,0) | x \text{ is an irrational number}\}$, let $X = A \cup B$, and give B the usual subspace topology $\mathcal{T}'$ of the plane. For each $a \in A$ where $a = (x,0)$ and each $\epsilon > 0$ let $T(a,\epsilon) = \{a\} \cup \{(x + k,h) \in X | \; |k| < |h| < \epsilon\}$. Then $\mathcal{T}' \cup \{T(a,\epsilon) | a \in A \text{ and } \epsilon > 0\}$ is a base for a topology $\mathcal{T}$ on X. It is easily verified that $(X,\mathcal{T})$ is a locally compact Moore space. Since A is a discrete closed subspace of $(X,\mathcal{T})$, it follows from Theorem 6.15 that every normal neighbornet of X contains a transitive neighbornet. Thus it suffices to establish that $(X,\mathcal{T})$ is not weakly orthocompact.

Set $\mathcal{C} = \{T(a,2) | a \in A\}$ and let $\mathcal{R}$ be an open refinement of $\mathcal{C}^F$. For each $a \in A$ let $R_a \in \mathcal{R}_a$ and note that there exists $\epsilon_a > 0$ such that if $a = (x,0)$, then $[(x - \epsilon_a, x + \epsilon_a) \times [0,\epsilon_a)] \cap R_a = T(a,\epsilon_a)$. By the Baire Category Theorem there exists an $m \in \mathbb{N}$ and $p, q \in \mathbb{R}$ such that the open interval $(p,q) \times \{0\}$ is a subset of the Euclidean closure of $\{a \in A | 1/m < \epsilon_a\}$. Since there are infinitely many rationals between p and q, there is a $k > m$ such that $q_k - 1/k$ is in the Euclidean interior of the Euclidean closure of $\{a \in A | 1/m < \epsilon_a\}$. Thus there is a natural number b such that for each natural number $j \geq b$, $(q_k + 1/j, 1/k) \notin \cap R_{(q_k, 1/k)}$. Hence R is not interior preserving. By Corollary 7.22, X is not quasi-metrizable.

5.35 EXAMPLE. The space X of Example 5.34 and Theorem 5.30 can be used to show that weak orthocompactness is not hereditary in Moore spaces. (In contrast we show in Theorem 7.26 that orthocompactness is hereditary in Moore spaces.) We can embed the space X as an open subspace in a Moore space Z that admits a perfect finite-to-one map onto a subspace of the Niemytzki plane (Example 5.17) as follows. Let $Z = X \cup (A \times \{0\})$, let X be an open subspace of Z, and for each $(x,0) \in A$, use the sets $\{(x,0,0)\} \cup (G(n,x) - T((x,0),2))$ (see §§5.17 and 5.34), with $n \in \mathbb{N}$ as a neighborhood base for $(x,0,0)$ in Z. Let f be the identity on X and define $f((x,0,0)) = (x,0)$ for each $(x,0) \in A$. Then f is a closed finite-to-one continuous map from Z onto the subspace of the Niemytzki plane corresponding to X.

4. PREORTHOCOMPACTNESS OF $X \times \beta X$

5.36 Although the paracompact spaces may be characterized as those Hausdorff spaces admitting a Lebesgue uniformity, paracompactness is an important topological property whose study is not subsumed by the study of uniform spaces. Similarly, we cannot insist that orthocompactness and preorthocompactness are quasi-uniform spaces properties; and, in the case of preorthocompactness, there is at present no characterization that is couched only in terms of quasi-uniform spaces. This last section of Chapter 5 is devoted to a discussion and proof of H. Junnila's analogue of Tamano's Theorem. Just as Tamano's Theorem is not a theorem of uniform spaces, Junnila's Theorem is somewhat beyond our purview; but it provides a method of constructing completely regular spaces that are not preorthocompact.

Before proving Junnila's Theorem we consider one question that this theorem suggests. Let X be a Tychonoff space. Tamano's Theorem states that X is paracompact if and only if $X \times \beta X$ is normal. Junnila's Theorem states that X is metacompact if and only if $X \times \beta X$ is preorthocompact. It is natural to ask which covering properties of a Tychonoff space X can be characterized in terms of a formally weaker covering property of $X \times \beta X$. From the viewpoint of quasi-uniform spaces, the following problem is of particular interest.

PROBLEM J. Does there exist a topological property P such that if X is a Tychonoff space then $X \times \beta X$ has property P if and only if X is (pre)orthocompact?

We now proceed to establish Junnila's Theorem.

5.37 Let $(X,\mathcal{T})$ be a topological space and let $\mathcal{R}$ and $\mathcal{C}$ be covers of X. We say that $\mathcal{R}$ is a W-*refinement* of $\mathcal{C}$ provided that for each $x \in X$ there is a finite subfamily $\mathcal{C}'$ of $\mathcal{C}$ such that if $R \in \mathcal{R}_x$ then there is a $C \in \mathcal{C}'$ such that $R \subset C$. If for each $x \in X$, $\mathcal{C}'$ may be chosen to consist of only one member of $\mathcal{C}$, then $\mathcal{R}$ is a *point-star refinement* of $\mathcal{C}$. Thus if $\mathcal{R}$ is either a point-star refinement or a point-finite refinement of $\mathcal{C}$, then $\mathcal{R}$ is a W-refinement of $\mathcal{C}$.

5.38 LEMMA. A topological space is metacompact if and only if every open cover has an open W-refinement.

Proof. Let $(X,\mathcal{T})$ be a topological space and suppose that every open cover of X has an open W-refinement. Let $\mathcal{C}_1$ be an open cover of X and for each $n \in \mathbb{N}$ let $\mathcal{C}_{n+1}$ be an open W-refinement of $\mathcal{C}_n$. Let $\mathcal{C}_1$ be well ordered as $\{C_\alpha | \alpha < \gamma\}$ where γ is an ordinal number. For each $C \in \cup_{n=1}^{\infty} \mathcal{C}_n$, let $\alpha(C)$ denote the least ordinal $\alpha < \gamma$ such that $C \subset C_\alpha$; note that if $C' \in \cup_{n=1}^{\infty} \mathcal{C}_n$ and $C \subset C'$ then $a(C) \le \alpha(C')$. For $n > 1$ set $\mathcal{W}_n = \{C \in \mathcal{C}_n | \alpha(C') = \alpha(C)$ whenever $C \subset C' \in \mathcal{C}_{n-1}\}$. For all $n > 1$ and $\alpha < \gamma$ set $V_{(\alpha,n)} = \cup\{C \in \mathcal{W}_n | \alpha(C) = \alpha\}$ and set $\mathcal{V}_n = \{V_{(\alpha,n)} | \alpha < \gamma\}$.

We first show that $\mathcal{V} = \cup_{n=2}^{\infty} \mathcal{V}_n$ is a σ-point-finite open refinement of $\mathcal{C}_1$. Since for each $n > 1$ and $\alpha < \gamma$, $V_{(\alpha,n)}$ is an open set that is contained in C_α, in order to show that $\mathcal{V}$ is an open refinement of $\mathcal{C}_1$ it suffices to show that $\cup_{n=2}^{\infty} \mathcal{W}_n$, and hence $\mathcal{V}$, covers X. Let $x \in X$. For each $n > 1$ set $\alpha_n = \sup \{\alpha(C) | C \in (\mathcal{C}_n)_x\}$; note that since $\mathcal{C}_n$ is a W-refinement of $\mathcal{C}_1$ we have that $\alpha_n < \gamma$. Moreover, since $\mathcal{C}_{n+1}$ is a refinement of $\mathcal{C}_n$, $\alpha_{n+1} \le \alpha_n$ for each $n \in \mathbb{N}$. Hence there is an ordinal $\beta < \gamma$ and a natural number $k > 2$ such that for each $n \ge k - 1$, $\alpha_n = \beta$. Since $\mathcal{C}_{k+1}$ is a W-refinement of C_k, there is a finite subfamily F of $(C_k)_x$ such that if $C \in (C_{k+1})_x$ then there is an $F \in F$ such that $C \subset F$. Let $F \in F$ be such that for each $F' \in F$ $\alpha(F') \leq \alpha(F)$. Evidently $\alpha_{k+1} \leq \alpha(F)$ and since $F \in C_k$ we have that $\alpha(F) \leq \alpha_k$. Since $\alpha_{k+1} = \alpha_k = \beta$, it follows that $\alpha(F) = \beta$. To show that $F \in W_k$ let $C \in C_{k-1}$ such that $F \subset C$. Since $x \in F$, $C \in (C_{k-1})_x$ so that $\alpha(C) \leq \beta = \alpha(F) \leq \alpha(C)$. Thus $\cup_{n=2}^{\infty} W_n$, and hence V, covers X.

In order to show that $V = \cup_{n=2}^{\infty} V_n$ is a σ-point-finite open refinement of C_1 it remains to establish that for each $n > 1$, V_n is a point-finite collection. Let $n \in N$, let $x \in X$, and let C' be a finite subcollection of C_n such that if $C \in (C_{n+1})_x$ then there is a $C' \in C'$ so that $C \subset C'$. In order to show that $(V_{n+1})_x$ is finite we show that card $((V_{n+1})_x) \leq$ card (C'). Let $A = \{\alpha(C') | C' \in C'\}$ and $B = \{\alpha < \gamma | x \in V_{(\alpha,n+1)}\}$. It suffices to establish that $B \subset A$. Let $\beta \in B$. Since $x \in V_{(\beta,n+1)}$ there exists $W \in W_{n+1}$ such that $\alpha(W) = \beta$ and $x \in W$. Then $W \in (C_{n+1})_x$ so that there is a $C' \in C'$ such that $W \subset C'$. Since $W \in W_{n+1}$, we have that $\alpha(C') = \alpha(W)$. Since $\alpha(W) = \beta$ and $\alpha(C') \in A$, $\beta \in A$. Thus $B \subset A$ and $(V_{n+1})_x$ is finite.

For each $n > 1$ set $R_n = \cup V_n$ and set $R = \{R_n | n \in N\}$. Let G be a W-refinement of R. Set $F_1 = \emptyset$ and for each $n > 1$ set $F_n = \{x | \text{st}(x,G) \subset \cup_{k=2}^{n} R_k\}$. For each $n > 1$ set $V'_n = \{V - F_{n-1} | V \in V_n\}$ and set $V' = \cup_{n=2}^{\infty} V'_n$. Let $x \in X$ and let j be the least natural number n so that $x \in R_n$. Then $x \in \cup V'_j$. Thus V' covers X and is an open refinement of C_1. Moreover as G is a W-refinement of R it follows easily that V' is a point-finite collection.

As is remarked in §5.37, every point-finite refinement is a W-refinement. Consequently it is evident that every open cover of a metacompact space has an open W-refinement. ■

5.39 LEMMA. Let (X,T) be a preorthocompact space. Then (X,T) is metacompact if and only if for each directed open cover C of X there is a neighbornet V of X so that $\{V^{-1}(x) | x \in X\}$ refines C.

Proof. Suppose first that (X,T) is metacompact. Let C be a directed open cover of X and let R be a point-finite open refinement of C. Set $V = \{(x,y) | y \in \cap R_x\}$. Let $x \in X$. Then $A = \{V(y) | x \in V(y)\}$ is a finite collection. For each $A \in A$ choose $C_A \in C$ such that $A \subset C_A$. Since C is a directed open cover, $V^{-1}(x) \subset \cup A \subset \cup\{C_A | A \in A\} \subset C$ for some $C \in C$.

Now suppose that for every directed open cover C of X there is a neighbornet V on X such that $\{V^{-1}(x) | x \in X\}$ refines C. To show that X is metacompact, in light of the previous lemma, it suffices to show that every open cover of X has an open W-refinement. Let $\mathcal{C}$ be an open cover of X. Since X is preorthocompact, there is an open cocushioned refinement $\mathcal{R}$ of $\mathcal{C}$. For each $R \in \mathcal{R}$ let $C_R \in \mathcal{C}$ such that for each subcollection $\mathcal{R}'$ of $\mathcal{R}$, $\bigcap\{R | R \in \mathcal{R}'\} \subset \text{int } (\bigcap\{C_R | R \in \mathcal{R}'\}$. By hypothesis there is a neighbornet V of X such that $\{V^{-1}(x) | x \in X\}$ is a refinement of $\mathcal{R}^F$. For each $x \in X$ set $D(x) = \text{int } [V(x) \cap \bigcap\{C_R | x \in R\}]$ and set $\mathcal{D} = \{D(x) | x \in X\}$. We assert that $\mathcal{D}$ is an open W-refinement of $\mathcal{C}$. Let $x \in X$. There is a finite subcollection $\{R_i | 1 \le i \le n\}$ of $\mathcal{R}$ such that $V^{-1}(x) \subset \cup_{i=1}^n R_i$. Let $D(y) \in \mathcal{D}_x$. There is a k with $1 \le k \le n$ such that $y \in R_k$ and so $D(y) \subset C_{R_k}$. Thus $\{C_{R_i} | 1 \le i \le n\}$ is a finite subcollection of $\mathcal{C}$ such that if $D(y) \in \mathcal{D}_x$ then $D(y)$ is a subset of some member of $\{C_{R_i} | 1 \le i \le n\}$. ∎

5.40 JUNNILA'S THEOREM. Let K be a compactification of a Tychonoff space X. Then $X \times K$ is preorthocompact if and only if X is metacompact.

Proof. If X is metacompact, by Corollary 5.31 we have that $X \times K$ is metacompact and hence preorthocompact.

Now suppose that $X \times K$ is preorthocompact. Let $\mathcal{C}$ be a directed open cover of X. By the previous lemma it suffices to establish the existence of a neighbornet A of X such that $\{A^{-1}(x) | x \in X\}$ refines $\mathcal{C}$. Since X is a subspace of K, there exists a family $\mathcal{V}$ of open subsets of K such that $\mathcal{C} = \{V \cap X | V \in \mathcal{V}\}$. Let $H = \cup\mathcal{V}$. Since K is a regular space, there exists a family $\mathcal{W}$ of open subsets of K such that $\cup\mathcal{W} = H$ and $\overline{W} \subset H$ for each $W \in \mathcal{W}$. It is easily seen that the family $\mathcal{G} = \{X \times H\} \cup \{(W \cap X) \times (K - \overline{W}) | W \in \mathcal{W}\}$ is an open cover of $X \times K$. There is a neighbornet R of $X \times K$ such that $\{R^2(p) | p \in X \times K\}$ refines $\mathcal{G}$. Set $A = \{(x,y) | (x,y) \in X \times X$ and $y \notin \pi_1(X \times K - \text{int } [R(\{x\} \times K)])\}$. Since π_1 is a closed mapping, A is a neighbornet of X. Consequently, we have only to establish that $\{A^{-1}(x) | x \in X\}$ is a refinement of $\mathcal{C}$. Let $x \in X$ and set $M = \{m \in K | (x,m) \notin \text{int } [R(\{x\} \times (K - H)]\}$; then M is a closed, and hence compact, subset of K. Evidently $M \subset H$ and since M is compact, M is covered by a finite subfamily of $\mathcal{V}$. Since $\mathcal{C}$ is a directed open cover there is a $C \in \mathcal{C}$ such that $M \cap X \subset C$. Thus we need only show that $A^{-1}(x) \subset M$. Suppose that $y \in A^{-1}(x) - M$. Since $y \notin M$, there exist $k \in K - H$ such that $(x,y) \in R((x,k))$. Since $x \in A(y)$, $x \notin \pi_1(X \times K - \text{int } R(\{y\} \times K))$ and so $\{x\} \times K \subset R(\{y\} \times K)$. Hence there is a $q \in K$ such that $(x,k) \in R((y,q))$ so that $(x,y) \in R((x,k)) \subset R^2((y,q))$. There is some $G \in \mathcal{G}$ so that $R^2((y,q)) \subset G$.

Since $R^2((y,q)) \nsubseteq X \times H$, there is a $W \in \mathcal{W}$ so that $R^2((y,q)) \subset (W \cap X) \times (K - \overline{W})$. Hence $\{(x,y), (y,q)\} \subset (W \cap X) \times (K - \overline{W})$ - a contradiction. Thus $A^{-1}(x) \subset M$ as was to be shown. ■

COROLLARY. A Tychonoff space X is metacompact if and only if $X \times \beta X$ is preorthocompact.

Chapter 6

TRANSITIVE SPACES

1. TRANSITIVE UNIFORMITIES

6.1 In the previous chapters we considered uniform spaces only to establish results that pertain to a wider class of quasi-uniformities, and in general, throughout this text we give the uniform spaces short shrift because an equitable allotment of time to them would prove overwhelming. Even when we restrict our attention to transitive quasi-uniformities the study of the uniformities among them leads from our province into such topics as dimension theory, valuation rings, and Boolean algebras. In this section, therefore, we only consider transitive uniformities in order to obtain insight into unresolved problems concerning transitive quasi-uniformities.

We have already seen that every topological space admits a transitive quasi-uniformity and that the collection of all transitive quasi-uniformities compatible with a given topology has a finest member. Our first proposition shows that, although a completely regular space need not admit a transitive uniformity, any space that does admit a transitive uniformity also admits a finest transitive uniformity.

PROPOSITION. Let $(X,\mathcal{T})$ be a topological space. There is a transitive uniformity compatible with $\mathcal{T}$ if and only if $(X,\mathcal{T})$ is a zero-dimensional space. If $(X,\mathcal{T})$ is a zero-dimensional space and $\mathcal{A}$ is the collection of all open partitions of X, then $\mathcal{U}_{\mathcal{A}}$ is the finest transitive uniformity compatible with $\mathcal{T}$.

Proof. Suppose that $\mathcal{U}$ is a transitive uniformity compatible with $\mathcal{T}$. Then there is a base $\mathcal{B}$ for $\mathcal{U}$ consisting of equivalence relations on X. For each $x \in X$ and $B \in \mathcal{B}$, $B(x)$ is both open and closed so that $\{B(x) \mid x \in X, B \in \mathcal{B}\}$ is a base for $\mathcal{T}$ consisting of closed sets.

Now suppose that $(X,\mathcal{T})$ is zero-dimensional and let $\mathcal{A}$ be the collection of all open partitions of X. It is evident that $\mathcal{A}$ satisfies the conditions of Theorem 2.6 so that $\mathcal{U}_{\mathcal{A}}$ is a transitive uniformity compatible with $\mathcal{T}$. Let $\mathcal{V}$ be a transitive uniformity compatible with $\mathcal{T}$ and let $\mathcal{B}$ be a base for $\mathcal{V}$ consisting of equivalence relations. For each $B \in \mathcal{B}$, $\{B(x) \mid x \in X\} \in \mathcal{A}$ and it follows that $B \in \mathcal{U}_{\mathcal{A}}$. Thus $\mathcal{V} \subset \mathcal{U}_{\mathcal{A}}$. ∎

6.2 In light of the previous proposition, it is natural to seek a characterization of those topological spaces that admit a coarsest transitive uniformity. If $(X,\mathcal{T})$ admits a coarsest transitive uniformity, this uniformity is the coarsest uniformity compatible with $\mathcal{T}$. This result can be established by showing that if X has a zero-dimensional compactification cX such that cX - X has more than one point, then by identifying two points of cX - X one obtains a smaller zero-dimensional compactification of X. It follows from the next proposition that a topological space admits a coarsest transitive uniformity if and only if it is locally compact and zero-dimensional.

PROPOSITION. Let $(X,\mathcal{T})$ be a locally compact zero-dimensional Hausdorff space. The coarsest quasi-uniformity compatible with $\mathcal{T}$ is a transitive quasi-uniformity.

Proof. Let δ be the quasi-proximity determined by the strong inclusion of Lemma 1.46. Then $\mathcal{U}_\delta$ is the coarsest quasi-uniformity compatible with $\mathcal{T}$ and so it suffices to show that $T(A,B)$ contains a transitive member of $\mathcal{U}_\delta$ whenever $A \not\delta B$. Let $B \neq \emptyset$ and let $A \not\delta B$. There is a compact set K and an open set G so that $A \subset K \subset G \subset X - B$. Since $(X,\mathcal{T})$ is zero-dimensional, we may assume without loss of generality that K is an open and closed set. Set $U = T(K,X - K)$. Then $U \in \mathcal{U}_\delta$, U is transitive and $U \subset T(A,B)$. ∎

6.3 LEMMA. Let $\mathcal{U}$ be a transitive quasi-uniformity on a set X. Then $\mathcal{U}_\omega$ is a transitive quasi-uniformity.

Proof. Let $\mathcal{V}$ be the quasi-uniformity on X for which $\{T(G,X - G) \mid G \not\delta_{\mathcal{U}} X - G\}$ is a subbase. Since $\mathcal{V}$ is a totally bounded transitive quasi-uniformity coarser than $\mathcal{U}$, it suffices to establish that $\delta_{\mathcal{V}} \subset \delta_{\mathcal{U}}$. Suppose that $A \not\delta_{\mathcal{U}} B$. Then there is a transitive entourage $U \in \mathcal{U}$ such that $A \times B \cap U = \emptyset$. It follows that $U \cap (U(A) \times X - U(A)) = \emptyset$ so that $T(U(A), X - U(A)) \in \mathcal{V}$ and $A \not\delta_{\mathcal{V}} B$. ■

6.4 THEOREM. Let $(X,\mathcal{T})$ be a completely regular space. Then the following statements are equivalent.

(a) If R is a cozero-set that contains a zero-set Z, then there is an open and closed set H such that $Z \subset H \subset R$.

(b) Every zero-set is the countable intersection of sets that are open and closed.

(c) Every countable cover of X by cozero-sets has an open refinement that is a partition of X.

(d) The finest uniformity compatible with $\mathcal{T}$ is a transitive uniformity.

(e) C^* is a transitive uniformity.

(f) C is a transitive uniformity.

Proof. We first establish that (a), (b) and (c) are equivalent.

(a) ⇒ (b) Let Z be a zero-set in X. There is a nested sequence $\langle R_m \rangle$ of cozero-sets so that $Z = \cap_{n=1}^{\infty} R_n$. By hypothesis for each $n \in \mathbb{N}$ there is an open and closed set H_n so that $Z \subset H_n \subset R_n$. Thus $Z = \cap_{n=1}^{\infty} H_n$.

(b) ⇒ (c) Let $\mathcal{R} = \{R_n | n \in \mathbb{N}\}$ be a countable cover of X by cozero-sets. For each $n \in \mathbb{N}$ there is a countable collection $\mathcal{R}_n$ of sets that are both open and closed such that $R_n = \cup\mathcal{R}_n$. Set $\mathcal{A} = \cup_{n=1}^{\infty} \mathcal{R}_n$, enumerate the members of $\mathcal{A}$ as $A_1, A_2, A_3, \ldots$, and set $A_0 = \emptyset$. Then $\{A_n - \cup_{i=0}^{n-1} A_i |$ $n \in N\}$ is an open partition of X that refines $\mathcal{R}$.

(c) ⇒ (a) Let R be a cozero-set that contains a zero set Z. There is an open partition $\mathcal{A}$ of X that refines $\{R, X - Z\}$. Thus $H = \cup\{A \in \mathcal{A} | Z \cap A \neq \emptyset\}$ is an open and closed set containing Z and contained in R.

(c) ⇒ (d) Let U be a member of the fine uniformity. It follows from Lemma 1.5 that there is a pseudo-metric d for X such that $\mathcal{T}(d) \subset \mathcal{T}$ and $\{(x,y) | d(x,y) < 1\} \subset U$. Let $\mathcal{A} = \{S(x,1/2) | x \in X\}$. Then $\mathcal{A}$ is $\mathcal{T}(d)$-open cover of the pseudo-metrizable space $(X,\mathcal{T}(d))$ so that $\mathcal{A}$ has a locally finite $\mathcal{T}(d)$-open refinement $\mathcal{R} = \{R_\lambda | \lambda \in \Lambda\}$, where the index set Λ is taken as a set well ordered by <. Since $(X,\mathcal{T}(d))$ is normal, there is a $\mathcal{T}(d)$-closed cover $\{Z_\lambda | \lambda \in \Lambda\}$ of X so that for each $\lambda \in \Lambda$, $Z_\lambda \subset R_\lambda$. With respect to $\mathcal{T}(d)$ each R_λ is a cozero-set and each Z_λ is a zero-set, and since $\mathcal{T}(d) \subset \mathcal{T}$, R_λ is a cozero-set in $(X,\mathcal{T})$ and Z_λ is a zero-set in $(X,\mathcal{T})$. Since we have already established the implication (c) ⇒ (a) it follows that for each $\lambda \in \Lambda$ there is an open and closed set H_λ so that $Z_\lambda \subset H_\lambda \subset R_\lambda$. For each $\lambda \in \Lambda$ set $H'_\lambda = H_\lambda - \cup\{H_\alpha | \alpha < \lambda\}$ and set $\mathcal{B} = \{H'_\lambda | \lambda \in \Lambda\}$. By Proposition 6.1, $U_{\mathcal{B}}$ is a member of the finest transitive uniformity compatible with $\mathcal{T}$; and since $U_{\mathcal{B}} \subset U$, U belongs to this transitive uniformity as well.

(d) ⇒ (e) This implication is an immediate consequence of Lemma 6.3.

(e) ⇒ (a) Let Z be a zero-set and let R be a cozero-set containing Z. There is a bounded real-valued continuous function f such that $f(Z) = 0$ and $f(X - R) = 1$. Since $U_{(1/2,f)} \in C^*$, there is an equivalence relation V on X such that $V \subset U_{(1/2,f)}$. Evidently, V(Z) is an open and closed set and $Z \subset V(Z) \subset R$.

(e) ⇔ (f) Let $U_{(\epsilon,f)}$ be a typical subbasic entourage of C and let $\mathcal{R} = \{(j\epsilon/3, (j+2)\epsilon/3) | j$ is an integer$\}$. Then $\mathcal{A} = \{f^{-1}(R) | R \in \mathcal{R}\}$ is a countable cover of X by cozero-sets so that there is a countable open partition $\{G_n | n \in \mathbb{N}\}$ of X that refines $\mathcal{A}$. Define $g : X \to \mathbb{R}$ by $g(x) = n$ if $x \in G_n$. Then $U_{(1,g)}$ is transitive and $U_{(1,g)} \subset U_{(\epsilon,f)}$. Thus (e) ⇒ (f) and the equivalence of conditions (e) and (f) now follows directly from Lemma 6.3. ∎

6.5 A completely regular space that satisfies any of the equivalent conditions of the previous theorem is called a *strongly zero-dimensional space*. In light of the previous theorem, it is obvious that every strongly zero-dimensional space is zero-dimensional and that every compact zero-dimensional space is strongly zero-dimensional. It is natural to seek less stringent conditions under which zero-dimensionality and strong zero-dimensionality coincide.

PROPOSITION. Every zero-dimensional Lindelöf space is strongly zero-dimensional.

Proof. Let $(X,\mathcal{T})$ be a zero-dimensional Lindelöf space, let Z be a zero-set, and let R be a cozero-set containing Z. For each $x \in X$ let G_x be an open and closed set that contains x such that if $G_x \cap Z \neq \emptyset$ then $G_x \subset R$. Then $\{G_x | x \in X\}$ has a countable subcover $\{G_{x_n} | n \in \mathbb{N}\}$. For each $n \in \mathbb{N}$ set $H_n = G_{x_n} - \cup_{i<n} G_{x_i}$ and set $H = \cup\{H_n | H_n \cap Z \neq \emptyset\}$. Then H is an open and closed set such that $Z \subset H \subset R$. ∎

6.6 COROLLARY. The fine uniformity of any countable regular space is transitive.

6.7 A topological space is a P-*space* provided that the intersection of any countable collection of open sets is open. It is easy to deduce from Theorem 6.4(b) that every completely regular P-space is strongly zero-dimensional. We give a proof of this result, which relies upon the characterization of strongly zero-dimensional spaces as those spaces whose fine uniformity is transitive. The advantage of our roundabout argument is that it can be extended to show that the finest quasi-uniformity on a P-space is transitive whereas the obvious argument, which relies upon open and closed subspaces, cannot.

PROPOSITION. Let $(X,\mathcal{T})$ be a completely regular P-space. Then $(X,\mathcal{T})$ is strongly zero-dimensional.

Proof. Let U be a member of the fine uniformity and let $\langle U_n \rangle$ be a normal sequence of symmetric entourages of the fine uniformity so that $U_1 \subset U$. Then $V = \cap_{n=1}^{\infty} U_n$ is an equivalence relation on X and for each $x \in X$, $V(x) = \cap_{n=1}^{\infty} U_n(x)$ is open. It follows that U is a member of the finest transitive uniformity compatible with $\mathcal{T}$. ■

6.8 It seems irresistible to conjecture that any topological space that admits both a transitive uniformity and a uniformity with a countable base also admits a uniformity with a countable transitive base. Indeed, the counterexample to this longstanding conjecture is both classic and formidable.

A quasi-metric d on a set X is a *non-Archimedean quasi-metric* provided that for all x, y, $z \in X$, $d(x,z) \le \max \{d(x,y), d(y,z)\}$.

THEOREM. Let $(X,\mathcal{T})$ be a T_0 topological space. The following statements are equivalent.

(a) $(X,\mathcal{T})$ is strongly zero-dimensional and metrizable.

(b) There is a uniformity $\mathcal{U}$ compatible with $\mathcal{T}$ that has a countable transitive base.

(c) $(X,\mathcal{T})$ is non-Archimedeanly metrizable.

Proof. (a) ⇒ (b) Let d be a metric on X compatible with $\mathcal{T}$. For each $n \in \mathbb{N}$ let $U_n = \{(x,y) | d(x,y) < 2^{-n}\}$. Since for each $n \in \mathbb{N}$, U_n is a member of the fine uniformity, which is transitive, for each $n \in \mathbb{N}$ there is an equivalence relation V_n in the fine uniformity so that $V_n \subset U_n$. Then $\{V_n | n \in \mathbb{N}\}$ is a countable transitive base for a uniformity that is compatible with $\mathcal{T}$.

(b) ⇒ (c) Let $\mathcal{U}$ be a uniformity compatible with $\mathcal{T}$ that has a countable transitive base $\{U_n | n \in \mathbb{N}\}$. Without loss of generality we assume that for each $n \in \mathbb{N}$, $U_{n+1} \subset U_n$ and U_n is symmetric. For each $x \in X$ define $d(x,x) = 0$. For $x \neq y$ set $d(x,y) = 2^{-n}$, where n is the least positive integer j for which $(x,y) \notin U_j$. Then d is a non-Archimedean metric compatible with $\mathcal{T}$.

(c) ⇒ (a) Let (X,d) be a non-Archimedean metric space and let F be a closed subset of X. For each $n \in \mathbb{N}$ set $V_n = \{(x,y) | d(x,y) < 1/n\}$. For each $n \in \mathbb{N}$, $V_n(F)$ is an open and closed set, and $F = \cap_{n=1}^{\infty} V_n(F)$. Thus condition (b) of Theorem 6.4 obtains. ■

6.9 As we indicated in the beginning of this section, our motivation for discussing transitive uniformities is to gain insight into problems concerning transitive quasi-uniformities. In particular in the next section we begin the study of the topological spaces for which $\mathcal{FT} = \mathcal{FINE}$. These spaces are, in a sense that Theorem 6.4 makes precise, the quasi-uniform analogues of strongly zero-dimensional spaces. Therefore, we end this section with a summary of those facts concerning strongly zero-dimensional spaces that appear related to the problems to be discussed in the next section. (See the notes.)

(a) There is a topological space X whose fine uniformity is transitive such that the fine uniformity of X × X is not transitive.

(b) There is a topological space that admits a transitive uniformity and a uniformity with a countable base, but does not admit a transitive uniformity with a countable base.

(c) If a topological space admits a uniformity with a countable transitive base, then its fine uniformity has a transitive base.

(d) Let $(X,\mathcal{T})$ be a zero-dimensional space that is a (compact, Lindelöf, countable, P-) space. The fine uniformity of $(X,\mathcal{T})$ is transitive.

(e) Let X be a topological space whose fine uniformity is transitive and let $f : X \to Y$ be a continuous function that is both an open map and a closed map. Then the fine uniformity of Y has a transitive base.

2. SUBSPACES AND MAPPINGS

6.10 A topological space $(X,\mathcal{T})$ is said to be a *transitive space* provided that $\mathcal{FT} = \mathcal{FINE}$; that is, provided the fine quasi-uniformity has a transitive base. In light of the importance of the strongly zero-dimensional spaces and the connection between these spaces and the transitive spaces, it is not surprising that the transitive spaces occupy a prominent position in the study of quasi-uniform spaces. A principal reason that the class of transitive spaces is important in the study of quasi-uniform spaces is that several of the properties we consider coincide in this class of spaces. We have already noted one such occurrence in Problem B, and the following proposition records another obvious instance.

PROPOSITION. A transitive space X is orthocompact if and only if every open cover of X is a quasi-normal cover.

6.11 Although, in general, it is difficult to determine whether a given topological space is a transitive space, the proof of Proposition 6.7 shows that the intersection of the members of a normal sequence of neighbornets of a P-space is a transitive neighbornet. Thus we have established the following proposition.

PROPOSITION. Every P-space is a transitive space.

6.12 EXAMPLE. A normal transitive space that is not preorthocompact. Let X be a normal P-space that is not metacompact. (See the notes.) Let $\mathcal{G}(X)$ be the space constructed from X that is described in §5.14. Then $\mathcal{G}(X)$ is a normal transitive space that is not preorthocompact.

6.13 PROPOSITION. Let X be a set and let $\mathcal{U}$ be a quasi-uniformity on X with a base $\mathcal{B}$ that is linearly ordered by inclusion. Then either $\mathcal{U}$ has a countable base or $\mathcal{U}$ is closed under countable intersections and has a transitive base.

Proof. Suppose that $\mathcal{U}$ has no countable base. Let $\mathcal{C} = \{U_n \mid n \in \mathbb{N}\}$ be a countable subset of $\mathcal{U}$ and for each $n \in \mathbb{N}$ let $B_n \in \mathcal{B}$ such that $B_n \subset U_n$. Since $\{B_n \mid n \in \mathbb{N}\}$ not a base for $\mathcal{U}$, there is an entourage $B \in \mathcal{B}$ such that for each $n \in \mathbb{N}$, $B_n \not\subset B$; and, since $\mathcal{B}$ is linearly ordered, $B \subset \bigcap_{n=1}^{\infty} B_n \subset \bigcap\mathcal{C}$. Hence $\bigcap\mathcal{C} \in \mathcal{U}$.

For each $B \in \mathcal{B}$ choose a normal sequence $\langle B_n \rangle$ of entourages in $\mathcal{B}$ such that $B_1 \subset B$ and set $D(B) = \bigcap_{n=1}^{\infty} B_n$. Set $\mathcal{S} = \{D(B) \mid B \in \mathcal{B}\}$. Then as above $\mathcal{S} \subset \mathcal{U}$ and it follows readily that $\mathcal{S}$ is a transitive base for $\mathcal{U}$. ∎

COROLLARY. Let X be a set and let $\mathcal{U}$ be a quasi-uniformity on X with a base that is linearly ordered by inclusion. If $\mathcal{U}$ has no countable base, then $(X,\mathcal{T}(\mathcal{U}))$ is a P-space.

Proof. Let $\{G_n | n \in \mathbb{N}\}$ be a countable collection of $\mathcal{T}(\mathcal{U})$-open sets and let $x \in \cap_{n=1}^{\infty} G_n$. For each $n \in \mathbb{N}$ there is a $U_n \in \mathcal{U}$ such that $U_n(x) \subset G_n$. Set $U = \cap_{n=1}^{\infty} U_n$. Then $U \in \mathcal{U}$ and $U(x) \subset \cap_{n=1}^{\infty} G_n$. Thus $\cap_{n=1}^{\infty} G_n \in \mathcal{T}(\mathcal{U})$. ∎

6.14 Although the following proposition is an immediate consequence of Corollary 2.19, it is so useful that it warrants a formal statement.

PROPOSITION. Let $(X,\mathcal{T})$ be a transitive space. Every subspace of X that is the intersection of an open set and a closed set is a transitive space.

COROLLARY. Every locally compact subspace of a Hausdorff transitive space is a transitive space.

6.15 We now consider conditions under which transitivity of open or closed subspaces of a space X forces transitivity of X itself. We make use of the following definition: A topological space is an F_σ-*discrete space* provided that it is the countable union of closed discrete subspaces.

THEOREM. Every countable union of closed transitive spaces is transitive.

Proof. Let $X = \cup_{n=1}^{\infty} F_n$, where each F_n is a closed and transitive subspace of X. To show that X is transitive, let U be a normal neighbornet and let $\langle U_n \rangle$ be a normal sequence of neighbornets of X such that $U_1^4 \subset U$. For each $n \in \mathbb{N}$, there exists a transitive neighbornet V_n of the subspace F_n such that $V_n \subset U_n$. By Proposition 2.19 there is for each $n \in \mathbb{N}$, a $W_n \in \mathcal{FT}_X$ such that $W_n \cap F_n \times F_n = V_n$. For each $x \in X$ let $m(x)$ denote the least element of the set $\{n \in \mathbb{N} | x \in F_n\}$. Define a neighbornet W of X by setting for each $x \in X$

$$W(x) = (W_{m(x)} \cap U_{m(x)})(x) - \cup\{F_n | n < m(x)\}$$

We show that the transitive neighbornet W^∞ is contained in U. Let $(x,y) \in W^\infty$ and let k be the least $n \in \mathbb{N}$ such that $(x,y) \in W^n$. Then there exists a finite sequence $x = x_1, x_2, x_3, \ldots, x_k, x_{k+1} = y$ such that for each $i \le k$, $(x_i,x_{i+1}) \in W$. Note that $m(x_i) \le m(x_{i+1})$ for all $i \le k$. Suppose there is an $i < k$ such that $m(x_i) = m(x_{i+1}) = m(x_{i+2}) = m$. Then

$x_{i+1} \in W_m(x_i) \cap F_m = V_m(x_i)$ and similarly $x_{i+2} \in V_m(x_{i+1})$. Thus $x_{i+2} \in V_m(x_i) \subset W_m(x_i) \cap U_m(x_i)$ so that $(x_i, x_{i+2}) \in W$ and $(x,y) \in W^{k-1}$ - a contradiction. Hence for each $i < k$, $m(x_i) < m(x_{i+2})$ and so $(x,y) \in U_{m(x_1)} \circ U_{m(x_2)} \circ U_{m(x_3)} \circ \ldots \circ U_{m(x_k)} \subset U_1 \circ U_1 \circ U_2 \circ U_2 \circ \ldots \circ U_k \circ U_k \subset U_1^4 \subset U$. ■

COROLLARY. Every F_σ-discrete space is a transitive space.

6.16 LEMMA. Let $(X,\mathcal{T})$ be a topological space that is the union of two transitive subspaces one of which is either open or closed. Then $(X,\mathcal{T})$ is a transitive space.

Proof. Without loss of generality, we assume that $X = A \cup B$ where A is open and transitive, B is closed and transitive, and $A \cap B = \emptyset$. Let U be a normal neighbornet of X and let U_1 be a normal neighbornet of X such that $U_1^3 \subset U$. Choose a transitive neighbornet V of A such that $V \subset U_1 \cap A \times A$ and choose a transitive neighbornet W of B such that $W \subset U_1 \cap B \times B$. As in Proposition 2.19 set $V^\dagger = V \cup (B \times X)$ and $W^\dagger = W \cup (X \times A)$. Set $Q = W^\dagger \cap V^\dagger \cap U_1 = W \cup V \cup (B \times A \cap U_1)$. Then $Q^\infty = Q^3 \subset U$. ■

6.17 It is evident from Lemma 6.16 that the union of finitely many open (or closed) transitive spaces is transitive. We can also use this lemma to establish an analogue of Theorem 6.15 for open transitive spaces.

THEOREM. Every countable union of open transitive spaces is transitive.

Proof. Let $X = \cup_{n=1}^\infty G_n$ where each G_n is an open transitive subspace of X. Set $G_0 = \emptyset$. In light of Lemma 6.16 we assume without loss of generality that for each $n \in \mathbb{N}$, $G_n \subset G_{n+1}$. For each $x \in X$ let $m(x)$ be that natural number n such that $x \in G_n - G_{n-1}$. Let U be a normal neighbornet of X and let $\langle U_n \rangle$ be a normal sequence such that $U_1^2 \subset U$. For each $n \in \mathbb{N}$, $U_n \cap G_n \times G_n$ is a normal neighbornet of the transitive space G_n so that there exists a transitive neighbornet V_n of G_n such that $V_n \subset U_n$. Let $V = \cup_{n=1}^\infty V_n \cap (G_n - G_{n-1}) \times G_n$. We assert that $V^\infty \subset U$. Let $(x,y) \in V^\infty$ and let k be the least positive integer such that $(x,y) \in V^k$. There is a finite sequence $x = x_1, x_2, \ldots, x_k, x_{k+1} = y$ such that for $1 \le i \le k$, $(x_i, x_{i+1}) \in V$. Evidently for each $i < k$, $m(x_i) \ge m(x_{i+1})$ and $(x_i, x_{i+2}) \in V$ whenever $m(x_i) = m(x_{i+1})$. Thus for each $i < k$, $m(x_i) > m(x_{i+1})$ so that $(x,y) \in U_1 \circ U_2 \circ \ldots \circ U_{k-1} \circ U_k \subset U_1^2 \subset U$. ■

It follows from the previous theorem that the Michael Line is a transitive space; by Corollary 6.21 and the previous theorem, the space Y of Example 5.25 is a transitive space.

6.18 A space $(X,\mathcal{T})$ is *locally transitive* provided that for each $x \in X$ there is an open transitive subspace containing x. Note that since every open subspace of a transitive space is transitive, this definition is equivalent to the condition that each x in X has a transitive neighborhood. Although we can present no counterexample, it seems unlikely that every locally transitive space is transitive. We show, however, that this result does hold in the class of θ-refinable spaces.

PROPOSITION. Let $(X,\mathcal{T})$ be a topological space and let $\mathcal{C} = \{C_i \mid i \in I\}$ be a point-finite open cover of X such that each member of $\mathcal{C}$ is a transitive subspace. Then $(X,\mathcal{T})$ is transitive.

Proof. Let V be a normal neighbornet of X. For each $i \in I$ there is a transitive neighbornet V_i of C_i such that for each $x \in C_i$, $V_i(x) \subset V(x)$. Let $W = \{(x,y) \mid y \in V_i(x)$ whenever $x \in C_i\}$. Then W is a transitive neighbornet of X and $W \subset V$. ■

COROLLARY. The topological sum of transitive spaces is a transitive space.

6.19 THEOREM. Every locally transitive θ-refinable space is a transitive space.

Proof. Let $(X,\mathcal{T})$ be a locally transitive θ-refinable space and for each $x \in X$ let G_x be an open transitive subspace containing x. Let $\langle\mathcal{R}_n\rangle$ be a sequence of open covers of X each of which refines $\{G_x \mid x \in X\}$ so that for each $x \in X$ there is an $n \in \mathbb{N}$ such that $\mathcal{R}_n$ is point finite at x. For each $(n,m) \in \mathbb{N} \times \mathbb{N}$ set $A_{n,m} = \{x \in X \mid$ at most m members of $\mathcal{R}_n$ contain $x\}$. As $A_{n,m}$ is a closed subspace of X, for each $R \in \mathcal{R}_n$, $A_{n,m} \cap R$ is a closed subset of the transitive space R. It follows from Proposition 6.18 that for each $(n,m) \in \mathbb{N} \times \mathbb{N}$, $A_{n,m}$ is a transitive subspace of X so that by Theorem 6.15, X is a transitive space. ■

6.20 Although we will show that the open continuous image of a transitive space need not be a transitive space, we do have the following positive result.

PROPOSITION. The closed continuous image of a transitive space is transitive.

Proof. Let $(X,\mathcal{T})$ be a transitive space and let f be a continuous closed map from X onto Y. Let U be a normal neighbornet of Y. There exists a normal sequence $\langle U_n \rangle$ of neighbornets of Y such that $U_1 = U$. For every $n \in \mathbb{N}$, define a neighbornet V_n of X by setting $V_n(x) = f^{-1}(U_n(f(x)))$ for every $x \in X$. It is easily verified that the sequence $\langle V_n \rangle$ is a normal sequence so that, in particular, V_1 is a normal neighbornet of X. Let W be a transitive neighbornet of X such that $W \subset V_1$ and define a neighbornet Q of Y by setting $Q(y) = Y - f(X - W(f^{-1}(y)))$ for every $y \in Y$. Since $W \subset V_1$, $Q \subset U_1$. It remains to show that Q is transitive. Let $v \in Q(y)$. Then $f^{-1}(v) \subset W(f^{-1}(y))$ and $W(f^{-1}(v)) \subset W^2(f^{-1}(y)) = W(f^{-1}(y))$. Consequently $Q(v) \subset Q(y)$. ∎

COROLLARY. Let X be a topological space. If X has a σ-locally finite cover by closed transitive spaces, then X is transitive.

One consequence of the preceding corollary is that the product of a discrete space and a transitive space is transitive. It follows that the product of the Michael line and a transitive space is transitive; in particular every finite product of Michael lines is transitive.

3. PROPERTIES THAT IMPLY TRANSITIVITY

6.21 Although we have established several properties of transitive spaces, it does not even follow from our previous results that the real line is a transitive space. In this section we establish that two classes of topological spaces, both of which include the real line, are transitive spaces. The arguments given to show that these two classes are contained in the class of transitive spaces actually show that if X belongs to either class and V is a neighbornet of X, then V^3 contains a transitive neighbornet. Curiously, although there are two different ways to establish that the real line is a transitive space, every known argument establishes the stronger conclusion that the cube of every neighbornet of the real line contains a transitive neighbornet.

A space $(X,\mathcal{T})$ is *semistratifiable* provided that to each closed set E of X one can assign a sequence $\langle E_n \rangle$ of open subsets of X so that $\bigcap_{n=1}^{\infty} E_n = E$ and for each $n \in \mathbb{N}$, $E_n \subset F_n$ whenever $E \subset F$. Thus semistratifiability is the condition that the closed sets of a space are uniform G_δ-sets, and so every semistratifiable space is countably metacompact. It is easy to show that a space X is semistratifiable if and only if there is a decreasing sequence $\langle V_n \rangle$ of neighbornets of X (called a *semistratification* of X) such that if for each $n \in \mathbb{N}$, $p \in V_n(x_n)$ then p is a cluster point of $\langle x_n \rangle$. (It is equivalent to require that $\langle x_n \rangle \to p$.) The diagonal of a space X is a G_δ-subset of $X \times X$ if and only if there is a decreasing sequence $\langle V_n \rangle$ of neighbornets of X such that if $\langle x_n \rangle$ is a sequence in X then $\cap\{V_n(x_n) \mid n \in \mathbb{N}\}$ contains at most one point. In this case $\langle V_n \rangle$ is called a G_δ-*diagonal sequence* and X is said to *have a* G_δ-*diagonal*. Every semistratifiable Hausdorff space has a G_δ-diagonal since in a Hausdorff space every semistratification is a G_δ-diagonal sequence.

LEMMA. Let U be a neighbornet of a semistratifiable space X.

(a) If X is preorthocompact, there exists a neighbornet V of X such that for each $x \in X$ there exist y, $z \in X$ so that $V^6(x) \subset U(y)$ and $\{x,y\} \subset U(z) \cap U^{-1}(z)$.

(b) If X is orthocompact, there exists $V \in \mathcal{FT}$ such that for each $x \in X$ there exist y, $z \in X$ so that $V(x) \subset U(y)$ and $\{x,y\} \subset U(z) \cap U^{-1}(z)$.

(c) If X is metacompact, there exists $V \in \mathcal{PF}$ such that for each $x \in X$ there exist y, $z \in X$ so that $V(x) \subset U(y)$ and $\{x,y\} \subset U(z) \cap U^{-1}(z)$.

Proof. Let $\langle U_n \rangle$ be a semistratification of X such that for each $n \in \mathbb{N}$, $U_n \subset U$. For each $n \in \mathbb{N}$ let $H_n = \{x \in X \mid U_n^{-1}(x) \subset U(x)\}$; then $\cup_{n=1}^{\infty} H_n = X$. For each $x \in X$ let $k_x = \min\{n \in \mathbb{N} \mid x \in \overline{H}_n\}$. Let $\langle V_n \rangle$ be a decreasing sequence of neighbornets such that for each $n \in \mathbb{N}$, $\{V_n^6(x) \mid x \in X\}$ refines $\{U_n(X) \mid x \in X\}$. Let $V = \{(x,y) \mid y \in V_{k_x}(x) - \cup_{n<k_x} \overline{H}_n\}$. Let $x \in X$ and let $z \in H_{k_x} \cap (U_{k_x} \cap V_{k_x})(x)$. There exists $y \in X$ such that $V_{k_x}^6(x) \subset U_{k_x}(y) \subset U(y)$. Clearly $\{x,y\} \subset U^{-1}(z)$. Since $z \in H_{k_x}$, $U_{k_x}^{-1}(z) \subset U(z)$ and $x \in U_{k_x}^{-1}(z) \subset U(z)$. Further $y \in U_{k_x}^{-1}(z) \subset U(z)$ so that $\{x,y\} \subset U(z)$.

In order to establish part (a), it remains to verify that $V^6(x) \subset U(y)$. Since $\langle V_n \rangle$ is a decreasing sequence of neighbornets, $V^6(x) \subset V_{k_x}^6(x) \subset U(y)$.

In order to establish part (b), we verify that if each term of the sequence $\langle V_n \rangle$ given above is transitive then the neighbornet V defined above is also transitive. Suppose that $c \in V(b)$ and $b \in V(a)$. Then $k_c \geq k_b \geq k_a$ so that $c \in V^2_{k_a}(a) - \cup_{n<k_a} \overline{H}_n = V_{k_a}(x) - \cup_{n<k_a} \overline{H}_n = V(a)$.

In order to establish part (c), we suppose that for each $n \in \mathbb{N}$ there is a point-finite open cover C_n so that $V_n = \{(x,y) | y \in \cap (C_n)_x\}$ and show that $V \in PF$. Let $p \in X$. Then $\{V(x) | p \in V(x)\} \subset \{V_{k_x}(x) - \cup_{n<k_x} \overline{H}_n | k_x \leq k_p$ and $p \in V_{k_x}(x)\}$, which is finite. ∎

THEOREM. Let X be a semistratifiable space.

(a) If X is preorthocompact and U is a neighbornet of X, then U^3 is a normal neighbornet.

(b) If X is preorthocompact, every open cover of X is a quasi-normal cover.

(c) If X is orthocompact, X is a transitive space.

(d) If X is metacompact, $PF = FINE$.

Proof. (a) Let U be a neighbornet of X. It follows from part (a) of the preceding lemma that there is a neighbornet V of X such that $V^6 \subset U^3$. By Proposition 1.6, U^3 is a normal neighbornet.

(b) Let C be an open cover of X. Well order C and for each $x \in X$, let C_x be the least member of C_x. Let $U = \cup\{\{x\} \times C_x | x \in X\}$. Evidently $\{U(x) | x \in X\}$ refines C. Let V be defined as in part (a) of the preceding lemma. In order to show that U is normal, it suffices in light of part (a) to establish that $V^6 \subset U$. Let $x \in X$. By part (a) of the preceding lemma, there exist $y, z \in X$ such that $V^6(x) \subset U(y)$, $x \in U(z)$, $z \in U(x)$, $y \in U(z)$ and $z \in U(y)$. Thus $U(y) = C_y = C_z = C_x = U(x)$ and $V^6(x) \subset U(y) = U(x)$.

(c) Let W be a normal neighbornet and let $U^3 \subset W$. By part (b) of the preceding lemma there is an entourage $V \in FT$ such that $V \subset U^3$; hence $W \in FT$.

(d) This implication follows from part (c) of the preceding lemma. ∎

COROLLARY. Every metrizable space is a transitive space.

6.22 The method of proof of Theorem 6.21 part (a) yields the following corollary. (See the notes.)

COROLLARY. If X is a preorthocompact developable space and U is a neighbornet of X, then U^2 is a normal neighbornet.

6.23 As we have observed previously it follows directly from Lemma 6.21 that if U is a neighbornet of an orthocompact semistratifiable space then U^3 contains a transitive neighbornet. The following lemma is used to show that there is a neighbornet U of a paracompact semistratifiable space X such that U^2 contains no transitive neighbornet of X.

LEMMA. Let D be a dense subset of a semistratifiable space X and let V be a neighbornet of X. Then $E = \{x \in X \mid x \notin \overline{D \cap V^{-1}(x)}\}$ is of the first category in X.

Proof. Let $\langle U_n \rangle$ be a semistratification of X such that for each $n \in \mathbb{N}$, U_n is an open neighbornet contained in V. For each $n \in \mathbb{N}$ set $F_n = X - U_n(D)$. Then $E \subset \cup_{n=1}^{\infty} F_n$, and so E is the countable union of nowhere dense sets. ■

EXAMPLE. A neighbornet U of an orthocompact semistratifiable space such that U^2 is not a normal neighbornet. As usual let $Q = \{q_n \mid n \in \mathbb{N}\}$ denote the set of all rational numbers and let J denote the set of all irrational numbers. Let $X = Q \cup J$. For each $x \in J$ and $n \in \mathbb{N}$ let $B_n(x) = \{y \in X \mid |x - y| < 1/n\}$. Let $D = \{d_n \mid n \in \mathbb{N}\}$ be a dense subset of J and for each $n \in \mathbb{N}$ choose $r_n \in Q \cap B_n(d_n)$. For each $x \in J$ choose as follows a sequence $\langle a_n(x) \rangle$ of rational numbers converging to x in the Euclidean topology: If $x \in B_n(d_n)$ set $a_n(x) = r_n$, otherwise choose $a_n(x) \in Q \cap B_n(x)$. For each $x \in J$ set $A_x = \{a_n(x) \mid n \in \mathbb{N}\}$. For each $n \in \mathbb{N}$ set $W_n = \{(x,y) \mid$ if $x \in Q$, $y = x$ and if $x \in J$ then $y \in B_n(x) - (A_x \cup \{q_i \mid i \leq n\})\}$ and let $\mathcal{T}$ be the topology on X for which $\{W_n(x) \mid n \in \mathbb{N}, x \in X\}$ is a base. Evidently, $\langle W_n \rangle$ is a semistratification for $(X,\mathcal{T})$. Indeed, one can show that $\mathcal{T}$ has a σ-closure-preserving base so that $(X,\mathcal{T})$ is paracompact.

We define the required neighbornet U as follows: $U = \{(x,y) \mid$ if $x \in Q$, $x = y$, if $x \in J - D$, $y \in X - A_x$ and if $x = d_n \in D$, $y \in W_n(d_n)\}$. Note that for each $n \in \mathbb{N}$, $r_n \notin U^2(d_n)$. Since X is a transitive space, in order to show that U^2 is not a normal neighbornet it suffices to show that U^2 contains no transitive neighbornet. Suppose that V is a transitive neighbornet contained in U^2 and let $H = \{x \in J \mid x \in \overline{D \cap V^{-1}(x)}\}$. For each $x \in H$, $x \in \overline{\{d \in D \mid V(d) = V(x)\}}$ because V is a transitive neighbornet. The topology that J inherits from X is the Euclidean topology on J, and so by the preceding lemma H is uncountable. For each $d \in D$ set $H_d = \{x \in H \mid V(x) = V(d)\}$. Since $H = \bigcup\{H_d \mid d \in D\}$, there is a $z \in D$ so that H_z is uncountable. Since the topology of J has a countable base, we can find $x, y \in H_z - D$ and an $n \in \mathbb{N}$ such that $x \neq y$ and $y \in W_n(x) \subset V(z)$. As $y \neq x$, y does not belong to the Euclidean closure of A_x; consequently there is a $k \in \mathbb{N}$ such that $B_k(y) \subset W_n(x)$. Since $y \in \overline{\{d \in D | V(d) = V(y)\}}$ and $y \notin D$, there exists $i \geq 2k$ such that $d_i \in B_{2k}(y)$ and $V(d_i) = V(y)$. Thus $r_i \in B_i(d_i) \subset B_k(y) \subset W_n(x) \subset V(z)$. Since $V(d_i) = V(y) = V(z)$, $r_i \in V(z) \subset U^2(d_i)$ - a contradiction.

6.24 One problem that is related to the results of section 6.21 is the problem of determining which topological spaces have the property that each of their neighbornets is a normal neighbornet. Since every F_σ-discrete space is semistratifiable, Lemma 6.21 enables us to find a large class of spaces with this property.

PROPOSITION. Every neighbornet of a preorthocompact F_σ-discrete space belongs to $\mathcal{FT}$.

Proof. Let $X = \cup_{n=1}^{\infty} F_n$ where, for each $n \in \mathbb{N}$, F_n is a closed discrete subspace of X. For each $x \in X$ let n(x) be the least natural number n so that $x \in F_n$ and choose $G_x \in \eta_x$ such that $G_x \cap F_{n(x)} = \{x\}$. Let W be a neighbornet of X and set $U = W \cap \{(x,y) | y \in G_x - \cup_{n<n(x)} F_n\}$. By part (a) of Lemma 6.21 there is a neighbornet V of X such that for each $x \in X$ there exist $y, z \in X$ such that $V^6(x) \subset U(y)$ and $\{x,y\} \subset U(z) \cap U^{-1}(z)$. Since $U(z) \cap U^{-1}(z) = \{z\}$, $V^6 \subset U$. By part (a) of Theorem 6.21, W is a normal neighbornet, and so by Corollary 6.15, $W \in \mathcal{FT}$. ∎

COROLLARY. Every neighbornet of a metacompact F_σ-discrete space belongs to $\mathcal{PF}$. In particular, if X is a countable T_1 space, $\mathcal{PF} = \mathcal{FINE}$.

It is easy to see that in general $\mathcal{PF} \neq \mathcal{FINE}$. Indeed we established in Example 3.20 that if X is the Michael line then $\mathcal{PF}_X | Q \times Q \neq \mathcal{PF}_Q$, whereas it follows from Corollary 2.19 and the previous corollary that $\mathcal{FINE}_X | Q \times Q = \mathcal{FINE}_Q = \mathcal{PF}_Q$. Thus the Michael line is a space with a σ-point-finite base for which $\mathcal{PF} \neq \mathcal{FINE}$. It appears that one hypothesis of Theorem 6.21 deserves special attention, namely the condition implied by semistratifiability that closed sets are G_δ-sets.

6.25 PROPOSITION. Let $(X,\mathcal{T})$ be a topological space and suppose that there is a closure-preserving closed cover of X by finite sets. Then every neighbornet of X belongs to $\mathcal{PF}$.

Proof. Let $\mathcal{C}$ be a closure-preserving closed cover of X by finite sets. Let V be a neighbornet of X and let $W = \{(x,y) | y \notin \cup(\mathcal{C} - \mathcal{C}_x)\} \cap V$. Note that for each $p \in X$, $W^{-1}(p) \subset \cap\mathcal{C}_p$, which is finite. Thus $\mathcal{W} = \{\text{int } W(x) | x \in X\}$ is a point-finite open cover of X and since $U_{\mathcal{W}} \subset V$, $V \in \mathcal{PF}$. ∎

6.26 It follows from the previous proposition that every neighbornet of the one-point compactification of an uncountable discrete space is a normal neighbornet. Thus, even within the class of compact Hausdorff spaces, Proposition 6.24 does not provide a necessary condition in order that every neighbornet be a normal neighbornet. We can, however, give a simple characterization of the first countable compact Hausdorff spaces for which every neighbornet is normal.

LEMMA. Let $(X,\mathcal{T})$ be a first countable T_1 space for which every neighbornet is normal. Then $(X,\mathcal{T})$ is quasi-metrizable.

Proof. Since X is first countable, there is a sequence $\langle V_n \rangle$ of neighbornets of X such that for each $x \in X$, $\{V_n(x) | n \in \mathbb{N}\}$ is a base for η_x. For each $n \in \mathbb{N}$ let $\langle W_{m,n} \rangle$ be a normal sequence of neighbornets whose first term is V_n. Then $\{W_{m,n} | (m,n) \in \mathbb{N} \times \mathbb{N}\}$ is a subbase for a quasi-uniformity compatible with $\mathcal{T}$. ∎

It follows from the preceding lemma and Corollary 7.22 that, although Example 5.34 is an F_σ-discrete space, it has a neighbornet that is not a normal neighbornet. Thus preorthocompactness cannot be omitted from the hypothesis of Proposition 6.24.

LEMMA. Let X be a first countable space with G_δ-diagonal that has no isolated points. If every neighbornet of X is a normal neighbornet, X is of first category in itself.

Proof. Suppose that every neighbornet of X is normal and let $\langle W_n \rangle$ be a G_δ-diagonal sequence for X. There is a collection $\{X_n | n \in \mathbb{N}\}$ of pairwise disjoint dense subsets of X such that $X = \cup_{n=1}^{\infty} X_n$ (see the notes). Define $U = \{(x,y) | y \in W_n(x) \text{ whenever } x \in X_n\}$ and let V be a neighbornet of X such that $V^2 \subset U$. For each $n \in \mathbb{N}$ set $A_n = \{x \in X | \text{for each } y \in X, V(x) - W_n(y) \neq \emptyset\}$. Suppose X is of second category in itself. Since $X = \cup_{n=1}^{\infty} A_n$, there is an $m \in \mathbb{N}$ so that $\text{int}(\bar{A}_m) \neq \emptyset$. Let $z \in X_m \cap \text{int}(\bar{A}_m)$ and let $p \in V(z) \cap A_m$. Then $V(p) \subset V^2(z) \subset U(z) = W_m(z)$ - a contradiction. ∎

PROPOSITION. Let X be a compact first countable Hausdorff space. Every neighbornet of X is normal if and only if X is countable.

Proof. Suppose that every neighbornet of X is a normal neighbornet. By the first lemma, X is quasi-metrizable so that X is metrizable by Corollary 2.30. Since the weight of a compact scattered Hausdorff space equals its cardinality, in order to show that X is countable, we show that X is scattered. (See the notes.) Let F be a nonempty closed subset of X and let G be a nonempty open subset of F. Since G is an open subspace of a Baire space, G is of second category, and by the second lemma, G contains an isolated point of F. Thus X is scattered.

The remaining implication follows immediately from Corollary 6.24. ∎

The method of proof of the previous proposition establishes the following corollary.

COROLLARY. Let X be a locally compact Hausdorff space with a G_δ-diagonal in which every neighbornet is normal. Then X is scattered and the set of isolated points of X is dense in X.

PROBLEM L. What familiar topological properties hold in those topological spaces in which every neighbornet is normal? In particular are these spaces necessarily either orthocompact or transitive?

6.27 We have commented about the behavior of the Michael line in §§3.20 and 6.24. We now reconsider the Michael line in order to construct a space that has a neighbornet V with the property that for each $n \in \mathbb{N}$, V^n is not a normal neighbornet.

PROPOSITION. Every neighbornet of the Michael line contains a transitive neighbornet.

Proof. As usual let J denote the set of all irrational numbers, let $Q = \{q_n | n \in \mathbb{N}\}$ denote the set of all rational numbers, and let M denote the Michael line. Let U be a neighbornet of M. For each $n \in \mathbb{N}$ choose an open interval G_n that contains q_n and has irrational endpoints such that $G_n \subset U(q_n) - \{q_i | i < n\}$ and for each $i < n$, G_n is either contained in, or disjoint from, G_i. Set $V = \{(x,y) | x = y \text{ if } x \in J \text{ and } y \in G_n \text{ if } x = q_n\}$. Then V is a transitive neighbornet and $V \subset U$. ∎

Although each neighbornet of M is normal, the following theorem shows that no power of M retains this well-behavior.

THEOREM. For each $n \in \mathbb{N}$ there is a neighbornet U_n of M^n such that U_n^{n-1} is not normal.

Proof. For each irrational $t \in M$ let $\epsilon(t) = 1$ and for each rational $t \in M$ let $\epsilon(t) = 1/q$ where q is the least positive denominator of t. For each $n \in \mathbb{N}$ define a neighbornet U_n of M^n in the following way. For $x = \langle x_1, \ldots, x_n \rangle \in M^n$ let $U_n(x)$ be the product of the following neighborhoods of the coordinates x_i of x: If x_i is irrational, take $\{x_i\}$ as the required neighborhood; if x_i is rational, take the neighborhood to be the largest open interval (r,s) containing x_i such that if t is a rational number in (r,s), other than x_i, then $\epsilon(t) < \min\{\epsilon(x_1), \ldots, \epsilon(x_n)\}$.

We show by induction on n that for any neighbornet W of M^n there exists an element a of M^n with all coordinates rational and an element z of M^n with all coordinates irrational such that $z \in W^n(a) - U_n^{n-1}(a)$. The case n = 1 holds trivially since, by definition, $U_1^0 = \Delta$. Assume that the inductive hypothesis holds for n - 1. For each irrational j and each $x' = \langle x_1, \ldots, x_{n-1} \rangle \in M^{n-1}$ we have that $U_n(\langle x',j \rangle) = U_{n-1}(x') \times \{j\}$. Let W be a neighbornet of M^n. By the inductive hypothesis for each irrational number j there exist $a'(j) \in M^{n-1}$ with all coordinates rational and $z'(j) \in M^{n-1}$ with all coordinates irrational such that $\langle z'(j),j \rangle \in W^{n-1}(\langle a'(j),j \rangle) - U_n^{n-2}(\langle a'(j),j \rangle)$. By the Baire Category Theorem there is an $\epsilon > 0$, an open interval (u,v), and a set D of irrational numbers that is Euclidean dense in (u,v) such that all $j \in D$ have a common $a'(j) = a'$ and such that for each $j \in D$, $\min\{|(z'(j))_i - (a')_i| \mid 1 \le i \le n-1\} > \epsilon$. Choose a rational number q in (u,v) such that $\epsilon(q) < \epsilon$ and let $a = \langle a',q \rangle$. Choose $d \in D$ such that $\langle a',d \rangle \in W(a)$ and let $z = \langle z'(d),d \rangle$. Then $z \in W^{n-1}(\langle a',d \rangle)$ and so $z \in W^n(a)$.

It remains to show that $z \notin U_n^{n-1}(a)$. Suppose that $z \in U_n^{n-1}(a)$. Then there is an element y of $U_n(a)$ such that $z \in U_n^{n-2}(y)$. Since for each i, with $1 \le i \le n$, $\epsilon(y_i) \le \epsilon(a_i)$ if y_i is a rational number, it follows from the definition of the neighbornet U_n that $U_n(y) \subset U_n(a)$ unless some coordinate of y is an irrational number. If $U_n(y) \subset U_n(a)$, then $z \in U_n^{n-2}(y) \subset U_n^{n-2}(a)$ - a contradiction. Let i be the least positive integer such that y_i is irrational and suppose that $1 \le i \le n-1$. Then $z_i = y_i$ is an irrational number and a_i is a rational number. Let (r,s) denote $\pi_i(U_n(a))$.

Note that if the length of (r,a_i) or (a_i,s) exceeds $\epsilon(q)$, one of these intervals contains a rational number t such that $\epsilon(t) \geq \epsilon(q)$ - a contradiction. Since $y_i \in (r,s)$, $|y_i - a_i| \leq \epsilon(q) < \epsilon$, whereas $|z_i - a_i| \geq \epsilon$ - a contradiction.

Therefore $y_1, y_2, \ldots, y_{n-1}$ are rational and $y_n = d$ is irrational. Since $y \in U_n(a)$, it follows that $U_n(y) \subset U_n(\langle a',d\rangle)$. Thus $z \in U_n^{n-2}(y) \subset U_n^{n-2}(\langle a',d\rangle)$ - a contradiction. This contradiction establishes that $z \notin U_n^{n-1}(a)$, as is required. ■

6.28 EXAMPLE. It is an immediate consequence of the preceding theorem that M^{ω_0} has a neighbornet V such that for each $n \in \mathbb{N}$, V^n is not normal.

6.29 As we have seen in the previous theorem, no finite product of Michael lines is as well behaved as the Michael line itself. Nonetheless, as we observed in §6.20, every finite product of Michael lines is a transitive space. It is an attractive conjecture that every finite product of GO-spaces is a transitive space. While this conjecture is supported by little more than wishful thinking, we can at least establish that every GO-space is transitive.

In order to establish that every GO space is a transitive space, we introduce the following notation. For each neighbornet U of a topological space $U^+ = \{(x,y) | y \in \cap\{U(G) | G \in \eta_x\}\}$. It is easily verified that if U is a neighbornet of a topological space $(X,\mathcal{T})$ and $G \in \mathcal{T}$ then $U^+(G) = U(G)$ and $U \subset U^+ \subset U^2$. If $(X,\mathcal{T},\leq)$ is a GO space then $\mathcal{T}^{\rightarrow}$ denotes the topology generated by $\mathcal{T} \cup \{[x,\rightarrow) | x \in X\}$ and $\mathcal{T}^{\leftarrow}$ denotes the topology generated by $\mathcal{T} \cup \{(\leftarrow,x] | x \in X\}$.

LEMMA. Let $(X,\mathcal{T},\leq)$ be a GO space and let U be a $\mathcal{T}$ neighbornet of X. There is a transitive $\mathcal{T}^{\rightarrow}$ neighbornet V such that $V \subset U^+$, where U^+ is taken with respect to the topology $\mathcal{T}^{\rightarrow}$.

Proof. Well-order X as $\{r_\alpha | \alpha \in \lambda\}$ where as usual the ordinal λ is the collection of all ordinals less than λ. For each $\alpha \in \lambda$ define R_α inductively as follows:

$$R_\alpha = \{x \in X | x \in \operatorname{int}_{\mathcal{T}^{\rightarrow}}[x,r_\alpha] \quad \text{and} \quad [x,r_\alpha] \subset (U^+(x) - \cup_{\beta\in\alpha} R_\beta)\}$$

We first show that $X = \operatorname{cl}_{\mathcal{T}^{\rightarrow}}(\cup_{\alpha\in\lambda} R_\alpha)$. Suppose that $z \in X - \operatorname{cl}_{\mathcal{T}^{\rightarrow}}(\cup_{\alpha\in\lambda} R_\alpha)$.

Then there is a $y \in X$ and a $\beta \in \lambda$ so that $y = r_\beta$, $z \in \mathrm{int}_{\mathcal{T}^{\rightarrow}}[z,y]$, and $[z,y] \subset U^+(z) - \cup_{\alpha\in\lambda} R_\alpha$. By definition of R_β, $z \in R_\beta$ - a contradiction. We now show that $X = \cup_{\alpha\in\lambda} R_\alpha$. Let $x \in X$ and let γ be the least ordinal in $\lambda \cup \{\lambda\}$ such that $x \in \mathrm{cl}_{\mathcal{T}^{\rightarrow}}(\cup_{\alpha\in\gamma} R_\alpha)$. We show that $x \in \cup_{\alpha\in\gamma} R_\alpha$. If γ is a successor, say $\gamma = \beta + 1$, then $x \in \mathrm{cl}_{\mathcal{T}^{\rightarrow}} R_\beta$. We show that $x \in R_\beta$. Let G be a $\mathcal{T}^{\rightarrow}$ open set containing x such that $G \subset [x,\rightarrow)$. Then $(x,r_\beta] \subset \cup\{[y,r_\beta] \mid y \in G \cap R_\beta\} \subset (U^+(G)) - \cup_{\beta'\in\beta} R_{\beta'} = U(G) - \cup_{\beta'\in\beta} R_{\beta'}$. Thus $(x,r_\beta] \subset U^+(x)$ and $(x,r_\beta] \cap (\cup_{\beta'\in\beta} R_{\beta'}) = \emptyset$. It follows from the definition of γ that $x \notin \cup_{\beta'\in\beta} R_{\beta'}$. Thus $x \in R_\beta$.

Suppose that γ is a limit ordinal. There is a $\beta \in \gamma$ and a $y \in R_\beta$ such that $[x,y] \subset U(x)$, because $x \in \mathrm{cl}_{\mathcal{T}^{\rightarrow}}(\cup_{\alpha\in\gamma} R_\alpha)$. Then $x \in \mathrm{int}_{\mathcal{T}^{\rightarrow}}([x,y] - \cup_{\alpha\in\beta+1} R_\alpha)$ and so there is an ordinal μ such that $\beta \in \mu \in \lambda$ and there is a point $z \in R_\mu$ such that $x \in \mathrm{int}_{\mathcal{T}^{\rightarrow}}[x,z]$ and $[x,z] \subset [x,y] - \cup_{\alpha\in\beta+1} R_\alpha$. We assume that μ is the least ordinal with the properties stated above. Then $[x,z] \subset [x,y] - \cup_{\alpha\in\mu} R_\alpha$. Since $z \in R_\mu$, $[z,r_\mu] \subset X - \cup_{\alpha\in\mu} R_\alpha$; consequently $[x,r_\mu] \subset X - \cup_{\alpha\in\mu} R_\alpha$. Since $\beta \in \mu$, $[x,r_\mu] \cap R_\beta = \emptyset$ and $y \notin [x,r_\mu]$. Thus $[x,r_\mu] \subset [x,y] \subset U(x)$. By the definition of R_μ, $x \in R_\mu$ and so $X = \cup_{\alpha\in\lambda} R_\alpha$.

For each $x \in X$ we may define $\alpha(x)$ to be the least $\alpha \in \lambda$ such that $x \in R_\alpha$. The desired $\mathcal{T}^{\rightarrow}$ neighbornet $V = \{(x,y) \mid y \in \mathrm{int}_{\mathcal{T}^{\rightarrow}}[x,r_{\alpha(x)}]\}$. Evidently V is a $\mathcal{T}^{\rightarrow}$ neighbornet contained in U^+. To show that V is transitive let $x \in X$ and let $y \in V(x)$. Then $y \in \mathrm{int}_{\mathcal{T}^{\rightarrow}}[x,r_{\alpha(x)}]$ and since $[x,r_{\alpha(x)}] \cap \cup_{\beta<\alpha(x)} R_\beta = \emptyset$, $\alpha(x) \in \alpha(y)$ or $\alpha(x) = \alpha(y)$. Suppose first that $\alpha(x) \in \alpha(y)$. Then $[y,r_{\alpha(x)}] \not\subset U^+(y)$ since $y \notin R_{\alpha(x)}$. As $[y,r_{\alpha(y)}] \subset U^+(y)$, $r_{\alpha(y)} < r_{\alpha(x)}$ and $V(y) \subset V(x)$. Now suppose that $\alpha(x) = \alpha(y)$. Then $V(y) = \mathrm{int}_{\mathcal{T}^{\rightarrow}}[y,r_{\alpha(x)}] \subset V(x)$. ■

THEOREM. Let $(X,\mathcal{T},\leq)$ be a GO space such that $\mathcal{T} = \mathcal{T}^{\rightarrow}$ and let U be a neighbornet of X. Then U^2 contains a transitive neighbornet.

Since the Sorgenfrey line is the prototype of those GO spaces $(X,\mathcal{T},\leq)$ for which $\mathcal{T} = \mathcal{T}^{\rightarrow}$, we note that it follows from the second lemma of §6.26 that the Sorgenfrey line has a neighbornet that is not a normal neighbornet. We note also that the Michael line does not satisfy the hypothesis of the preceding theorem, but it does satisfy the conclusion.

6.30 In the following theorem if U is a neighbornet of X, $U^{\rightarrow} = \{(x,y) \in U | x \leq y\}$ and $U^{\leftarrow} = \{(x,y) \in U | x \geq y\}$.

THEOREM. Let $(X,\mathcal{T},\leq)$ be a GO space and let U be a neighbornet of X. Then U^3 contains a transitive neighbornet.

Proof. Without loss of generality we assume that U is an open neighbornet. By the preceding lemma there is a transitive $\mathcal{T}^{\leftarrow}$ neighbornet L such that $L = L^{\leftarrow} \subset U^+$ and a transitive $\mathcal{T}^{\rightarrow}$ neighbornet R such that $R = R^{\rightarrow} \subset U^+$. Moreover, by the proof of the previous lemma we may assume that for each $t \in X$, $R(t)$ and $L(t)$ are order convex. Set $W = (R \cup L)^2$. When $W^+ \subset (U^+)^{2^+} = U^{2^+} \subset U^3$. In order to show that W^+ contains a transitive $\mathcal{T}$ neighbornet, it suffices to find transitive $\mathcal{T}$ neighbornets V and V' such that $V^{\rightarrow} \cup (V')^{\leftarrow} \subset W^+$. (Then $V \cap V'$ is a transitive $\mathcal{T}$ neighbornet contained in W^+.) We prove the existence of V, and the existence of V' follows by a "symmetric" argument.

As in the preceding lemma, let X be well-ordered as $\{r_\alpha | \alpha \in \lambda\}$, let $R_\alpha = \{x \in X | x \in \mathrm{int}_{\mathcal{T}^{\rightarrow}}[x,r_\alpha]$ and $[x,r_\alpha] \subset (W^+(x) - \cup_{\beta\in\alpha} R_\beta)\}$ and let $\alpha(x)$ be the least $\alpha \in \lambda$ such that $x \in R_\alpha$. Set $Q = \{(x,y) | y \in \mathrm{int}_{\mathcal{T}^{\rightarrow}}(\leftarrow,r_{\alpha(x)}]\}$. By the proof of the lemma, $Q^{\rightarrow}$ is a transitive $\mathcal{T}^{\rightarrow}$ neighbornet of X. Set $V = \{(x,y) \in Q | Q(y) \subset Q(x)\}$. Note that V is transitive and $V^{\rightarrow} \subset Q^{\rightarrow} \subset W^+$. Since $Q^{\rightarrow}$ is transitive, $Q^{\rightarrow} \subset V$ and V is a $\mathcal{T}^{\rightarrow}$ neighbornet. Thus to show that V is a $\mathcal{T}$ neighbornet of X, it suffices to show that $L \subset V$. Let $(x,y) \in L$ with $x \neq y$. If $r_{\alpha(y)} \leq r_{\alpha(x)}$, then $Q(y) \subset Q(x)$ so that $(x,y) \in V$. Assume that $r_{\alpha(y)} > r_{\alpha(x)}$. Then since $y \in L(x) = L^{\leftarrow}(x)$, $x \in \mathrm{int}_{\mathcal{T}^{\rightarrow}}[y,r_{\alpha(y)}]$ and since $[y,r_{(y)}] \subset X - \cup_{\beta\in\alpha(y)} R_\beta$, $\alpha(y) \in \alpha(x)$. Thus $x \notin R_{\alpha(y)}(y)$ and so $[x,r_{\alpha(y)}] \not\subset W^+(x)$. Since $y \neq x$ and $y \in L(x)$, $y \in \mathrm{int}_{\mathcal{T}}L(x)$ so that $W^+(y) \subset W(L(x))$. Since L and R are transitive, $W(L(x)) = (R \cup L)^2(L(x)) = LRL(x)$. Recall that for each $t \in X$, $R(t)$ is order convex so that $[x,\rightarrow) \cap LRL(x) = [x,\rightarrow) \cap LR(x) \subset W(x)$. Combining some of the previous inclusions, we have that $[x,\rightarrow) \cap W^+(y) \subset W^+(x)$. As $x \geq y$, $[x,r_{\alpha(y)}] = [x,\rightarrow) \cap [y,r_{\alpha(y)}] \subset [x,\rightarrow) \cap W^+(y) \subset W^+(x)$ - a contradiction. Hence $r_{\alpha(y)} \leq r_{\alpha(x)}$ and as shown above $(x,y) \in V$. ■

6.31 There is an interesting parallel between the transitivity of orthocompact semistratifiable spaces and GO spaces, which the following example furthers.

EXAMPLE. *The Engelking-Lutzer Line.* A GO space $(X,\mathcal{T},\leq)$ and a neighbornet U of X such that U^2 does not contain a transitive neighbornet. Let X denote the set of all real numbers, let J denote the set of all irrational numbers, and let $Q = \{q_n | n \in \mathbb{N}\}$ denote the set of all rational numbers. Define a topology on X by taking basic neighborhoods of a point x to be an interval of the form $[x, x + \epsilon)$ if $x \in Q$ and $(x - \epsilon, x]$ if $x \in J$. For each $n \in \mathbb{N}$, choose $\epsilon_n \leq 1/n$ such that for each $i < n$, $[q_n, q_n + \epsilon_n)$ is either disjoint from or contained in $[q_i, q_i + \epsilon_i)$. Let $U = \{(x,y) | y \in (x - 1, x]$ if $x \in J$ and $y \in [q_n, q_n + \epsilon_n)$ if $x = q_n\}$. Suppose that there exists a transitive neighbornet V such that $V \subset U^2$. For each $n \in \mathbb{N}$ let $A_n = \{x \in X | [x, x + 1/n) \subset V(x)\}$ and let $B_n = \{x \in X - \cup_{i=1}^{\infty} A_i | (x - 1/n, x] \subset V(x)\}$. Then $X = \cup_{n=1}^{\infty} A_n \cup \cup_{n=1}^{\infty} B_n$. Suppose that the Euclidean closure of B_n contains an interval. There exist $q \in Q$ and $m > n$ such that $[q, q + 1/m) \subset V(q)$ and $B_n \cap [q, q + 1/m) \neq \emptyset$. Let $x \in B_n \cap [q, q + 1/m)$. Then $x - 1/n < q \leq x < q + 1/m$ and so $V(q) \subset V(x)$. Choose $k \in \mathbb{N}$ such that $x + 1/k < q + 1/m$. Then $[x, x + 1/k) \subset V(x)$, and so $x \in A_k$ - a contradiction. Now suppose that the Euclidean closure of A_n contains an interval. For some $m > n$ there exists $q_m \in Q$ such that $A_n \cap [q_m, q_m + 1/m) \cap V(q_m) \neq \emptyset$. Choose $x \in A_n \cap [q_m, q_m + 1/m) \cap V(q_m)$. Then $[x, x + 1/n) \subset V(x) \cap [q_m,\to) \subset V(q_m) \cap [q_m,\to) \subset U^2(q_m) \cap [q_m,\to)$. Since $U^2(q_m) \subset (q_m - 1, q_m + 1/m)$, $[x, x + 1/n) \subset [q_m, q_m + 1/m)$. Thus X is the countable union of nowhere dense sets - a contradiction.

PROBLEM M. Although it is no picnic to prove that the GO spaces and the orthocompact semistratifiable spaces are transitive, the problem of finding topological spaces that are not transitive appears more difficult still. Indeed, we have no counterexample to refute the seemingly ridiculous conjecture that every compact space is transitive. By Propositions 6.14 and 6.20, if there exists a compact nontransitive Hausdorff space of weight λ, then $\{0,1\}^\lambda$ is not transitive. Besides the question of transitivity of all compact spaces, the results of this chapter leave unanswered another basic and obviously related question: Is every subspace of a transitive space transitive?

6.32 The following example is of paramount importance in the study of transitive spaces because every known example of a space that is not transitive is based upon it.

EXAMPLE. *The Kofner Plane.* A realcompact hereditarily orthocompact perfect subparacompact completely regular space that is not a transitive space. Let X be the plane. For each $x \in X$ and each $\epsilon > 0$ let $C(x,\epsilon)$ be the circle of radius ϵ lying above the horizontal line through x and tangent to this line at x. For each nonnegative integer n let $U_n = \{(x,y) \mid y = x$ or y is inside $C(x,2^{-n})\}$ and for each positive integer m let $U_{m,n} = U_n \cap \{(x,y) \mid \pi_2(y) - \pi_2(x) < 2^{-m}\}$. Figure 2 indicates that $U_n = U_{n+1}^2$ for each $n \in \mathbb{N}$ so that $\langle U_n \rangle$ is a base for a quasi-uniformity $\mathcal{U}$. Indeed, for all j, $n \in \mathbb{N}$, $(U_{j,n+1})^2 \subset U_{j,n}$. The Kofner plane is the set X endowed with the topology $\mathcal{T}(\mathcal{U})$. We establish in Example 7.7 that the Kofner plane is completely regular.

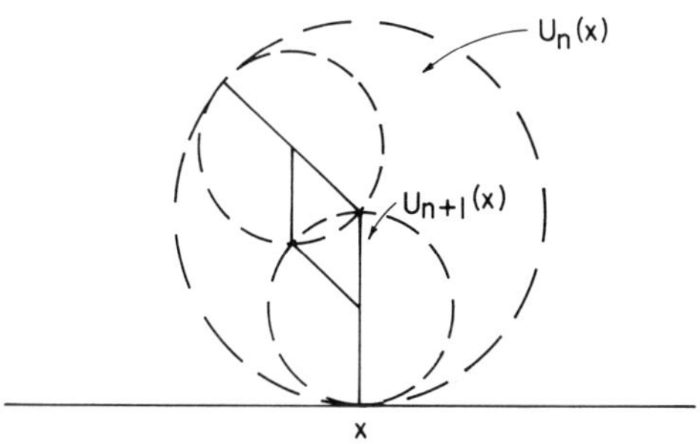

Figure 2. The Kofner plane

<u>X is not transitive</u>. To see that X is not a transitive space suppose that U_1 contains a transitive neighbornet W. For each $m,n \in \mathbb{N}$ set $A(m,n) = \{x \mid U_{m,n}(x) \subset W(x)$ and for each $j \in \mathbb{N}$, $U_{j,n-1}(x) \not\subset W(x)\}$. Since $X = \bigcup\{A(m,n) \mid m, n \in \mathbb{N}\}$, it follows from the Baire Category Theorem that there is a Euclidean open set G and m, $n \in \mathbb{N}$ so that $G \subset \overline{A(m,n)}$ (throughout $\overline{}$ denotes the Euclidean closure). Let $x \in A(m,n) \cap G$. Since G is a Euclidean open set, there is a $j > m$ so that $U_{j,n}(x) \subset G$. Set $B = A(m,n) \cap U_{j,n}(x)$. Then $U_{j,n}(x) \subset \overline{B}$ and $B \subset U_{j,n}(x) \subset U_{m,n}(x) \subset W(x)$. We show that $U_{j,n-1}(x) \subset U_{j,n}(B)$. Let $z \in U_{j,n-1}(x)$. There is a $p \in U_{j,n}(x) \subset \overline{B}$ such that $z \in U_{j,n}(p)$. Without loss of generality we assume that $p \neq z$. For each $t \in X$ let $S(t,\epsilon)$ denote the Euclidean ϵ-sphere about t. Choose $\epsilon > 0$ so that $S(z,\epsilon) \subset U_{j,n}(p)$. Since $p \in \overline{B}$, there is a $y \in S(p,\epsilon) \cap B$; it follows that $z \in U_{j,n}(y)$. Since $B \subset A(m,n)$, $U_{m,n}(B) \subset W(B) \subset W(W(x)) = W(x)$. Thus $U_{j,n-1}(x) \subset W(x)$ - a contradiction.

X is perfect and subparacompact. In order to show that X is perfect and subparacompact we show that if $G \in T(U)$ and C is an open cover of G then C has a σ-discrete closed refinement in X. Let $G \in T(U)$ and let C be an open cover of G. Let C' denote the collection of all Euclidean interiors of members of C, and let $G' = \cup C'$. Then C' has a closed σ-discrete refinement in the Euclidean topology and hence in $T(U)$. Let $F = G - G'$ and for each $n \in \mathbb{N}$ let $F_n = \{x \in F | U_n(x) \subset C$ for some $C \in C\}$. Since $F = \cup_{n=1}^{\infty} F_n$, we need only show that each F_n is σ-discrete in X.

Let $n \in \mathbb{N}$. In order to show that F_n is σ-discrete we introduce the topology T' on the plane for which $\{U_n(x) \cup U_n^{-1}(x) | x \in X\}$ is a base. The topology T' is coarser than $T(U)$ and finer than the Euclidean topology on X, and (X,T') is a first countable semistratifiable space. Therefore $\{U_n(x) \cup U_n^{-1}(x) | x \in F_n\}$ has a σ-discrete closed refinement $R = \cup_{k=1}^{\infty} R_k$ in (X,T'). For each $k \in \mathbb{N}$ and $x \in X$ there is an $m \in \mathbb{N}$ so that $U_m(x)$ intersects at most one member of R_k. Therefore, since R is a closed family in $(X,T(U))$, in order to show that the trace of R on F_n is a closed σ-discrete collection, we have only to show that for each $x \in F_n$, $[U_n(x) \cup U_n^{-1}(x)] \cap F_n = \{x\}$. Suppose that x and y are two different points of F_n and that $y \in U_n(x) \cup U_n^{-1}(x)$. Then either $y \in G'$ or $x \in G'$ - a contradiction.

X is hereditarily orthocompact. We need only show that X is σ-orthocompact, since every perfect σ-orthocompact space is hereditarily orthocompact. Let C be an open cover of X. Let C' be the collection of all Euclidean interiors of C. Then C' has a locally finite open refinement in the Euclidean topology and hence in $T(U)$. Let $F = X - \cup C'$ and for each $n \in \mathbb{N}$ let $F_n = \{x \in F | U_n(x) \subset C$ for some $C \in C\}$. Since $F = \cup_{n=1}^{\infty} F_n$, we need only show that each $n \in \mathbb{N}$, $\{U_n(x) | x \in F_n\}$ has an open interior-preserving refinement.

Let $n \in \mathbb{N}$. Let B be a uniform base for the Euclidean topology such that $B = \cup_{k=1}^{\infty} B_k$, where for each $k \in \mathbb{N}$, B_k is point finite and the diameter of each member of B_k is less than $1/k$. For each $x \in F_n$ and $x' \in U_{n+1}(x)$ let $k(x,x')$ be the smallest $k \in \mathbb{N}$ such that for some $B \in B_k$, $x' \in B \subset U_n(x)$. For $x \in F_n$ and $x' \in U_{n+1}(x)$ let $B(x,x') \in B_{k(x,x')}$ such that $x' \in B(x,x') \subset U_n(x)$. Suppose that $\langle x_j \rangle$ is a sequence in F_n and for each $j \in \mathbb{N}$, $x_j' \in U_{n+1}(x_j)$ and $x_j' \neq x_j$. If $\lim_{j\to\infty} \pi_2(x_j') - \pi_2(x_j) = 0$, then $\lim_{j\to\infty} \pi_1(x_j') - \pi_1(x_j) = 0$. Therefore if $\lim_{j\to\infty} k(x_j,x_j') = \infty$, then $\lim_{j\to\infty} \rho(x_j,x_j') = 0$, where ρ denotes the Euclidean metric of the plane. For each $x \in F_n$ set $B(x) = \cup\{B(x,x') | x' \in U_{n+1}(x)\} \cup \{x\}$. Evidently, $B(x) \subset U_n(x)$ for each $x \in F_n$.

Suppose that $\{B(x)|x \in F_n\}$ is not interior preserving and let $p \in X$ such that $\cap\{B(x)|p \in B(x), x \in F_n\}$ is not a neighborhood of p. There are sequences $\langle x_j \rangle$ and $\langle x_j' \rangle$ such that for each $j \in \mathbb{N}$, $x_j \in F_n$ and $x_j' \in U_{n+1}(x_j)$ and $p \in B(x_j, x_j')$. Thus $\{B(x_j,x_j')|j \in \mathbb{N}\}$ is a base for the Euclidean neighborhood base of p. Then $\lim_{j\to\infty} k(x_j,x_j') = \infty$ and $\lim_{j\to\infty} \rho(x_j,x_j') = 0$. Thus $\lim_{j\to\infty} \rho(x_j',p) = 0$ so that $p \in F$. Let $j \in \mathbb{N}$ such that $x_j \neq p$. Then $p \in B(x_j) - \{x_j\} \subset U_n(x_j) - \{x_j\} \subset \cup C'$ - a contradiction.

X is hereditarily realcompact. Since the topology of X is finer than the Euclidean topology, X is hereditarily realcompact. ■

6.33 The method of proof used to prove that the Kofner plane is not transitive establishes the following proposition, which is used in Example 7.7.

PROPOSITION. Let X be the Kofner plane and let W be a transitive neighbornet of X. For each $n \in \mathbb{N}$, $\{x|W(x) \subset U_n(x)\}$ is a first category set in the Euclidean plane.

6.34 It is evident that the Kofner plane satisfies the first axiom of countability so that by Ponamarev and Hanai's Theorem the Kofner plane is the continuous open image of a metrizable space; it follows from Theorem 6.21(c) that the continuous open image of a transitive space need not be a transitive space. This negative result contrasts with Proposition 6.20.

Every first countable Hausdorff space is a closed continuous image of a scattered Hausdorff space. (See the notes.) It follows from Proposition 6.20 that there exists a scattered Hausdorff space that is not transitive.

6.35 We have already observed that the Kofner plane is realcompact and so can be embedded as a closed subspace of real lines. Since the real line is transitive and the Kofner plane is not, it follows from Proposition 6.14 that the arbitrary product of transitive spaces is not necessarily a transitive space.

PROBLEM N. Is the product of two transitive spaces necessarily a transitive space?

Chapter 7

QUASI-METRIZABLE SPACES

1. THE QUASI-METRIZATION PROBLEM

7.1 As we observed in Chapter 1, a widely known result in the study of quasi-uniform spaces is the theorem of A. H. Frink that a T_1 space admits a quasi-uniformity with a countable base if and only if it is quasi-metrizable. We begin this section by considering the problem of finding necessary and sufficient topological conditions for a topological space to admit a quasi-metric. Unlike the corresponding metrization problem, no completely satisfactory solution to the quasi-metrization problem has yet been found; nonetheless there is a purely topological characterization of the non-Archimedeanly quasi-metrizable spaces.

THEOREM. Let $(X,\mathcal{T})$ be a T_1 space. The following statements are equivalent.

(a) There is a σ-interior-preserving base for $\mathcal{T}$.

(b) $(X,\mathcal{T})$ admits a non-Archimedean quasi-metric.

(c) $(X,\mathcal{T})$ admits a transitive quasi-uniformity with a countable base.

Proof. (a) $\Rightarrow$ (b) For each $n \in \mathbb{N}$ let $\mathcal{B}_n$ be an interior-preserving open collection and suppose that $\mathcal{B} = \cup\{\mathcal{B}_n | n \in \mathbb{N}\}$ is a base for $\mathcal{T}$. For $x, y \in X$ with $x \neq y$, let $d(x,y) = 1/n$ where n is the least natural number such that $y \notin \cap(\mathcal{B}_n)_x$. Define $d(x,x) = 0$ for each $x \in X$. If $d(x,y) \leq 1/n$ and $d(y,z) \leq 1/n$, then for $1 \leq k < n$, $y \in \cap(\mathcal{B}_k)_x$ and $z \in \cap(\mathcal{B}_k)_y$ so that $z \in \cap(\mathcal{B}_k)_x$ and $d(x,z) \leq 1/n$. Thus d is a non-Archimedean quasi-metric and $\mathcal{T}(d) = \mathcal{T}$ since for each $x \in X$ and $n \in \mathbb{N}$, $\cap_{i=1}^{n} (\cap(\mathcal{B}_j)_x) \subset \{y | d(x,y) < 1/n\} \subset \cap(\mathcal{B}_n)_x$.

(b) ⇒ (c) For each $n \in \mathbb{N}$ set $U_n = \{(x,y) | d(x,y) < 1/n\}$. Then $\{U_n | n \in \mathbb{N}\}$ is a transitive base for a quasi-uniformity compatible with $\mathcal{T}$.

(c) ⇒ (a) Let $\{U_n | n \in \mathbb{N}\}$ be a transitive base for a quasi-uniformity compatible with $\mathcal{T}$ and for each $n \in \mathbb{N}$ set $\mathcal{B}_n = \{U_n(x) | x \in X\}$. Then $\cup\{\mathcal{B}_n | n \in \mathbb{N}\}$ is a σ-interior-preserving base for $\mathcal{T}$. ∎

COROLLARY. Let $(X,\mathcal{T})$ be a T_1 topological space. If $\mathcal{T}$ has a σ-point-finite base, $(X,\mathcal{T})$ is quasi-metrizable.

7.2 Curiously the problem of characterizing those topological spaces that admit a totally bounded quasi-metric is easily resolved and its solution brings to mind the solution of the corresponding problem for metrizable spaces.

PROPOSITION. Let $(X,\mathcal{T})$ be a T_1 topological space. The following statements are equivalent.

(a) $(X,\mathcal{T})$ satisfies the second axiom of countability.

(b) $(X,\mathcal{T})$ admits a totally bounded non-Archimedean quasi-metric.

(c) $(X,\mathcal{T})$ admits a totally bounded quasi-uniformity with a countable base.

Proof. (a) ⇒ (b) Let $\{B_n | n \in \mathbb{N}\}$ be a countable base for $\mathcal{T}$. Let $\mathcal{U}$ be the quasi-uniformity for which $\{T(B_n, X - B_n) | n \in \mathbb{N}\}$ is a subbase. Since $\mathcal{U}$ is totally bounded, the implication follows from the previous theorem.

(b) ⇒ (c) This implication is evident.

(c) ⇒ (a) Let $\langle U_n \rangle$ be a normal sequence of neighbornets that is a base for a quasi-uniformity compatible with $\mathcal{T}$. For each $n \in \mathbb{N}$ let $\{A(n,j) | 1 \le j \le k_n\}$ be a cover of X such that $\cup_{j=1}^{k_n} A(n,j) \times A(n,j) \subset U_n$, choose $x_{n,j} \in A(n,j)$ for $1 \le j \le k_n$, and set $\mathcal{B} = \{U_n(x_{n,j}) | n \in \mathbb{N}, 1 \le j \le k_n\}$. Let $G \in \mathcal{T}$, $x \in G$ and let $m, n \in \mathbb{N}$ such that $U_n(x) \subset G$ and $x \in A(n+1,m)$. Suppose that $y \in U_{n+1}(x_{n+1,m})$. Since $(x,x_{n+1,m}) \in A(n+1,m) \times A(n+1,m) \subset U_{n+1}$, $y \in U_{n+1}^2(x)$ so that $x \in A(n+1,m) \subset U_{n+1}(x_{n+1,m}) \subset U_{n+1}^2(x) \subset U_n(x) \subset G$. Thus $\mathcal{B}$ is a countable base for $\mathcal{T}$. ∎

7.3 It is easily seen that the restriction of the Sorgenfrey line to the closed unit interval provides an example of a precompact quasi-metric space that is not a totally bounded quasi-metric space. We note one consequence of Proposition 7.2: Every topological space that admits a totally bounded quasi-metric also admits a totally bounded non-Archimedean quasi-metric. The following proposition provides a more important instance in which quasi-metrizable spaces and non-Archimedeanly quasi-metrizable spaces coincide.

PROPOSITION. Let $(X,\mathcal{T})$ be a transitive space. The following statements are equivalent.

(a) $(X,\mathcal{T})$ is quasi-metrizable.

(b) $(X,\mathcal{T})$ is non-Archimedeanly quasi-metrizable.

(c) $(X,\mathcal{T})$ is a T_1 space and $\mathcal{T}$ has a σ-interior-preserving base.

Proof. By Theorem 7.1 it suffices to show that if $(X,\mathcal{T})$ is quasi-metrizable $(X,\mathcal{T})$ admits a transitive quasi-uniformity with a countable subbase. Suppose that $(X,\mathcal{T})$ is quasi-metrizable and let $\langle U_n \rangle$ be a normal sequence that is a base for a quasi-uniformity $\mathcal{U}$ compatible with $\mathcal{T}$. Since $\mathcal{U} \subset \mathcal{FT}$, for each $n \in \mathbb{N}$ there is a transitive entourage $V_n \in \mathcal{FT}$ such that $V_n \subset U_n$. Evidently, $\{V_n | n \in \mathbb{N}\}$ is a subbase for a quasi-uniformity compatible with $\mathcal{T}$. ∎

The method of proof of the previous proposition establishes the following corollary. As a consequence of this corollary we see that for the Sorgenfrey line $\mathcal{PF} \neq \mathcal{FINE}$.

COROLLARY. Let $(X,\mathcal{T})$ be a quasi-metrizable space. If $\mathcal{PF} = \mathcal{FINE}$, then $\mathcal{T}$ has a σ-point-finite base and if $\mathcal{LF} = \mathcal{FINE}$, then $\mathcal{T}$ has a σ-locally-finite base.

PROBLEM 0. (The quasi-metrization problem) Determine necessary and sufficient topological conditions in order that a topological space admit a quasi-metric.

7.4 One reason that the quasi-metrization problem has proved less tractable than the metrization problem is that the class of quasi-metrizable spaces includes some of topology's most pathological counterexamples. The Michael line, the Sorgenfrey plane, and the Kofner plane are quasi-metrizable and it follows from Corollary 7.24 that the Niemytzki plane is quasi-metrizable as well. Although the examples listed above are known for their wanton behavior, each has a G_δ-diagonal. We now use Corollary 7.1 to establish that the example given below admits a non-Archimedean quasi-metric. The example is a LOTS that is not metrizable and therefore does not have a G_δ-diagonal. (See the notes.)

EXAMPLE. A quasi-metrizable LOTS that is not metrizable. Set $X = \{(x,n) \in \mathbb{R} \times \mathbb{Z} | n = 0$ if x is rational$\}$. Give X the order topology of the lexicographic order. As $\{(x,0) | x \in \mathbb{R}\}$ is a homeomorphic copy of the Michael line, X is not metrizable and as the order topology has a σ-disjoint base it follows from Corollary 7.1 that X is quasi-metrizable.

7.5 A quasi-metric d on a set X is *continuous* provided that for each $x \in X$ the function $d(x,\_) : (X,\mathcal{T}(d)) \to \mathbb{R}$ defined by $d(x,\_)(p) = d(x,p)$ is a continuous function. All the examples discussed in §7.4 admit a continuous quasi-metric and so are completely regular. Of course many of these spaces are obviously completely regular; indeed many are GO spaces. Nonetheless, the corollary given below provides an easy way of determining that the Kofner plane is completely regular.

PROPOSITION. Let (X,d) be a quasi-metric space and let $x \in X$. Then $d(x,\_)$ is a continuous function if and only if for each $\epsilon > 0$, $\overline{S(x,\epsilon)} \subset \{y | d(x,y) \leq \epsilon\}$.

Proof. Suppose that $d(x,\_)$ is continuous, let $\epsilon > 0$ and let $y \in \overline{S(x,\epsilon)}$. There is a sequence $\langle x_n \rangle$ in $S(x,\epsilon)$ such that $\langle x_n \rangle \to y$. For each $n \in \mathbb{N}$, $d(x,x_n) < \epsilon$ and $\langle d(x,x_n) \rangle \to d(x,y)$ so that $d(x,y) \leq \epsilon$.

Now suppose that for each $\epsilon > 0$, $\overline{S(x,\epsilon)} \subset \{y | d(x,y) \leq \epsilon\}$ and suppose that $d(x,\_)$ is not continuous, say at y. Let $\langle y_n \rangle$ be a sequence converging to y such that $\langle d(x,y_n) \rangle$ does not converge to $d(x,y)$. There is a $\delta > 0$ and a subsequence $\langle y_{j_n} \rangle$ of $\langle y_n \rangle$ such that for each $n \in \mathbb{N}$, $|d(x,y_{j_n}) - d(x,y)| > \delta$. Since $\langle y_n \rangle \to y$, there is an $m \in \mathbb{N}$ so that for each $n > m$, $d(y,y_n) < \delta$. Thus for each $n > m$, $d(x,y_n) \leq d(x,y) + d(y,y_n) < d(x,y) + \delta$, so that $d(x,y_n) - d(x,y) < \delta$. Since $|d(x,y_{j_n}) - d(x,y)| > \delta$, we have that $d(x,y) > \delta + d(x,y_{j_n})$ for $j_n > m$. Choose $\epsilon > 0$ so that for each $n \in \mathbb{N}$ with $j_n > m$, $d(x,y_{j_n}) < \epsilon < d(x,y)$. Then $\langle y_{j_n} \rangle \to y \in \overline{S(x,\epsilon)}$ so that by hypothesis $d(x,y) \leq \epsilon$ - a contradiction. ■

COROLLARY. Let (X,d) be a quasi-metric space such that for each $\epsilon > 0$ and $x \in X$, $\overline{S(x,\epsilon)} \subset \{y | d(x,y) \leq \epsilon\}$. Then $(X,\mathcal{T}(d))$ is a completely regular space.

7.6 PROPOSITION. Let (X,d) be a quasi-metric space. Then $d(\_,x)$ is an upper semi-continuous function for each $x \in X$ if and only if $\mathcal{T}(d^{-1}) \subset \mathcal{T}(d)$.

Proof. Suppose that $d(\_,x)$ is upper semi-continuous for each $x \in X$. For each $x \in X$ and $\epsilon > 0$, $S_{d^{-1}}(x,\epsilon) = \{y \mid d(y,x) < \epsilon\} \in \mathcal{T}(d)$. Thus $\mathcal{T}(d^{-1}) \subset \mathcal{T}(d)$.

Suppose that $\mathcal{T}(d^{-1}) \subset \mathcal{T}(d)$. Let $x \in X$, let $\epsilon > 0$, and let $p \in S_{d^{-1}}(x,\epsilon)$. There is a $\delta > 0$ so that $p \in S_d(x,\delta) \subset S_{d^{-1}}(x,\epsilon)$. Hence $S_{d^{-1}}(x,\epsilon) \in \mathcal{T}(d)$. ■

COROLLARY. Let (X,d) be a quasi-metric space and suppose that $d(\_,x)$ is an upper semi-continuous function for each $x \in X$. Then $(X,\mathcal{T}(d))$ is metrizable.

7.7 We return to the Kofner plane. As the reader probably already suspects, this space shows that Theorem 7.1 does not solve the quasi-metrization problem.

EXAMPLE. (The Kofner plane). A hereditarily orthocompact perfect subparacompact completely regular quasi-metric space whose topology does not have a σ-interior-preserving base. Let X be the plane. For each $x \in X$ and $\epsilon > 0$ let $C(x,\epsilon)$ be the circle of radius ϵ lying above the horizontal line through x and tangent to this line at x. For x, $y \in X$ define $d(x,y)$ as follows.

$$d(x,y) = \begin{cases} 1 \text{ if } y \text{ is in the exterior of } C(x,1) \\ r \text{ if } y \in C(x,r) \text{ and } y \neq x \text{ and } r \leq 1 \\ 0 \text{ if } x = y \end{cases}$$

For each $n \in \mathbb{N}$ if U_n is the entourage given in Example 6.32 then $U_n = \{(x,y) \mid d(x,y) < 2^{-n}\}$ so that the quasi-uniformity $\mathcal{U}$ of Example 6.32 is generated by d. It follows from Corollary 7.5 that $(X,\mathcal{T}(d))$ is completely regular. Suppose that $\langle W_k \rangle$ is a base for a transitive quasi-uniformity compatible with $\mathcal{T}(d)$. For each k, $n \in \mathbb{N}$ set $B(k,n) = \{x \mid W_k(x) \subset U_n(x)\}$. Then $X = \cup\{B(k,n) \mid k, n \in \mathbb{N}\}$ so that by Proposition 6.33 the Euclidean plane is of first category - a contradiction.

PROBLEM P. If $(X,\mathcal{T})$ is a topological space and $\mathcal{T}$ has a σ-interior preserving base, is $(X,\mathcal{T})$ a transitive space?

2. PRODUCTS AND MAPPINGS

7.8 We now turn to results that establish the behavior of quasi-metrizable spaces with regard to products and mappings.

PROPOSITION. Every countable product of quasi-metrizable spaces is quasi-metrizable.

Proof. Let $\{X_n | n \in \mathbb{N}\}$ be a collection of quasi-metrizable spaces and let $X = \Pi\{X_n | n \in \mathbb{N}\}$. For each $n \in \mathbb{N}$ let $\{V(m,n) | m \in \mathbb{N}\}$ be a base for a quasi-uniformity compatible with the topology of X_n. For each $m, n \in \mathbb{N}$ let $W(m,n) = \{(x,y) \in X \times X | (\pi_n(x), \pi_n(y)) \in V(m,n)\}$. Since the product quasi-uniformity induces the product topology, the collection $\{W(m,n) | m, n \in \mathbb{N}\}$ is a subbase for a quasi-uniformity compatible with the product topology. ∎

7.9 PROPOSITION. Every finite-to-one open image of a quasi-metrizable space is quasi-metrizable.

Proof. Let X be a quasi-metrizable space, let $f : X \to Y$ be a finite-to-one open map of X onto Y, and let $\langle U_n \rangle$ be a normal sequence of neighbornets of X such that for each $x \in X$, $\{U_n(x) | x \in X\}$ is a base for η_x. For each $n \in \mathbb{N}$ set $V_n = \{(y,z) | z \in f(U_n(x))$ whenever $f(x) = y\}$. As f is an open map, each V_n is a neighbornet of Y. We show that $\langle V_n \rangle$ is a normal sequence. Suppose that $(y,z) \in V_{n+1}^2$. There is a $q \in Y$ such that $(y,q) \in V_{n+1}$ and $(q,z) \in V_{n+1}$. Let $x \in f^{-1}(y)$. There is a point $p \in U_{n+1}(x) \cap f^{-1}(q)$, and so $z \in f(U_{n+1}(p)) \subset f(U_{n+1}^2(x)) \subset f(U_n(x))$. We have that $z \in f(U_n(x))$ whenever $x \in f^{-1}(y)$; that is $(y,z) \in V_n$. It is easy to show that the quasi-uniformity for which $\{V_n | n \in \mathbb{N}\}$ is a base is compatible with the topology of Y. ∎

7.10 Let X and Y be topological spaces. A map $f : X \to Y$ is a *compact* map provided that for each $y \in Y$, $f^{-1}(y)$ is compact; a map $f : X \to Y$ is a *pseudo-open* map provided that if G is a neighborhood of $f^{-1}(y)$ then $y \in \operatorname{int}[f(G)]$. We now consider the behavior of quasi-metrizable spaces under pseudo-open and compact open maps.

EXAMPLE. A pseudo-open two-to-one map from a quasi-metrizable Moore space X onto a space Y that is neither developable nor quasi-metrizable.

Let $X_1 = \mathbb{R} \times \mathbb{R} \times \{1\}$, $X_2 = \mathbb{R} \times \mathbb{R} \times \{2\}$, and let $Y = \mathbb{R} \times \mathbb{R} \times \{0\}$. For each $x \in \mathbb{R}$ and $\epsilon > 0$ set $U_\epsilon((x,0)) = \{(x_1,y_1) \in \mathbb{R} \times \mathbb{R} | (x - x_1)^2 + (y_1 + \epsilon)^2 < \epsilon^2\}$, set $V_\epsilon((x,0)) = \{(x_1,y_1) \in \mathbb{R} \times \mathbb{R} | (x - x_1)^2 + (y_1 - \epsilon)^2 < \epsilon^2\}$, and set $W_\epsilon((x,0)) = U_\epsilon((x,0)) \cup V_\epsilon((x,0))$. Let $p = (x,y,z) \in X_1 \cup X_2 \cup Y$. If $y \neq 0$ let p have the Euclidean neighborhoods inherited from E^3, if $y = 0$ and $z = 0$ let p have basic neighborhoods of the form $\{p\} \cup (W_\epsilon((x,0)) \times \{0\})$, $\epsilon > 0$, if $y = 0$ and $z = 1$ let p have basic neighborhoods of the form $\{p\} \cup (U_\epsilon((x,0)) \times \{0\})$, $\epsilon > 0$, and if $y = 0$ and $z = 2$, let p have basic neighborhoods of the form $\{p\} \cup (V_\epsilon((x,0)) \times \{0\})$, $\epsilon > 0$. Let $X = X_1 \cup X_2$. Since X_1 and X_2 are each homeomorphic to the Niemytzki plane (Example 5.17), it follows from Corollary 7.24 that X is a quasi-metrizable space; as it is well known that Y is a semistratifiable space that is not developable, it follows from Proposition 7.20 that Y is not quasi-metrizable. Evidently the projection map $\pi : X \to Y$ is a pseudo-open mapping. In light of the preceding proposition it is comforting to note that if $K = \{(0,y,0) | y = 0 \text{ or } y = 1/n \text{ for some } n > 1\}$ then K is compact but $\pi^{-1}(K)$ is not.

7.11 Although the following example is given to show that the compact open image of a quasi-metrizable space need not be quasi-metrizable, we note that the domain space of the example is an orthocompact Moore space and that the range space is a Moore space; hence it follows from Theorem 7.26 that the example may be substituted for Example 5.25.

EXAMPLE. A compact open map from a quasi-metrizable Moore space onto a Moore space that is not quasi-metrizable. We construct a separable quasi-metrizable Moore space X_0, a separable 0-dimensional Moore space Y_0 that is not quasi-metrizable, and a compact open map f from X_0 onto Y_0. Let $A = \{(x,0) | x \text{ is irrational}\}$ and let $B = \{(x,y) | x \text{ and } y \text{ are rational and } y > 0\}$. Set $Y_0 = A \cup B$ and give B the usual subspace topology $\mathcal{T}'$ of the plane. For each $a \in A$ where $a = (x,0)$ and each $\epsilon > 0$ let $\nabla(a,\epsilon) = \{a\} \cup \{(x + k,h) | |k| < |h| < \epsilon\}$. Then $\mathcal{T}' \cup \{\nabla(a,\epsilon) | a \in A \text{ and } \epsilon > 0\}$ is a base for a topology $\mathcal{T}$ on Y_0. It is easy to see that $(Y_0,\mathcal{T})$ is a 0-dimensional Moore space; and, as the space X of Example 5.34 is a subspace of Y_0, Y_0 is not quasi-metrizable.

Let $I = \{x | 0 \leq x \leq 1\}$ and let $X_0 = A \times \{0\} \cup B \times I$. The set $B \times I$ is open in X_0 and has the usual subspace topology of $\mathbb{R}^3$; for each $p \in B \times I$ and $n \in \mathbb{N}$ let $U_n(p)$ denote the intersection of $B \times I$ with the sphere of radius $1/n$ and center p. For each $p \in A \times \{0\}$ where $p = (x_0,0,0)$ and $n \in \mathbb{N}$, let $U_n(p)$ denote the set consisting of p together with all points of $B \times I$ that belong to the interior of the solid bounded by the cones

$$C_1 : (x - x_0)^2 + (1 + 1/n)z^2 = y^2$$

$$C_2 : (x - x_0)^2 + (1 - 1/n)z^2 = y^2$$

and the plane $y = 1/n$. For each $n \in \mathbb{N}$ set $\mathcal{G}_n = \{U_n(x) \mid x \in X_0\}$. The set X_0 is given the topology for which $\langle \mathcal{G}_n \rangle$ is a development.

The projection $f((x,y,z)) = (x,y)$ is a compact open map of X_0 onto Y_0 since $f(U_n((x,0,0))) = \nabla((x,0),1/n)$ and for each $y \in Y_0$, $f^{-1}(y) = \{(y,0)\}$ or $\{y\} \times I$.

In light of Corollary 7.24, in order to show that X_0 is quasi-metrizable we may show that for each $n \in \mathbb{N}$, $\mathcal{G}_n$ has an interior-preserving open refinement. The subspace $B \times I$ is metrizable so that for $n \in \mathbb{N}$, $\{U_n(x) \mid x \in B \times I\}$ has a locally finite open refinement $\mathcal{R}_n$.

Let $n \in \mathbb{N}$. For each $p \in A \times \{0\}$ and $q \in U_{n+1}(p) - \{p\}$, let $k(p,q)$ denote the least $k \in \mathbb{N}$ such that $st(q,\mathcal{R}_k) \subset U_n(p)$ and choose $G(p,q) \in \mathcal{R}_{k(p,q)}$ such that $q \in G(p,q)$. Set $V_n(p) = \{p\} \cup (\cup\{G(p,q) \mid q \in U_{n+1}(p)$ and $q \neq p\})$. Then $U_{n+1}(p) \subset V_n(p) \subset U_n(p)$. We show that $\{V_n(p) \mid p \in A \times \{0\}\}$ is interior preserving. Let $x \in \cap\{V_n(p_\lambda) \mid \lambda \in \Lambda\}$; without loss of generality we assume that for each $\lambda \in \Lambda$, $p_\lambda \neq x$. Thus for each $\lambda \in \Lambda$ there exists $q_\lambda \neq p_\lambda$ such that $q_\lambda \in U_{n+1}(p_\lambda)$ and $x \in G(p_\lambda,q_\lambda)$. If $\{k(p_\lambda,q_\lambda) \mid \lambda \in \Lambda\}$ has an upper bound, $\cap\{G(p_\lambda,q_\lambda) \mid \lambda \in \Lambda\}$ is open and so $\cap\{V_n(p_\lambda) \mid \lambda \in \Lambda\}$ is a neighborhood of x.

If $\{k(p_\lambda,q_\lambda) \mid \lambda \in \Lambda\}$ is not bounded, we may assume without loss of generality that $\mathbb{N} \subset \Lambda$ and that $\langle k(p_i,q_i) \rangle \to \infty$. For each $i \in \mathbb{N}$, $\{x,q_i\} \subset G(p_i,q_i) \in \mathcal{R}_{k(p_i,q_i)}$, and so $\langle q_i \rangle \to x$. Since $x \in B \times (0,1]$, there exists $\epsilon > 0$ such that for all $i \in \mathbb{N}$, $q_i \in B \times [\epsilon,1]$. Let δ denote the Euclidean distance between the sets $U_{n+1}(p) \cap B \times [\epsilon,1]$ and $B \times I - U_n(p)$. Then $\delta > 0$ and δ does not depend on $p \in A \times \{0\}$. Choose $m \in \mathbb{N}$ such that $1/m < \delta$. For each $i \in \mathbb{N}$, $st(q_i,\mathcal{R}_m) \subset U_n(p_i)$ so that $k(p_i,q_i) \leq m$ - a contradiction.

7.12 Although the example given above shows that quasi-metrizability is not preserved by compact open maps, the range space Y_0 does not belong to Arhangel'skii's class MOBI, the smallest class of spaces containing all metrizable spaces that is closed under compact open maps. We now give an example of a member of MOBI that is not quasi-metrizable.

Let A, B, $\nabla(a,\epsilon)$ and $U_n(b)$ have the same meaning as before. Let $\mathcal{A}$ be the collection of all countable infinite subsets of A and let $Y = A \cup (B \times \mathcal{A})$. For each $a \in A$ set $V_n(a) = \{a\} \cup [\nabla(a,1/n) \times \{\alpha \in \mathcal{A} | a \in \alpha\}]$. For each $n \in \mathbb{N}$ and each $(b,\alpha) \in B \times \mathcal{A}$ set $V_n((b,\alpha)) = U_n(b) \times \{\alpha\}$ and for each $n \in \mathbb{N}$ set $G_n = \{V_n(y) | y \in Y\}$. Let Y have the topology for which $<G_n>$ is a development.

To see that Y is not quasi-metrizable, suppose that there exists a quasi-metric d for Y. For each $(m,n) \in \mathbb{N} \times \mathbb{N}$ set $\alpha(m,n) = \{a \in A | V_n(a) \subset S(a,1/m)$ and $S(a,2/m) \subset V_1(a)\}$. Since $A = \cup\{\alpha(m,n) | (m,n) \in \mathbb{N} \times \mathbb{N}\}$, by the Baire Category Theorem some $\alpha(m,n)$ is dense in an open interval L of $\mathbb{R} \times \{0\}$ in the Euclidean topology. Let T be an isosceles right triangle that lies above L whose hypotenuse has length less than 1/n and lies on the interval L. Choose a point $b \in B$ inside T and a countable set $\alpha \subset \alpha(m,n) \cap L$ that is dense in L in the Euclidean topology. Let $Y_\alpha = B \times \{\alpha\}$. There exists $k \in \mathbb{N}$ such that $U_k(b) \times \{\alpha\} \subset S((b,\alpha), 1/m) \cap Y_\alpha$. Since α is dense in L in the Euclidean topology and the length of the hypotenuse of T does not exceed 1/n, there exists $a \in \alpha$ such that $b \in \nabla(a,1/n)$ and such that one side of $\nabla(a,1/n)$ is so close to b that $U_k(b) \not\subset \nabla(a,1)$. Choose $c \in U_k(b) - \nabla(a,1)$.

Since $a \in \alpha$ and $\alpha \subset \alpha(m,n)$, $(\nabla(a,1/n) \times \{\alpha\}) - \{(a,\alpha)\} = V_n(a) \cap Y_\alpha \subset S(a,1/m) \cap Y_\alpha \subset S(a,2/m) \cap Y_\alpha \subset V_1(a) \cap Y_\alpha = \nabla(a,1) \times \{\alpha\}$. We have that $d(a,(c,\alpha)) \geq 2/m$, because $S(a,2/m) \cap Y_\alpha \subset \nabla(a,1) \times \{\alpha\}$ and $c \notin \nabla(a,1)$. Since $(b,\alpha) \in \nabla(a,1/n) \times \{\alpha\} \subset S(a,1/m)$ and $(c,\alpha) \in U_k(b) \times \{\alpha\} \subset S((b,\alpha),1/m)$ we have that $d(a,(b,\alpha)) < 1/m$ and $d((b,\alpha), (c,\alpha)) < 1/m$. This contradiction establishes that Y is not quasi-metrizable.

In order to show that Y is a member of MOBI we construct a metacompact Moore space X and show that Y is the image of X under a compact open map. For each $\alpha \in \mathcal{A}$ let $\alpha^* = \alpha \cup \{\alpha\}$ and let $X_\alpha = Y_\alpha \times \alpha^*$. Set $X = A \cup (\cup\{X_\alpha | \alpha \in \mathcal{A}\})$. Let $n \in \mathbb{N}$. For each $a \in A$ set $W_n(a) = \{a\} \cup (\cup\{\nabla(a,1/n) \times \{a\} | a \in \alpha \text{ and } \alpha \in \mathcal{A}\})$. For each $(b,\alpha,a) \in X_\alpha$, and $a \in \alpha$ set $W_n(b,\alpha,a) = U_n(b) \times \{\alpha\} \times \{a\}$. For each $(b,\alpha,\alpha) \in X_\alpha$ where $\alpha = \{a_n | n \in \mathbb{N}\}$ set $W_n((b,\alpha,\alpha)) = \{(b,\alpha,\alpha)\} \cup (\cup_{m \geq n} U_m(b) \times \{\alpha\} \times \{a_m\})$. Set $G_n = \{W_n(x) | x \in X\}$. Let X have the topology for which $<G_n>$ is a development. Since X - A is the topological sum of open countable subspaces, X - A is an open metrizable subspace of X. If a and b are two points of A, then $W_n(a) \cap W_n(b) = \emptyset$. Hence X is metacompact.

Define f mapping X onto Y as follows. For $a \in A$ set $f(a) = a$ and for $(b,\alpha,x) \in X_\alpha$ define $f(b,\alpha,x) = (b,\alpha)$. It is easily verified that f is a compact open map.

7.13 THEOREM. A closed image of a quasi-metrizable space is quasi-metrizable if it is first countable.

Proof. Let $f : X \to Y$ be a closed map from a quasi-metrizable space X onto a first countable space Y. Let d be a quasi-metric that induces the topology of X; for each $n \in \mathbb{N}$ let $U_n = \{(x,y) \mid d(x,y) < 2^{-n}\}$. Let $\langle V_n \rangle$ be a decreasing sequence of neighbornets of Y such that for each $y \in Y$, $\{V_n(y) \mid n \in \mathbb{N}\}$ is a base for η_y. For each $n \in \mathbb{N}$ let $W_n = V_n \cap \{(y,z) \mid f^{-1}(z) \subset U_n(f^{-1}(y))\} = V_n - \{(y,z) \mid z \in f(X - U_n(f^{-1}(y)))\}$. Then $\langle W_n \rangle$ is a decreasing sequence of neighbornets of Y. For each $n \in \mathbb{N}$ let $\hat{W}_n = \cup\{W_{m_1} \circ W_{m_2} \circ \ldots \circ W_{m_k} \mid \Sigma_{i=1}^k 2^{-m_i} \le 2^{-n}\}$. Then $\langle \hat{W}_n \rangle$ is a normal sequence, and the proof may be completed by showing that for each $y \in Y$, $\{\hat{W}_n(y) \mid n \in \mathbb{N}\}$ is a base for η_y. Suppose not. Then there exist $y \in Y$, $H \in \eta_y$ and a sequence $\langle y_n \rangle$ such that $y_n \in \hat{W}_n(y) - H$ for each $n \in \mathbb{N}$. For each $n \in \mathbb{N}$ there is a finite sequence $y = y_n^0, y_n^1, y_n^2, \ldots, y_n^k = y_n$ and integers $m_1, m_2, \ldots, m_k$ (where k and each m_i depend on n) such that $\Sigma_{i=1}^k 2^{-m_i} \le 2^{-n}$ and $y_n^i \in W_{m_i}(y_n^{i-1})$ for $1 \le i \le k$. Without loss of generality, for all $n \in \mathbb{N}$, $y_n^1 \ne y$. For each $n \in \mathbb{N}$ choose $x_n = x_n^k \in f^{-1}(y_n^k)$ and for each i $(i = k-1, k-2, \ldots, 1)$ choose $x_n^i \in f^{-1}(y_n^i)$ such that $x_n^{i+1} \in U_{m_{i+1}}(x_n^i)$. Then $(x_n^1, x_n) \in U_{m_2} \circ U_{m_3} \circ \ldots \circ U_{m_k}$, and since $\Sigma_{i=2}^k 2^{-m_i} \le 2^{-n}$, $x_n \in U_n(x_n^1)$. For each $n \in \mathbb{N}$, $y_n^1 \in W_{m_1}(y) \subset V_n(y)$, and so $\langle y_n^1 \rangle \to y$. Since f is a closed map, the sequence $\langle x_n^1 \rangle$ has a cluster point $x \in f^{-1}(y)$. As x is a cluster point of $\langle x_n^1 \rangle$ and for each $n \in \mathbb{N}$, $x_n \in U_n(x_n^1)$, it follows that x is a cluster point of $\langle x_n \rangle$. Thus y is a cluster point of $\langle y_n \rangle$ - a contradiction. ■

COROLLARY. Every perfect image of a quasi-metrizable space is quasi-metrizable.

Proof. Let X be quasi-metrizable and let $f : X \times Y$ be a perfect map. By the previous theorem it suffices to show that Y is first countable. Let $\{U_n \mid n \in \mathbb{N}\}$ be a base for a quasi-uniformity compatible with the topology of X. For each $n \in \mathbb{N}$ set $W_n = \{(y,z) \mid z \notin f(X - U_n(f^{-1}(y)))\}$ and note that W_n is a neighbornet of Y. Let $y \in Y$ and let G be an open set containing y. By Proposition 5.1 there is an $m \in \mathbb{N}$ such that $U_m(f^{-1}(y)) \subset f^{-1}(G)$. It follows that $W_m(y) \subset G$ and so $\{W_n(y) \mid n \in \mathbb{N}\}$ is a base for η_y. ■

7.14 THEOREM. A topological space X is quasi-metrizable if it satisfies any of the following conditions.

(a) X has a countable open cover by quasi-metrizable spaces.

(b) X has a point-finite open cover by quasi-metrizable spaces.

(c) X has a locally finite closed cover by quasi-metrizable spaces.

Proof. (a) Let $\{G_n | n \in \mathbb{N}\}$ be a cover of X where each G_n is an open quasi-metrizable subspace of X. For each $n \in \mathbb{N}$ let $\{V_{m,n} | m \in \mathbb{N}\}$ be a base for a quasi-uniformity compatible with the topology of G_n. For each $m, n \in \mathbb{N}$ set $W_{m,n} = \{(x,y) \in X \times X | (x,y) \in V_{m,n}$ whenever $x \in G_n\}$. Then $\{W_{m,n} | m, n \in \mathbb{N}\}$ is a subbase for a quasi-uniformity compatible with the topology of X.

(b) Let $\{G_\alpha | \alpha \in A\}$ be a point-finite cover of X where each G_α is an open quasi-metrizable subspace of X. For each $\alpha \in A$ let $\langle V_{n,\alpha} \rangle$ be a normal sequence such that $\{V_{n,\alpha}(x) | x \in G_\alpha, n \in \mathbb{N}\}$ is a base for the topology of G_α. For each $n \in \mathbb{N}$ let $W_n = \{(x,y) | y \in V_{n,\alpha}(x)$ whenever $x \in G_\alpha\}$. Then $\langle W_n \rangle$ is a normal sequence and $\{W_n(x) | x \in X, n \in \mathbb{N}\}$ is a base for the topology of X.

(c) Since the disjoint sum of quasi-metrizable spaces is quasi-metrizable, it is a consequence of Corollary 7.13 that any space with a locally finite cover by closed quasi-metrizable spaces is quasi-metrizable. ■

COROLLARY. Every metacompact space that has an open cover by quasi-metrizable spaces is quasi-metrizable.

We note that the LOTS ω_1 demonstrates that an orthocompact first countable space may have an interior-preserving open cover by metrizable spaces yet fail to be quasi-metrizable.

3. γ-SPACES

7.15 If we use an arbitrary filter $\mathcal{U}$ of reflexive relations on a set X to define a topology on X in the same way that a topology is defined by a quasi-uniformity, it may happen that there is an $x \in X$ and an entourage $U \in \mathcal{U}$ such that $x \notin \operatorname{int}(U(x))$. What is required to avoid this unhappy circumstance is the additional condition given in the following definition. A filter $\mathcal{U}$ of reflexive relations on a set X is a *local quasi-uniformity* provided that for each $x \in X$ and $U \in \mathcal{U}$ there is a $V \in \mathcal{U}$ such that $V^2(x) \subset U(x)$. If $\mathcal{U}$ has a base of symmetric entourages, $\mathcal{U}$ is said to be a *local uniformity*.

THEOREM. A space $(X,\mathcal{T})$ admits a quasi-uniformity with a countable base if it admits a local quasi-uniformity $\mathcal{U}$ with a countable base such that $\mathcal{U}^{-1}$ is a local quasi-uniformity.

Proof. Let $(X,\mathcal{T})$ be a topological space and let $\langle V_n \rangle$ be a base for a local quasi-uniformity such that $\langle V_n^{-1} \rangle$ is a base for a local quasi-uniformity. Without loss of generality we assume that $V_{n+1} \subset V_n$ for each $n \in \mathbb{N}$. Inductively we define a sequence $\langle U_n \rangle$ of neighbornets of X as follows: $U_1 = V_1^2 = \cup\{V_1^{-1}(x) \times V_1(x) \mid x \in X\}$ and for each $n \in \mathbb{N}$, $U_{n+1} = \cup\{V_m^{-1}(x) \times V_m(x) \mid m \geq n + 1, x \in X$ and $V_m^{-3}(x) \times V_m^3(x) \subset U_n\}$. To see that $\langle U_n \rangle$ is a base for a quasi-uniformity let $n \in \mathbb{N}$ and let $(p,r) \in U_{n+1}^2$. There is a $q \in X$ so that $(p,q) \in U_{n+1}$ and $(q,r) \in U_{n+1}$; hence there are $x, y \in X$ and $k, m \in \mathbb{N}$ with $k \geq n + 1$ and $m \geq n + 1$ such that $(p,q) \in V_k^{-1}(x) \times V_k(x)$, $(q,r) \in V_m^{-1}(y) \times V_m(y)$, $V_k^{-3}(x) \times V_k^3(x) \subset U_n$, and $V_m^{-3}(y) \times V_m^3(y) \subset U_n$. If $k \geq m$, $p \in V_m^{-3}(y)$ and $r \in V_m(y)$ so that $(p,r) \in U_n$. If $k < m$, $p \in V_k^{-1}(x)$, and $r \in V_k^3(x)$ so that $(p,r) \in U_n$. Since $U_n \subset V_n^2$ for each $n \in \mathbb{N}$, $\{U_n \mid n \in \mathbb{N}\}$ is a base for a quasi-uniformity compatible with $\mathcal{T}$. ■

It is evident from the proof of the previous theorem that a T_1 space that admits a local uniformity with a countable base also admits a uniformity with a countable base. Thus, in light of Theorem 1.5, we have the following corollary.

COROLLARY. A T_1 space is metrizable if it admits a local uniformity with a countable base.

It is easy to see that a quasi-uniformity $\mathcal{U}$ is locally symmetric if and only if $\mathcal{U}$ contains a local uniformity compatible with $\mathcal{T}(\mathcal{U})$. Moreover, a quasi-uniformity $\mathcal{U}$ with a countable base is locally symmetric if and only if $\mathcal{U}$ contains a local uniformity compatible with $\mathcal{T}(\mathcal{U})$ that has a countable base. Therefore, as a consequence of the previous corollary, we have the result of Theorem 2.32 that a topological space that admits a locally symmetric quasi-metric is metrizable.

7.16 At first glance it appears that Corollary 7.15 is an insignificant generalization of Theorem 1.5; the ease with which we may now obtain one direction of the Nagata-Smirnov metrization theorem dispels this first impression.

THEOREM. (Nagata and Smirnov). Every regular T_1 space with a σ-locally finite base is metrizable.

Proof. Let $\mathcal{B} = \cup_{n=1}^{\infty} \mathcal{B}_n$ be a σ-locally finite base for a regular T_1 topology $\mathcal{T}$ on a set X. Without loss of generality we assume that each $\mathcal{B}_n$ is locally finite and contains ∅ and X and that for each $n \in \mathbb{N}$, $\mathcal{B}_n \subset \mathcal{B}_{n+1}$. For each $n \in \mathbb{N}$ and $x \in X$ set $G(n,x) = \cap\{X - \overline{B} | x \in X - \overline{B} \text{ and } B \in \mathcal{B}_n\}$ and set $V_n = \cup\{(\cap(\mathcal{B}_n)_x \cap G(n,x)) \times (\cap(\mathcal{B}_n)_x \cap G(n,x)) | x \in X\}$. Each V_n is an open symmetric neighbornet of X. To see that $<V_n>$ is a base for a local uniformity let $x \in X$ and let $n \in \mathbb{N}$. There is a $j > n$ and members B_1 and B_2 of $\mathcal{B}_j$ such that $x \in B_1 \subset \overline{B}_1 \subset B_2 \subset V_n(x)$. Note that $V_j(B_1) \subset B_2$; for if $b \in B_1$ and $q \in V_j(b)$ there is a $y \in B_2$ such that $\{b,q\} \subset \cap(\mathcal{B}_j)_y \cap G(j,y)$ and so $q \in \cap(\mathcal{B}_j)_y \subset B_2$. By the same argument there is an $m > j$ such that $V_m(x) \subset B_1$. Thus $V_m^2(x) \subset V_j(B_1) \subset B_2 \subset V_n(x)$. It is clear that the local uniformity generated by $<V_n>$ is compatible with $\mathcal{T}$. ■

7.17 The Nagata-Smirnov metrization theorem suggests that there may be an intrinsic characterization of those T_1 spaces that admit a local quasi-uniformity with a countable base. Let $(X,\mathcal{T})$ be a topological space. A subset ξ of $\mathcal{T} \times \mathcal{P}(X)$ is a *pairbase for* X provided for each $x \in X$ and each $G \in \eta_x$ there exists $(A,B) \in \xi$ such that $x \in A \subset B \subset G$. A collection ξ of ordered pairs of subsets of X is *cocushioned* provided that for each $\zeta \subset \xi$, $\cap\{A | (A,B) \in \zeta\} \subset \text{int}[\cap\{B | (A,B) \in \zeta\}]$. A pairbase ξ is a *σ-cocushioned pairbase* provided that $\xi = \cup_{n=1}^{\infty} \xi_n$, where each ξ_n is cocushioned.

THEOREM. A T_1 space X admits a local quasi-uniformity with a countable base if and only if there is a σ-cocushioned pairbase for X.

Proof. Suppose that X admits a local quasi-uniformity with a base $<U_n>$. For each $n \in \mathbb{N}$ let $\xi_n = \{(R,U_n(R)) | R \text{ is an open set}\}$. The proof of Proposition 5.10 shows that each ξ_n is cocushioned and so it is evident that $\cup_{n=1}^{\infty} \xi_n$ is a σ-cocushioned pairbase for X.

Now suppose that $\xi = \cup_{n=1}^{\infty} \xi_n$ is a σ-cocushioned pairbase for X, where without loss of generality we assume that $\xi_n \subset \xi_{n+1}$ for each $n \in \mathbb{N}$. For each $n \in \mathbb{N}$ set $U_n = \{(x,y) | y \in B$ whenever $x \in A$ and $(A,B) \in \xi_n\}$. Let $x \in X$ and let $G \in \eta_x$. There is an $n \in \mathbb{N}$, and (A,B) and $(C,D) \in \xi_n$ such that $x \in A \subset B \subset C \subset D \subset G$. Suppose that $z \in U_n^2(x)$. There is a y in X such that $(x,y) \in U_n$ and $(y,z) \in U_n$. Since $y \in U_n(x)$ and $x \in A$, $y \in C$. Since $z \in U_n(y)$ and $y \in C$, $z \in G$. ■

COROLLARY. Let X be a regular space that has a base $\mathcal{B} = \cup_{n=1}^{\infty} \mathcal{B}_n$ such that for each $n \in \mathbb{N}$, $\cap\{\bar{B} | x \in B$ and $B \in \mathcal{B}_n\}$ is a neighborhood of x. Then X admits a local quasi-uniformity with a countable base.

Proof. For each $n \in \mathbb{N}$ set $\xi_n = \{(B,\bar{B}) | B \in \mathcal{B}_n\}$. Then $\xi = \cup_{n=1}^{\infty} \xi_n$ is a σ-cocushioned pairbase. ■

Although the converse of the preceding corollary provides an attractive conjecture, Example 7.7 shows that this conjecture is false.

7.18 T_1 spaces that admit a local quasi-uniformity with a countable base have been studied extensively under various aliases. The following theorem provides several useful characterizations of these spaces.

THEOREM. Let $(X,\mathcal{T})$ be a T_1 topological space and let $\langle V_n \rangle$ be a decreasing sequence of neighbornets of X. The following statements are equivalent.

(a) If $x_n \in V_n(y_n)$ and $y_n \in V_n(p)$ for each $n \in \mathbb{N}$, then $\langle x_n \rangle \to p$.

(b) $\{V_n | n \in \mathbb{N}\}$ is a base for a local quasi-uniformity for $(X,\mathcal{T})$.

(c) If K is a compact subspace of X and G is an open subset of X that contains K, there is an $n \in \mathbb{N}$ such that $V_n(K) \subset G$.

(d) If p is a cluster point of $\langle y_n \rangle$ and for each $n \in \mathbb{N}$, $x_n \in V_n(y_n)$, then p is a cluster point of $\langle x_n \rangle$.

Proof. (a) ⇒ (b). Let $p \in X$ and $m \in \mathbb{N}$. Suppose that for each $n \in \mathbb{N}$, there is an $x_n \in V_n^2(p) - V_m(p)$. Then for each $n \in \mathbb{N}$ there exists $y_n \in V_n(p)$ such that $x_n \in V_n(y_n)$. It follows from condition (a) that p is a cluster point of $\langle x_n \rangle$ - a contradiction. Thus $\{V_n | n \in \mathbb{N}\}$ is a base for a local quasi-uniformity $\mathcal{U}$ on X, and it is evident that $\mathcal{U}$ is compatible with $\mathcal{T}$.

(b) ⇒ (c). Let K be a compact set and let G be an open set containing K. For each $x \in K$ let $n(x) \in \mathbb{N}$ such that $V_{n(x)}^2(x) \subset G$. There is a finite subset F of K so that $K \subset \cup\{V_{n(x)}(x) | x \in F\}$. Let $n = \max\{n(x) | x \in F\}$. If $p \in V_n(K)$ there is a $y \in K$ and an $x \in F$ such that $p \in V_n(y)$ and $y \in V_{n(x)}(x)$ so that $p \in V_n^2(x) \subset G$.

(c) ⇒ (d). For each $n \in \mathbb{N}$, let $x_n \in V_n(y_n)$ and let p be a cluster point of $\langle y_n \rangle$. To show that p is a cluster point of $\langle x_n \rangle$, let $G \in \eta_p$ and let $m' \in \mathbb{N}$. Since X is first countable, $\langle y_n \rangle$ has a subsequence $\langle y_{j_n} \rangle$ that converges to p. Hence there exists $m'' \geq m'$ such that $y_{j_n} \in G$ whenever $n \geq m''$. Set $K = \{y_{j_n} | n \geq m''\}$. Then $K \subset G$ so that for some $m \geq m''$, $V_m(K) \subset G$. It follows that $x_{j_m} \in V_m(y_{j_m}) \subset V_m(K) \subset G$.

(d) ⇒ (a). Let $p \in X$. Considering the constant sequence $\langle p \rangle$, we see that it follows from condition (d) that if $y_n \in V_n(p)$ for each $n \in \mathbb{N}$ then p is a cluster point of $\langle y_n \rangle$. Consequently if, as in the hypothesis of condition (a), $x_n \in V_n(y_n)$ and $y_n \in V_n(p)$ for each $n \in \mathbb{N}$, then p is a cluster point of $\langle x_n \rangle$. Suppose that $\langle x_n \rangle$ does not converge to p. Then there is an open set G about p and a subsequence $\langle x_{j_n} \rangle$ of $\langle x_n \rangle$ such that for all $n \in \mathbb{N}$, $x_{j_n} \notin G$. For each $n \in \mathbb{N}$, $x_{j_n} \in V_{j_n}(y_{j_n}) \subset V_n(y_{j_n})$ and $y_{j_n} \in V_{j_n}(p) \subset V_n(p)$. Therefore p is a cluster point of $\langle x_{j_n} \rangle$ - a contradiction. ∎

In the literature, spaces that have a sequence of neighbornets satisfying (a) are called γ-spaces; spaces that have a sequence of neighbornets satisfying (c) are called coNagata spaces; spaces that have a sequence of neighbornets satisfying (d) are said to admit a coconvergent open neighborhood assignment. Hereafter, we call a T_1 space that has a sequence of neighbornets that satisfies any of the conditions of Theorem 7.18 a *γ-space*, and we call a sequence of neighbornets that satisfies condition (a) of Theorem 7.18 a *γ-sequence*.

7.19 A comparison of Theorem 1.5 and Corollary 7.15 suggests the conjecture that every γ-space is quasi-metrizable. We first establish an important class of spaces in which this conjecture obtains.

PROPOSITION. Let $(X,\mathcal{T})$ be a γ-space and suppose that there is an $m \in \mathbb{N}$ so that for each neighbornet U of X, U^m is a normal neighbornet of X. Then $(X,\mathcal{T})$ is quasi-metrizable.

Proof. Let $\langle U_n \rangle$ be a countable nested base for a local quasi-uniformity on X compatible with $\mathcal{T}$ and let m be a natural number such that for each $n \in \mathbb{N}$, U_n^m is a normal neighbornet. For each $k \in \mathbb{N}$ let $\{V_{n,k} | n \in \mathbb{N}\}$ be a normal sequence whose first term is U_k^m. Then $\{V_{n,k} | n, k \in \mathbb{N}\}$ is a subbase for a quasi-uniformity on X that is compatible with $\mathcal{T}$. ∎

The following corollary is a consequence of Theorem 6.30 and Proposition 7.3.

COROLLARY. Any GO space that is a γ-space is non-Archimedeanly quasi-metrizable.

7.20 We now characterize those developable spaces that are quasi-metrizable. Note that our characterization provides an important case in which γ-spaces and quasi-metrizable spaces coincide.

PROPOSITION. A semistratifiable γ-space is both quasi-metrizable and developable.

Proof. Let X be a semistratifiable γ-space, let $\langle U_n \rangle$ be a γ-sequence for X, and let $\langle V_n \rangle$ be a semistratification of X. For each $n \in \mathbb{N}$ set $W_n = U_n \cap V_n$ and suppose that $\{p, x_n\} \subset W_n(y_n)$ for each $n \in \mathbb{N}$. Since $\langle W_n \rangle$ is a semistratification of X, p is a cluster point of $\langle y_n \rangle$ and hence of $\langle x_n \rangle$. Thus, if for each $n \in \mathbb{N}$ we set $\mathcal{W}_n = \{\text{int } W_n(x) | x \in X\}$, $\{\mathcal{W}_n | n \in \mathbb{N}\}$ is a development for X. (See the notes.)

Since X is countably metacompact, it is an immediate consequence of Proposition 7.22 that X is preorthocompact; hence by Theorem 6.21(a) and Proposition 7.19, X is quasi-metrizable. ∎

7.21 We know from Proposition 7.19 that if there is a γ-space that is not quasi-metrizable there is a γ-space X and a neighbornet U of X with the property that no finite power of U is a normal neighbornet. The space M^{ω_0} of Example 6.28 has such a neighbornet; and, although this space is itself quasi-metrizable, it can be used to build a γ-space that is not quasi-metrizable.

EXAMPLE. Let $X = M^{\omega_0}$ and let V be a neighbornet of X with the property that for each $n \in \mathbb{N}$, V^n is not a normal neighbornet. Since X is the countable product of quasi-metrizable spaces, it follows from Proposition 7.8 that X is quasi-metrizable. Let $\langle V_n \rangle$ be a γ-sequence for X such that for each $n \in \mathbb{N}$, $V_{n+1} \subset V_n \subset V$. Let $Y = \cup_{\alpha \leq \omega_0} X^\alpha$. For each $n \in \mathbb{N}$, $\alpha < \omega_0$ and $x \in X^\alpha$ define

$$U_n(x) = \{x_1\} \times \{x_2\} \times \ldots \times \{x_{\alpha-1}\} \times V_n(x_\alpha)$$

For each $n \in \mathbb{N}$ and $x \in X^{\omega_0}$ define

$$U_n(x) = \{x\} \cup \cup_{i=n}^{\infty} \{x_1\} \times \{x_2\} \times \ldots \times \{x_{i-1}\} \times V_n^{i-n+1}(x_i)$$

We show that $\{U_n | n \in \mathbb{N}\}$ is a base for a local quasi-uniformity $\mathcal{U}$ on Y and that $(Y,\mathcal{T}(\mathcal{U}))$ is a γ-space that is not quasi-metrizable. Let α be a natural number and let $x \in X^{\alpha}$. Then $U_m^2(x) \subset U_n(x)$ whenever $V_m^2(x_\alpha) \subset V_n(x_\alpha)$. For each $x \in X^{\omega_0}$, $U_{n+1}^2(x) \subset U_n(x)$. Thus $\langle U_n \rangle$ is a γ-sequence for Y.

Now suppose that $\{W_n | n \in \mathbb{N}\}$ is a base for a quasi-uniformity on Y that is compatible with $\mathcal{T}(\mathcal{U})$. Inductively we define a sequence $\langle (x_n, y_n) \rangle$ in $X \times X$ such that $(x_n,y_n) \notin V^n$ and $((x_1, x_2, \ldots, x_n), (x_1, x_2, \ldots, x_{n-1}, y_n)) \in W_n$. Since V cannot contain a normal neighbornet we may choose $(x_1,y_1) \in W|X \times X - V$. Assume that $\{(x_i,y_i)|1 \leq i < n\}$ has been defined and set $Y_n = \{x_1\} \times \{x_2\} \times \ldots \times \{x_{n-1}\} \times X$. In order to simplify notation we identify Y_n with its canonically homeomorphic image, X. Since $W_n|Y_n \times Y_n$ is a normal neighbornet of Y_n and V^n is not a normal neighbornet of Y_n, there exists $((x_1, x_2, \ldots, x_n), (x_1, x_2, \ldots, x_{n-1}, y_n)) \in W_n$ such that $(x_n,y_n) \notin V^n$. Let x be the sequence whose n^{th} term is x_n and for each $n \in \mathbb{N}$ define points p_n and q_n of Y_n by $p_n = (x_1, x_2, \ldots, x_n)$ and $q_n = (x_1, x_2, \ldots, x_{n-1}, y_n)$. Then $q_n \in W_n(p_n)$ and x is a cluster point of $\langle p_n \rangle$ so that, by Theorem 7.18(d), x is a cluster point of $\langle q_n \rangle$. Since $V_1 \subset V$, $U_1(x) \subseteq \{x\} \cup \cup_{i=1}^{\infty} \{x_1\} \times \{x_2\} \times \ldots \times \{x_{i-1}\} \times V^i(x_i)$. Let $n \in \mathbb{N}$. As $y_n \notin V^n(x_n)$, $q_n \notin U_1(x)$. Thus no term of $\langle q_n \rangle$ belongs to $U_1(x)$ - a contradiction.

7.22 Paracompactness is important in the study of uniform spaces because it is a property shared by metrizable spaces and compact spaces and has a natural characterization in terms of uniform covers. The range space Y of Example 5.25 shows that the fine quasi-uniformity of a quasi-metrizable space need not be a Lebesgue quasi-uniformity; indeed the example shows that a quasi-metrizable space need not be preorthocompact. The following characterization of paracompactness suggests an appropriate generalization of paracompactness and quasi-metrizability: A regular T_1 space X is paracompact if and only if for each open cover $\mathcal{C}$ of X there is a sequence $\langle V_n \rangle$ of symmetric neighbornets of X such that for each $x \in X$ there is an $n \in \mathbb{N}$ and a $C \in \mathcal{C}$ such that $V_n^2(x) \subset C$.

A topological space X is γ-*refinable* provided that for each open cover $\mathcal{C}$ of X there is a sequence $\langle V_n \rangle$ of neighbornets of X such that for each $x \in X$ there is an $n \in \mathbb{N}$ and a $C \in \mathcal{C}$ such that $V_n^2(x) \subset C$. Evidently a space that is either a γ-space or a preorthocompact space is γ-refinable; and, in spite of the example mentioned above, we have the following result.

PROPOSITION. Every countably metacompact γ-refinable space is preorthocompact.

Proof. Let X be a countably metacompact γ-refinable space and let $\mathcal{C}$ be an open cover of X. There is a sequence $\langle V_n \rangle$ of neighbornets of X such that for each $x \in X$ there is an $m \in \mathbb{N}$ and a $C \in \mathcal{C}$ so that $V_m^3(x) \subset C$. We assume without loss of generality that $V_{n+1} \subset V_n$ for each $n \in \mathbb{N}$. For each $n \in \mathbb{N}$ let $G_n = \text{int}\,\{x \in X \mid \text{for some } C \in \mathcal{C},\ V_n^2(x) \subset C\}$. To see that $\{G_n \mid n \in \mathbb{N}\}$ is an open cover of X let $x \in X$ and let $m \in \mathbb{N}$ such that $V_m^3(x) \subset C$ for some $C \in \mathcal{C}$. Then $x \in G_m$, since $V_m^2(V_m(x)) = V_m^3(x)$. Let $\mathcal{H} = \{H_n \mid n \in \mathbb{N}\}$ be a point-finite open cover of X such that $H_n \subset G_n$ for each $n \in \mathbb{N}$. For each $x \in X$ let n_x be the largest natural number such that $x \in H_n$ and set $V = \{(x,y) \in X \times X \mid y \in V_{n_x}(x) \cap \bigcap \mathcal{H}_x\}$. It is easy to check that V is a neighbornet of X such that $\{V^2(x) \mid x \in X\}$ refines $\mathcal{C}$. ∎

COROLLARY. Every open cover of a countably metacompact quasi-metrizable space is a quasi-normal cover.

Proof. Let X be a countably metacompact quasi-metric space and let $\mathcal{C}$ be an open cover of X. We may now assume that the sequence $\langle V_n \rangle$ of neighbornets of the preceding proposition is a normal sequence of neighbornets. For each $n \in \mathbb{N}$ set $W_n = \{(x,y) \mid y \in V_{n_x+n-1}(x) \cap \bigcap \mathcal{H}_x\}$. Since W_1 is the neighbornet V of the preceding proposition and each W_n is a neighbornet of X, we need only show that $W_{n+1}^2 \subset W_n$ for each $n \in \mathbb{N}$. Let $n \in \mathbb{N}$, let $x \in X$, and let $z \in W_{n+1}^2(x)$. There is a $y \in W_{n+1}(x)$ such that $z \in W_{n+1}(y)$. Since $\bigcap \mathcal{H}_z \subset \bigcap \mathcal{H}_y \subset \bigcap \mathcal{H}_x$, $n_z \ge n_y \ge n_x$. Thus $y \in W_{n+1}(x) \subset V_{n_x+n}(x)$ and $z \in W_{n+1}(y) \subset V_{n_y+n}(y) \subset V_{n_x+n}(y)$ so that $z \in V_{n_x+n}^2(x) \subset V_{n_x+n-1}(x)$. Thus $z \in W_n(x)$. ∎

The method of proof of the preceding corollary could be used to establish that every countably metacompact σ-orthocompact space is orthocompact. A comparison of the proof of this result given in §5.9 and the proof indicated above provides an example of the clarity afforded by using neighbornets in place of covers.

It follows from Proposition 7.20 and the previous corollary that the locally compact Moore space of Example 5.34 is not a γ-space.

7.23 A sequence $\langle \mathcal{G}_n \rangle$ of open covers of a topological space X is a *subdevelopment* for X provided that for each $x \in X$ and each sequence $\langle G_n \rangle$ with $x \in G_n \in \mathcal{G}_n$, $\{G_n | n \in \mathbb{N}\}$ is a subbase for η_p. A Tychonoff space has a subdevelopment if and only if it has a G_δ-diagonal and is a G_δ-subset of its Stone-Čech compactification. A paracompact Hausdorff space that has a subdevelopment is metrizable.

THEOREM. Let X be a T_1 space that has a subdevelopment.

(a) If X is γ-refinable, X is a γ-space.

(b) If X admits a Lebesgue quasi-uniformity, X is quasi-metrizable.

(c) If X is σ-orthocompact, X is non-Archimedeanly quasi-metrizable.

Proof. We only establish part (a) since its proof is representative. Assume that X is γ-refinable and let $\langle \mathcal{G}_n \rangle$ be a subdevelopment for X. For each $n \in \mathbb{N}$ let $\{V_{m,n} | m \in \mathbb{N}\}$ be a sequence of neighbornets of X such that for each $x \in X$ there is an $m \in \mathbb{N}$ and a $G \in \mathcal{G}_n$ so that $V^2_{m,n}(x) \subset G$. For each $n \in \mathbb{N}$ set $W_n = \cap\{V_{i,j} | i + j \le n\}$. To see that $\langle W_n \rangle$ is a γ-sequence, for each $n \in \mathbb{N}$ let $x_n \in W_n(y_n)$ and let $y_n \in W_n(p)$. For each $n \in \mathbb{N}$ let $m_n \in \mathbb{N}$ and $G_n \in \mathcal{G}_n$ such that $V^2_{m_n}(p) \subset G_n$. Let $j_1 = m_1 + 1$ and for $n > 1$ let $j_n = m_n + j_{n-1}$. Note that for each $n \in \mathbb{N}$, $x_{j_n} \in W^2_{j_n}(p) = \cap^n_{i=1} W^2_{j_i}(p) \subset \cap^n_{i=1} G_i$. Thus $\{p, x_{j_n}\} \subset \cap^n_{i=1} G_i$ so that p is a cluster point of $\langle x_{j_n} \rangle$ and hence of $\langle x_n \rangle$. ∎

7.24 THEOREM. A topological space X admits a point-symmetric quasi-metric if and only if X is developable and every open cover is a quasi-normal cover.

Proof. Suppose that X is developable and that every open cover of X is a quasi-normal cover. Let $\langle \mathcal{G}_n \rangle$ be a development for X. There is a normal sequence $\langle U_n \rangle$ such that for each $n \in \mathbb{N}$, $\{U_n(x) | x \in X\}$ refines $\mathcal{G}_n$. Let $\mathcal{U}$ be the quasi-uniformity generated by $\{U_n | n \in \mathbb{N}\}$. It is easily established, as in the previous theorem, that $\mathcal{U}$ is compatible with $\mathcal{T}$. To see that $\mathcal{U}$ is point symmetric let $n \in \mathbb{N}$ and let $p \in X$. There is an $m \in \mathbb{N}$ such that $st(p,\mathcal{G}_m) \subset U_n(p)$. Let $y \in U^{-1}_m(p)$. Then $y \in st(p,\mathcal{G}_m) \subset U_n(p)$.

Now suppose that X admits a point-symmetric quasi-metric. By Proposition 2.28, X is developable and it follows from Corollary 7.22 that every open cover of X is a quasi-normal cover. ■

COROLLARY. A topological space admits a point-symmetric non-Archimedean quasi-metric if and only if it is an orthocompact developable space.

Proof. The proof of Theorem 7.24 may be easily modified to show that every orthocompact developable space admits a point-symmetric non-Archimedean quasi-metric. The remaining implication is an immediate consequence of Proposition 2.28, Theorem 7.1, and Proposition 5.9. ■

7.25 Since it is an unanswered question whether every preorthocompact developable space is orthocompact, we are unable to say whether there is any difference between Theorem 7.24 and its corollary. We are at least able to give an example of a non-orthocompact space X for which every open cover of X is a quasi-normal cover.

EXAMPLE. A completely regular non-orthocompact quasi-metric space X such that every open cover of X is a quasi-normal cover. Let (K,d) be the Kofner plane, let $D = \{0\} \cup \{1/n | n \in \mathbb{N}\}$ and let $X = K \times D$. We define a quasi-uniformity on X as follows. For each $m, n \in \mathbb{N}$ set $U_m((x,1/n)) = S_d(x,1/mn) \times \{1/n\}$ and for $m \in \mathbb{N}$ set $U_m((x,0)) = \{(x,0)\} \cup (\cup\{U_m((x,1/n)) | n > m\}$. Then $\{U_m | m \in \mathbb{N}\}$ is a base for a quasi-uniformity $\mathcal{U}$ on X, and $(X,\mathcal{T}(\mathcal{U}))$ is the desired quasi-metrizable space. Since the quasi-metric d on K is continuous, the quasi-metric generated by $\mathcal{U}$ is also continuous so that $(X,\mathcal{U})$ is completely regular. Note that $(X,\mathcal{T}(\mathcal{U}))$ is the countable union of pairwise disjoint closed and open subspaces homeomorphic to K and a discrete subspace. It follows that $(X,\mathcal{T}(\mathcal{U}))$ is perfect and hence countably metacompact. By Corollary 7.22 every open cover of X is a quasi-normal cover.

It remains to show that X is not orthocompact. Suppose that $\mathcal{R}$ is an interior-preserving open refinement of the open cover $\{U_1((x,0)) | x \in K\}$ and for each $n \in \mathbb{N}$ let $f_n : K \to X$ be defined by $f_n(x) = (x,1/n)$. Then for each $n \in \mathbb{N}$, $\mathcal{A}_n = \{f_n^{-1}(R) | R \in \mathcal{R}, n \in \mathbb{N}\}$ is an interior-preserving open collection in K. To show that $\cup\mathcal{A}_n$ is a base for the topology of K, let $x \in K$ and let G be an open subset of K containing x. There is an $n \in \mathbb{N}$ such that $S_d((x,1/n)) \subset G$ and $R \in \mathcal{R}$ such that $(x,0) \in R \subset U_1((x,0))$. Thus there is an $m \geq n$ such that $x \in f_m^{-1}(R) \subset f_m^{-1}(U_1((x,0))) = S_d((x,1/m)) \subset G$. It follows that $\cup_{n=1}^{\infty} A_n$ is a base for the topology of the Kofner plane - a contradiction. Thus X is not orthocompact.

PROBLEM Q. Is every open cover of a preorthocompact space a quasi-normal cover?

7.26 We end our discussion of quasi-metrizable developable spaces with a theorem that is merely an attempt to cover under one roof the results concerning these spaces that are strewn throughout the preceding sections.

THEOREM. Let X be a developable space. The first six of the following statements are equivalent, the remaining statements are equivalent, and if X is a transitive space all the following statements are equivalent.

(a) Every open cover of X is a quasi-normal cover.
(b) X admits a point-symmetric quasi-metric.
(c) X is quasi-metrizable.
(d) X is a γ-space.
(e) X is γ-refinable.
(f) X is preorthocompact.
(g) X admits a point-symmetric non-Archimedean quasi-metric.
(h) X admits a non-Archimedean quasi-metric.
(i) X is hereditarily orthocompact.
(j) X is orthocompact.
(k) The topology of X has a σ-interior-preserving base.

Proof. The equivalence of (a) and (b) is given in Theorem 7.24; the implications (b) $\Rightarrow$ (c) $\Rightarrow$ (d) $\Rightarrow$ (e) are evident; and, by Proposition 7.22, (e) $\Rightarrow$ (f). By Theorem 6.21(b), (f) $\Rightarrow$ (a). The equivalence of (g) and (j) has been given in Corollary 7.24 and the equivalence of (h) and (k) has been given in Theorem 7.1. Evidently, (g) $\Rightarrow$ (h). By Theorem 7.1, if (h) holds then every subspace of X has a σ-interior-preserving base so that, by Proposition 5.9, (h) $\Rightarrow$ (i) $\Rightarrow$ (j). If X is a transitive space, conditions (c) and (h) are equivalent by Proposition 7.3. ∎

PROBLEM R. Theorem 7.26 suggests the following questions. Is every preorthocompact semistratifiable space orthocompact? Is every semistratifiable space a transitive space? These questions are unresolved even when in their formulation the term "semistratifiable" is replaced by the term "Moore".

4. QUASI-METRIZABLE SUBTOPOLOGIES

7.27 Following Arhangelśkii, we say that a topological space X can be *condensed* onto a topological space Y provided there is a one-to-one continuous mapping from X onto Y. A space $(X,\mathcal{T})$ can be condensed onto a space Y with property P if and only if there is a topology $\mathcal{T}'$ coarser than $\mathcal{T}$ (called a *subtopology* of $\mathcal{T}$) such that $(X,\mathcal{T}')$ has property P. In this section we consider those spaces that can be condensed onto a quasi-metrizable space. We begin by characterizing a seemingly more restrictive property, which was introduced in the literature by F. Siwiec and J. Nagata. A cover $\mathcal{C}$ of a set X is a *point-separating* cover provided that for each two points x and y of X there exists a $C \in \mathcal{C}$ such that $x \in C$ and $y \notin C$. A space with a σ-closure-preserving point-separating closed cover is called a *σ#-space*. As with non-Archimedeanly quasi-metrizable spaces, countable products, perfect images and subspaces of σ#-spaces are σ#-spaces.

PROPOSITION. Let X be a topological space. The following statements are equivalent.

(a) X is a σ#-space.

(b) There is a sequence $\langle V_n \rangle$ of transitive neighbornets of X such that $\bigcap_{n=1}^{\infty} V_n = \Delta$.

(c) X can be condensed onto a non-Archimedeanly quasi-metrizable space.

Proof. (a) ⇒ (b). Let $\mathcal{H} = \bigcup\{\mathcal{H}_i \mid i \in \mathbb{N}\}$ be a σ-closure-preserving point-separating closed cover of X. For each $n \in \mathbb{N}$ let $V_n = \{(x,y) \mid$ for each $i \leq n$ and each $H \in \mathcal{H}_i$, $x \in H$ if $y \in H\}$. The sequence $\langle V_n \rangle$ satisfies the conditions of (b).

(b) ⇔ (c). These implications are immediate consequences of Theorem 7.1.

(b) ⇒ (a). Let $\langle V_n \rangle$ be a sequence of transitive neighbornets of X such that $\bigcap_{n=1}^{\infty} V_n = \Delta$. For each $n \in \mathbb{N}$ set $\mathcal{H}_n = \{X - V_n(x) \mid x \in X\}$. It is easily verified that $\mathcal{H} = \bigcup\{\mathcal{H}_n \mid n \in \mathbb{N}\}$ is a σ-closure-preserving point-separating closed cover of X. ■

COROLLARY. A topological space can be condensed onto a quasi-metrizable space if and only if there is a normal sequence $\langle V_n \rangle$ of neighbornets of X such that $\cap_{n=1}^{\infty} V_n = \Delta$.

COROLLARY. A transitive space can be condensed onto a quasi-metrizable space if and only if it is a $\sigma\#$-space.

7.28 If X is a countable set and $\mathcal{T}$ is the cofinite topology on X, then $(X,\mathcal{T})$ is non-Archimedeanly quasi-metrizable. Therefore, every countable T_1 space is a $\sigma\#$-space, and so, in general, a topological space can be condensed onto topologically distinct quasi-metrizable spaces. The following two contrasting results indicate that for some topological spaces there is no shortage of quasi-metrizable Hausdorff subtopologies while other topological spaces may have exactly one Hausdorff subtopology that is quasi-metrizable.

PROPOSITION. Let $(X,\mathcal{T})$ be a topological space and let $\mathcal{T}_1$ be a topology strictly coarser $\mathcal{T}$ such that $(X,\mathcal{T}_1)$ is quasi-metrizable. There is a topology $\mathcal{T}_2$ on X such that $(X,\mathcal{T}_2)$ is quasi-metrizable and $\mathcal{T}_1 \subsetneq \mathcal{T}_2 \subset \mathcal{T}$.

Proof. Let $\langle U_n \rangle$ be a base for a quasi-uniformity compatible with $\mathcal{T}_1$ and let $G \in \mathcal{T} - \mathcal{T}_1$. Set $V = G \times G \cup X - G \times X$ and for each $n \in \mathbb{N}$ set $V_n = U_n \cap V$. Then $\langle V_n \rangle$ is a base for a quasi-uniformity $\mathcal{V}$ on X and $\mathcal{T}(\mathcal{V})$ is the required topology $\mathcal{T}_2$. ∎

THEOREM. A countably compact space has at most one Hausdorff quasi-metrizable subtopology.

Proof. Let $(X,\mathcal{T})$ be a countably compact space and suppose that $\mathcal{T}_1$ and $\mathcal{T}_2$ are Hausdorff quasi-metrizable subtopologies of $\mathcal{T}$. Without loss of generality we may assume that $\mathcal{T}_1 \subset \mathcal{T}_2$. By Corollary 2.29, $\mathcal{T}_1$ and $\mathcal{T}_2$ are compact so that the identity map on X is a $\mathcal{T}_2 - \mathcal{T}_1$ continuous map from a compact space onto a Hausdorff space. Since the identity is a homeomorphism, $\mathcal{T}_1 = \mathcal{T}_2$. ∎

PROBLEM S. Are any two of the following statements equivalent? (a) X can be condensed onto a γ-space. (b) X can be condensed onto a quasi-metrizable space. (c) X is a $\sigma\#$-space. Note that although the Kofner plane is quasi-metrizable and not non-Archimedeanly quasi-metrizable, it can be condensed onto a separable metric space, the Euclidean plane.

7.29 The following theorem constitutes an overreaction to Problem S, since it establishes considerably more than is required to show that the questions raised in Problem S have affirmative answers in the class of linearly ordered topological spaces.

THEOREM. Let $(X,\mathcal{T})$ be a LOTS. Then $(X,\mathcal{T})$ can be condensed onto a γ-space if and only if $(X,\mathcal{T})$ is non-Archimedeanly quasi-metrizable.

Proof. Let $\langle V_n \rangle$ be a γ-sequence for a topology $\mathcal{T}'$ coarser than $\mathcal{T}$ and without loss of generality assume for each $x \in X$ and $n \in \mathbb{N}$ that $V_n(x)$ is an open (but possibly degenerate) interval about x. By Corollary 7.19, it suffices to show that $\langle V_n \rangle$ is a γ-sequence for $\mathcal{T}$. Suppose that for each $n \in \mathbb{N}$, $x_n \in V_n(y_n)$, $y_n \in V_n(p)$ and p is not a cluster point of $\langle x_n \rangle$. There is an open interval (a,b) about p and an $m \in \mathbb{N}$ such that $x_n \notin (a,b)$ and $y_n \in (a,b)$ whenever $n > m$. Without loss of generality we assume there is a subsequence $\langle x_{j_n} \rangle$ of $\langle x_n \rangle$ such that for all $n \in \mathbb{N}$, $x_{j_n} \le a$. Let $K = \{p\} \cup \{y_{j_n} | n > m\}$. We may assume that $\langle y_{j_n} \rangle$ converges to p in $(X,\mathcal{T}')$, so that K is a compact subset of $(X,\mathcal{T}')$. By Theorem 7.18(c) there is an $n > m$ such that $a \notin V_n(K)$. But $a \in [x_{j_n}, y_{j_n}] \subset V_{j_n}(y_{j_n}) \subset V_n(K)$ - a contradiction. ∎

7.30 Since every GO space is a transitive space, it may be supposed that the previous theorem would also hold in the class of GO spaces. While this apparently reasonable conjecture is incorrect, the questions raised in Problem S have affirmative answers in the class of GO spaces.

PROPOSITION. Every GO space that can be condensed onto a γ-space is a $\sigma\#$-space.

Proof. Let $(X,\mathcal{T})$ be a GO space that can be condensed onto a γ-space and let $\langle V_n \rangle$ be a base for a local quasi-uniformity compatible with a subtopology of $\mathcal{T}$. Then $\cap\{V_n^3 | n \in \mathbb{N}\} = \Delta$. By Theorem 6.30, for each $n \in \mathbb{N}$ there is a transitive neighbornet W_n such that $W_n \subset V_n^3$ and so by Proposition 7.27, X is a $\sigma\#$-space. ∎

EXAMPLE. A GO space that can be condensed onto $\mathbb{R}$ and is not a γ-space. Let $(X,\mathcal{T})$ be the Engelking-Lutzer line. This space has already been used in §6.31 as an example of a GO space with a neighbornet U such that U^2 does not contain a transitive neighbornet. Evidently, the usual topology of $\mathbb{R}$ is a subtopology of $\mathcal{T}$, so that it only remains to establish that $(X,\mathcal{T})$ is not a γ-space. To see that $(X,\mathcal{T})$ is not a γ-space suppose that there exists a γ-sequence $\langle V_n \rangle$ for X and suppose without loss of generality that for each $n \in \mathbb{N}$ and $x \in X$, $V_n(x)$ is a basic neighborhood of x. For each $m, n \in \mathbb{N}$ let $J(m,n) = \{y \in J \mid V_m^2(y) \subset V_1(y) \text{ and } (y - 1/n, y] \subset V_m(y)\}$. Since $J = \cup\{J(m,n) \mid m, n \in \mathbb{N}\}$, there exist $M, N \in \mathbb{N}$ and an open interval (a,b) such that (a,b) is a subset of the Euclidean closure of $J(M,N)$. Choose $x \in Q \cap (a,b)$ and choose y and z in $V_M(x) \cap (x, x + 1/N) \cap J(M,N)$ such that $y < z$. Because $y \in J(M,N)$, $x \in (y - 1/N, y] \subset V_M(y)$ and so $z \in V_M(x) \subset V_M^2(y) \subset V_1(y)$ - a contradiction.

7.31 It is known that every semistratifiable space is a $\sigma\#$-space and that every σ-orthocompact space with a G_δ-diagonal is a $\sigma\#$-space. In order to provide a unified treatment of these disparate propositions, we are content to prove the second of them in a slightly less general setting.

PROPOSITION. Let X be a topological space that has a G_δ-diagonal and suppose that for each open cover $\mathcal{C}$ of X, the cover $\{st(x,\mathcal{C}) \mid x \in X\}$ has a σ-closure-preserving closed refinement. Then X is a $\sigma\#$-space.

Proof. Let $\langle \mathcal{G}_n \rangle$ be a G_δ-diagonal sequence for X. For each $n \in \mathbb{N}$ let $\{\mathcal{H}_{m,n} \mid m \in \mathbb{N}\}$ be a family of closure-preserving closed collections such that $\cup_{m=1}^{\infty} \mathcal{H}_{m,n}$ refines $\{st(x,\mathcal{G}_n) \mid x \in X\}$. To show that $\cup\{\mathcal{H}_{m,n} \mid m, n \in \mathbb{N}\}$ is a point-separating cover, let x and y be two points of X. Let $n \in \mathbb{N}$ such that $x \notin st(y,\mathcal{G}_n)$ and let $m \in \mathbb{N}$ and $H \in \mathcal{H}_{m,n}$ such that $y \in H$. Then $H \subset st(y,\mathcal{G}_n)$ so that $y \in H$ and $x \notin H$. ■

COROLLARY. Let X be a topological space with a G_δ-diagonal. If X is either θ-refinable or orthocompact, X is a $\sigma\#$-space.

PROBLEM T. Can every topological space with a G_δ-diagonal be condensed onto a γ-space? Is every space with a G_δ-diagonal a $\sigma\#$-space?

7.32 EXAMPLE. A separable Moore space (hence a $\sigma\#$-space) that cannot be condensed onto a Hausdorff γ-space. Let J denote the set of all irrational numbers, let $A_1 = J \times \{0\}$, $A_2 = J \times \{-1\}$, $E = \{(r,s) | r$ and s are rational numbers and $s > 0\}$, $X = A_1 \cup A_2 \cup E$ and give E the usual subspace topology $\mathcal{T}'$ of the plane. For each $a \in J$ and each $n \in \mathbb{N}$ let $U_n(a,0) = \{(a,0)\} \cup \{(r,s) \in E | a < r < (s + an)/n, s < 1/n\}$ and let $U_n(a,-1) = \{(a,-1)\} \cup \{(r,s) \in E | (-s + an)/n < r < a, s < 1/n\}$. Then $\{U_n(a,i) | a \in J; i = 0, -1; n \in \mathbb{N}\} \cup \mathcal{T}'$ is a base for a topology $\mathcal{T}$ on X, and X is a separable completely regular Moore space.

Suppose that $(X,\mathcal{T})$ can be condensed onto a Hausdorff γ-space. There is a Hausdorff topology $\mathcal{H} \subset \mathcal{T}$ and a γ-sequence $\langle V_n \rangle$ for $(X,\mathcal{H})$. For each $n \in \mathbb{N}$ set $J_n = \{x \in J | U_n(x,0) \times U_n(x,-1) \cap V_n = \emptyset\}$. A straightforward argument, which uses the Baire Category Theorem, establishes that $J \neq \cup_{n=1}^{\infty} J_n$. Let $z \in J - \cup_{n=1}^{\infty} J_n$ and let H_0 and H_1 be disjoint members of $\mathcal{H}$ such that $(z,0) \in H_0$ and $(z,-1) \in H_1$. Choose $k \in \mathbb{N}$ so that $V_k^2((z,0)) \subset H_0$ and choose $n > k$ so that $U_n(z,-1) \subset H_1$ and $U_n(z,0) \subset V_k((z,0))$. Then $z \in J_n$ - a contradiction. ∎

5. COMPLETE QUASI-METRIC SPACES

7.33 Let $(X,\mathcal{U})$ be a quasi-uniform space. A sequence $\langle x_n \rangle$ in $(X,\mathcal{U})$ is a *Cauchy sequence* provided that for each $U \in \mathcal{U}$ there exists $n \in \mathbb{N}$ and $x \in X$ such that $\{x_i | i \geq n\} \subset U(x)$. A sequence $\langle x_n \rangle$ is a Cauchy sequence if and only if the elementary filter associated with $\langle x_n \rangle$ is a Cauchy filter. It follows easily that if $(X,\mathcal{U})$ is (convergence) complete, every Cauchy sequence in $(X,\mathcal{U})$ has a cluster point (converges).

THEOREM. Let $(X,\mathcal{U})$ be a quasi-metrizable quasi-uniform space. The following are equivalent.

(a) $(X,\mathcal{U})$ is complete.
(b) Every countable Cauchy filter base has a cluster point.
(c) Every Cauchy sequence has a cluster point.

Proof. The implications (a) ⇒ (b) and (b) ⇔ (c) are obvious. To see that (b) implies (a), let $\{U_n | n \in \mathbb{N}\}$ be a base for $\mathcal{U}$ such that for each $n \in \mathbb{N}$ and each $x \in X$, $U_n(x) \in \mathcal{T}(\mathcal{U})$. Let $\mathcal{F}$ be a Cauchy filter and for each $n \in \mathbb{N}$ choose $F_n \in \mathcal{F} \cap \{U_n(x) | x \in X\}$. Set $K = \cap_{n=1}^{\infty} \overline{F}_n$. In order to show that K is compact, it suffices, by Corollary 2.29, to prove that K is countably compact. Let $\{H_n | n \in \mathbb{N}\}$ be a filter base on K. For each $n \in \mathbb{N}$, $\overline{F_n \cap H_n} = \overline{H}_n$ so that $\{F_n \cap H_n | n \in \mathbb{N}\}$ is a Cauchy filter base and $\emptyset \neq \cap_{n=1}^{\infty} \overline{F_n \cap H_n} \subset K \cap \cap_{n=1}^{\infty} \overline{H}_n$. Thus K is compact so that in order to show that $\mathcal{F}$ has a cluster point it suffices to show that for each $F \in \mathcal{F}$, $\overline{F} \cap K \neq \emptyset$. Let $F \in \mathcal{F}$. Then $\{F \cap F_n | n \in \mathbb{N}\}$ is a Cauchy filter base and so $\emptyset \neq \cap_{n=1}^{\infty} \overline{F \cap F_n} \subset \overline{F} \cap K$. ■

7.34 The unique uniformity compatible with the topology of the space of all countable ordinals with the order topology is an equinormal uniformity that is not complete. S. G. Mrowka has shown that every equinormal metric is a complete metric. A modification of Mrowka's proof establishes a generalization of this result for quasi-metric spaces.

THEOREM. Every equinormal quasi-metric space is complete.

Proof. Let (X,d) be an equinormal quasi-metric space and let $\{U_n | n \in \mathbb{N}\}$ be a base for the quasi-uniformity generated by d. Suppose that $\langle x_n \rangle$ is a Cauchy sequence that has no cluster point. For each $n \in \mathbb{N}$, there exists a $y_n \in X$ and an integer $j_n \geq n$ such that $T_n = \{x_i | i \geq j_n\} \subset U_n(y_n)$. For each $n \in \mathbb{N}$ let $t_n = x_{j_n}$. We note $t_n \in U_n(y_n)$ for each $n \in \mathbb{N}$. Therefore, if $\langle y_n \rangle$ has a cluster point, by Theorem 7.18(d), $\langle t_n \rangle$ also has a cluster point. Consequently $A = \{t_n | n \in \mathbb{N}\}$ and $B = \{y_n | n \in \mathbb{N}\}$ are infinite discrete sets. Choose subsequences of $\langle t_n \rangle$ and $\langle y_n \rangle$ as follows: Set $f(1) = 1$. Let $g(1)$ be the least positive integer j such that $y_j \notin \{t_1\}$. Let $f(2)$ be the least positive integer j such that $t_j \notin \{y_{g(1)}\}$; in general, let $f(n)$ be the least positive integer j such that $t_j \notin \{y_{g(1)}, y_{g(2)}, \ldots, y_{g(n-1)}\}$ and let $g(n)$ be the least positive integer j such that $y_j \notin \{t_{f(1)}, t_{f(2)}, \ldots, t_{f(n)}\}$. Let $A' = \{t_{f(n)} | n \in \mathbb{N}\}$ and $B' = \{y_{g(n)} | n \in \mathbb{N}\}$. Then A' and B' are disjoint closed sets, and there is an $n \in \mathbb{N}$ such that $A' \cap U_n(B') = \emptyset$. Let k be the least integer j greater than n such that $y_j \in B'$. There is an integer $m \geq k$ such that $T_m \subset U_k(y_k) \subset U_n(B')$ so that $A' \cap U_n(B') \neq \emptyset$ - a contradiction. ■

7.35 We have already observed that a topological space is compact if and only if each admissible quasi-uniformity is complete. We now establish the corresponding result for quasi-metrizable spaces.

LEMMA. Let $(X,\mathcal{U})$ be a quasi-uniform space, let $\mathcal{B}$ be a base for $\mathcal{U}$, and let $\mathcal{F}$ be a filter base on X that has no $\mathcal{T}(\mathcal{U})$-cluster point. Then

$\{U \cup U^{-1}(F) \times U(F) \mid U \in \mathcal{B} \text{ and } F \in \mathcal{F}\}$ is a base for a quasi-uniformity $\mathcal{V}$ that is compatible with $\mathcal{T}(\mathcal{U})$, and $\mathcal{F}$ is a $\mathcal{V}$-Cauchy filter.

Proof. Let $F \in \mathcal{F}$, let $V \in \mathcal{B}$, and $U \in \mathcal{B}$ such that $U^2 \subset V$. Then $[U \cup (U^{-1}(F) \times U(F))]^2 \subset V \cup (V^{-1}(F) \times V(F))$ so that $\mathcal{V}$ is a quasi-uniformity on X. To see that $\mathcal{V}$ is compatible with $\mathcal{T}(\mathcal{U})$, it suffices to prove that $\mathcal{T}(\mathcal{U}) \subset \mathcal{T}(\mathcal{V})$. Let $G \in \mathcal{T}(\mathcal{U})$ and let $x \in G$. Choose $U \in \mathcal{U}$ and $F \in \mathcal{F}$ such that $U(x) \subset G$ and $U(x) \cap F = \emptyset$. Then $[U \cup (U^{-1}(F) \times U(F))](x) = U(x)$. It is clear that $\mathcal{F}$ is a $\mathcal{V}$-Cauchy filter. ∎

THEOREM. A quasi-metrizable space X is (countably) compact if and only if every quasi-metric on X is complete.

7.36 Evidently the quasi-uniformity $\mathcal{V}$ constructed in Lemma 7.35 is a uniformity whenever $\mathcal{U}$ is a uniformity. Hence the following well-known theorem of Niemytzki and Tychonoff also follows immediately from Lemma 7.35.

THEOREM. A metrizable space X is compact if and only if every metric on X is complete.

7.37 Although every metric space has a metric completion, we now show that a quasi-metric space may fail to have a completion - quasi-metric or not - and consider the related problems of characterizing those quasi-metric spaces that have (quasi-metric) completions and those topological spaces that admit a complete quasi-metric. Let $(X,\mathcal{T})$ be a topological space. A base $\mathcal{B}$ for $\mathcal{T}$ is a *base of countable order* if whenever $\langle B_n\rangle$ is a strictly decreasing sequence of members of $\mathcal{B}$ and $p \in X$ such that for each $n \in \mathbb{N}$, $p \in B_n$, then $\{B_n \mid n \in \mathbb{N}\}$ is a base for η_p. A topological space is an *Aronszajn space* if it is a regular T_1 space whose topology has a base of countable order. A regular T_1 space $(X,\mathcal{T})$ is a *complete Aronszajn space* provided there is a sequence $\langle \mathcal{G}_n\rangle$ of bases for $\mathcal{T}$ such that if for each $n \in \mathbb{N}$, $G_{n+1} \subset G_n$ and $G_n \in \mathcal{G}_n$, then $\{\overline{G}_n \mid n \in \mathbb{N}\}$ is a convergent filter base.

THEOREM. Every regular complete quasi-metric space is a complete Aronszajn space.

Proof. Let (X,d) be a regular complete quasi-metric space, let $\mathcal{U}$ be the quasi-uniformity generated by d, and let $\{U_n \mid n \in \mathbb{N}\}$ be a base for $\mathcal{U}$. For each $n \in \mathbb{N}$ set $\mathcal{G}_n = \{G \in \mathcal{T}(d) \mid \text{for some } x \in G, \overline{G} \subset U_n(x)\}$. Suppose that for each $n \in \mathbb{N}$, $G_{n+1} \subset G_n$ and $G_n \in \mathcal{G}_n$. For each $n \in \mathbb{N}$ choose $y_n \in G_n$ such that $\overline{G}_n \subset U_n(y_n)$. Since $\langle y_n \rangle$ is a Cauchy sequence, it has a cluster point p. Suppose that $\{\overline{G}_n | n \in \mathbb{N}\}$ does not converge to p. There exists $H \in \eta_p$ such that for each $n \in \mathbb{N}$, there exists $x_n \in \overline{G}_n - H$. For each $n \in \mathbb{N}$, $x_n \in U_n(y_n)$ and p is a cluster point of $\langle y_n \rangle$ so that by Theorem 7.18(d), p is a cluster point of $\langle x_n \rangle$ - a contradiction. ∎

7.38 The example given below was constructed by C. Pixley and P. Roy in their study of uncompletable Moore spaces. Since our use of their example closely parallels its original purpose, we only outline the argument required to establish its pertinent properties.

EXAMPLE. A quasi-metric Moore space that has a completion but has no regular quasi-metric completion. Let Λ be the collection of all nonempty finite subsets of $\mathbb{R}$. Topologize Λ by using all sets of the form $[F,G] = \{F' \in \Lambda | F \subset F' \subset G\}$ as a neighborhood base at $F \in \Lambda$, where G is allowed to be any open subset of $\mathbb{R}$ that contains F. The space $(\Lambda,\mathcal{T})$ thus defined is called the *Pixley-Roy space*. It is well known that the Pixley-Roy space is a metacompact Moore space that contains no dense metrizable subspace. As it can be shown that any subspace of a complete Aronszajn space contains a dense metrizable subspace, (see the notes) it follows from Corollaries 3.45 and 7.24 that the Pixley-Roy space admits a quasi-metrizable quasi-uniformity that has a (quasi-uniform) completion but no regular quasi-metric completion.

PROBLEM U. Characterize the class of quasi-metric spaces that have quasi-metric completions. Theorem 7.37 suggests that this problem is likely to be difficult.

7.39 The following proposition indicates that it is demanding to ask that a quasi-metric space have a completion - even one that is not necessarily a quasi-metric completion.

PROPOSITION. Let $(X,\mathcal{U})$ be a quasi-metrizable quasi-uniform space that has a completion. Then X has a subdevelopment.

Proof. Let $\{U_n | n \in \mathbb{N}\}$ be a base for $\mathcal{U}$ and suppose that for each $n \in \mathbb{N}$ and $x \in X$, $U_n(x)$ is an open set. For each $n \in \mathbb{N}$ set

$G_n = \{U_n(x) | x \in X\}$. We assert that $\langle G_n \rangle$ is a subdevelopment for X. Let $p \in X$ and let $\langle y_n \rangle$ be a sequence in X such that $p \in U_n(y_n)$ for each $n \in \mathbb{N}$. Suppose that $\{U_n(y_n) | n \in \mathbb{N}\}$ is not a subbase for η_p. There exists an open set G about p and a sequence $\langle x_n \rangle$ such that for each $n \in \mathbb{N}$, $x_n \in \cap_{i=1}^{n} U_i(y_i) - G$. Let F be the elementary filter associated with $\langle y_n \rangle$. Then F is a U-Cauchy filter and p is a $T(U^{-1})$ cluster point of F so that by Theorem 3.43(b), p is a $T(U)$-cluster point of F. For each $n \in \mathbb{N}$, $x_n \in U_n(y_n)$ and p is a $T(U)$-cluster point of $\langle y_n \rangle$ so that by Theorem 7.18(d), p is a $T(U)$-cluster point of $\langle x_n \rangle$ - a contradiction. ∎

COROLLARY. If U is a quasi-uniformity with a countable base and U is compatible with the topology of either the Michael line or the Sorgenfrey line, then $(\mathbb{R}, U)$ has no completion. In particular, $(\mathbb{R}, M)$ and $(\mathbb{R}, Z)$ have no completion.

Proof. If a paracompact Hausdorff space has a subdevelopment, it is metrizable. ∎

7.40 A development $\langle G_n \rangle$ for a space X is *complete* provided that if $\langle T_n \rangle$ is a decreasing sequence of closed sets and for each $n \in \mathbb{N}$ there exists $G_n \in G_n$ such that $T_n \subset \overline{G}_n$ then $\cap_{n=1}^{\infty} T_n \neq \emptyset$. A Moore space is *complete* provided there exists a complete development for X. Every complete Moore space is a complete Aronszajn space; however, there is a Moore space that is not complete but is a complete Aronszajn space.

THEOREM. A regular quasi-metrizable space is a complete Moore space if and only if it is θ-refinable and admits a complete quasi-metric.

Proof. Let (X,T) be a regular θ-refinable space and let $\{U_n | n \in \mathbb{N}\}$ be a base for a complete quasi-uniformity compatible with T. By Theorem 7.37, T has a base of countable order. Since (X,T) is θ-refinable, it has a development $\langle G_n \rangle$. For each $n \in \mathbb{N}$ let H_n be an open cover of X such that $\{\overline{H} | H \in H_n\}$ refines $\{U_n(x) \cap G | G \in G_n$ and $x \in X\}$. Then $\langle H_n \rangle$ is a complete development for (X,T).

Now suppose that $\langle G_n \rangle$ is a complete development for (X,T). By Corollary 7.22 there is a normal sequence $\langle U_n \rangle$ of neighbornets such that for each $n \in \mathbb{N}$, $\{U_n(x) | x \in X\}$ refines G_n. It is easy to show that $\{U_n | n \in \mathbb{N}\}$ is a base for a quasi-uniformity U compatible with T. To see that U is complete, let $\{B_n | n \in \mathbb{N}\}$ be a decreasing Cauchy filter base. For each

$n \in \mathbb{N}$ there exists $x_n \in X$ and $j_n \geq n$ such that $B_{j_n} \subset U_n(x_n)$. For each $n \in \mathbb{N}$ there exists $G_n \in \mathcal{G}_n$ such that $\overline{B}_{j_n} \subset \overline{G}_n$; since $\mathcal{G}_n$ is a complete development, $\bigcap_{n=1}^{\infty} \overline{B_n} \neq \emptyset$. ■

PROBLEM V. Example 5.34 is a locally compact (and hence complete) Moore space that is not quasi-metrizable, and as Example 7.41 indicates it it also possible that a complete quasi-metric space fail to be developable. Is every regular complete quasi-metric space quasi-developable?

7.41 EXAMPLE. A complete quasi-metric space that is not developable. Let I denote the interval $[0,1]$ of $\mathbb{R}$ and let $\{H_\alpha | \alpha \in A\}$ be a family of subsets of I maximal under the partial order of containment with respect to the following property: The family is an almost disjoint family of convergent sequences in I each of which is without its limit point. Set $X = I \cup A$ and for each $\alpha \in A$ let x_α denote the limit point (with respect to the Euclidean topology) of H_α. For each $n \in \mathbb{N}$ set $U_n = \Delta \cup (\{(\alpha,x) \in A \times I | |x - x_\alpha| \leq 1/n\})$ and let $\mathcal{U}$ be the quasi-uniformity for which $\langle U_n \rangle$ is a base. Since I is not a F_σ-subset of $(X,\mathcal{T}(\mathcal{U}))$, X is not developable, and, since every Cauchy sequence in X is either eventually constant or converges to a point of A, $\mathcal{U}$ is complete.

NOTES

These notes comment on the historical development of our subject. References to well-known results are usually omitted.

CHAPTER 1

§1.1. The study of quasi-uniformities begins with Nachbin [116], who calls these structures semi-uniform structures. The term quasi-uniformity is suggested by Császár, [33], p. 66. Alternate characterizations of quasi-uniform spaces in terms of families of quasi-pseudo-metrics and in terms of families of covers are given by Gantner and Steinlage [58] and Császár [33].

§1.5. Asymmetric distance functions are considered by Hausdorff [64], pp. 145-146, Niemytzki [124], and Wilson [176]. The proof of Lemma 1.5 given by Kelley [85], pp. 185-186, is based upon Frink [55]. The Michael line and the Sorgenfrey line are defined in [109] and [161].

§1.6. The term "neighbornet" and the concept of an unsymmetric neighbornet are due to Junnila [79].

§1.7. Junnila [79], Lemma 3.3. Neighbornets satisfying the hypothesis of Proposition 1.7 are called semi-neighborhoods of the diagonal by Michael [108].

§1.9. The concept of an R_0 space is due to Shanin [153], who calls these spaces weakly regular spaces. The terminology "R_0-space" is introduced by Davis [34].

§1.10. Niemytzki [124], Hilfsatz p. 670.

§1.20. Nachbin [116], p. 774. Corollary 1.20 is due to Weil [173], Theorem 4.

§1.21. Salbany [146], Theorem 4.7, p. 76 and Example p. 77.

§1.22. The term quasi-proximity appears first in Pervin [130] and Steiner [162]. A discussion of the historical development of proximities may be found in [118].

§1.24. Quasi-proximities are considered in terms of strong inclusion by Dowker [37] and as part of a more general study of Császár [33]. In the terminology of [33], Chapter 7, a relation << satisfying 1.24 (a)-(f) is a simple syntopogenous structure or a topogenous structure.

§1.33. Theorem 1.33 is established by Hunsaker and Lindgren [74]. A proof of the corresponding result for totally bounded syntopogenous structures is given by Császár [33], (19.40). Császár attributes the result to Czipszer [33], Footnote 1, p. 339. The equivalence of proximities and totally bounded uniformities is shown by Smirnov [160], Alfsen and Fenstad [4], and Gál [56].

§1.40. Alfsen and Njastad [5], Lemma p. 236.

§1.44. Blatter and Seever [12], 5.8 (Translation Lemma).

§1.45. The corollary is established in greater generality by Lambrinos [94], Theorem 1.

§1.46. The proof of the Lemma 1.46 is based upon an argument of Gál given in [56], Example 1, p. 316. The equivalence of (b') and (c) of Theorem 1.46 is due to Samuel [147], Theorem XIV, p. 129.

§§1.53 and 1.54. Dowker [37] and Isbell [76], Problem 12, p. 34; J. Czipszer in Császár [33], (19.18)-(19.20), (19.34).

§1.55. Alfsen and Njastad [5], Corollary 2 of Theorem 6.

§1.56. In the terminology of Murdeshwar and Naimpally [113], the members of $Q(\mathcal{U})$ are said to be quasi-uniformly lower semi-continuous.

§1.57. The lemma is the asymmetric analogue of Smirnov [160], Lemma 3, p. 23. The analogue for syntopogenous structures is given by Császár [33], (12.41). (See Footnote 3, p. 169). The results of this section appear in Hunsaker and Lindgren [75], and the corollary is proved in Brümmer [15], Teorema 4.

§1.58. Dowker [37].

§1.59. The theorem is proved by Hunsaker and Lindgren [75], Theorem 5.5. The corollary is Theorem 19 of Smirnov [160], p. 30, and the examples are given by Hunsaker and Lindgren [75], Examples 5.1 and 5.4.

§1.60. The special case of Proposition 1.60 that a mapping between metric spaces is uniformly continuous if and only if it is p-continuous is proved by Efremovič [38] and is one of the earliest results in the theory of proximities.

CHAPTER 2

§2.1. Krishnan [93] shows that every topological space admits a quasi-uniformity; subsequent proofs of this result were obtained by Császár [33] and Pervin [129]. The proof given here is Pervin's. A comparison of the constructions of Császár and Pervin is given by Votaw [170], Section 5.

§2.2. Pervin [130] defines δ_p; this quasi-proximity had been defined previously in terms of strong inclusion by Császár [33], Section 7.19. The weight of the literature dictates our nomenclature, Pervin quasi-uniformity and Pervin quasi-proximity. The pretty example, which reappears in Sections 4.15 and 4.30 is due to Brümmer (unpublished).

§2.3. The characterization of *FINE* in terms of quasi-pseudo-metrics follows from Császár [33], Section 13.51, Brümmer [14], Proposition 2.1, and Reilly [139], Corollary 1.

§2.4. Hunsaker and Lindgren [74], Theorem 3.

§2.5. The term interior preserving is due to Junnila [79].

§2.6. Fletcher [46], Theorem 1.

§2.8. Fletcher and Lindgren [48], Lemma 4.1.

§2.9. Krishnan [93] and Nielsen and Sloyer [122] Propositions 1.1 and 1.2.

§2.10. The semi-continuous quasi-uniformity is defined in [9] and [48] in terms of upper semi-continuous functions and is denoted by *USC*. Proposition 2.10 is proved in Hunsaker and Lindgren [74], Theorem 2, Brümmer [15], Stelling 3, and Votaw [170], Section 5.

§2.12. The concept of a spectrum is due to Leader [95]. Theorem 2.12 is Theorem 2.1 of Fletcher and Lindgren [48].

§2.21. The terminology point-symmetric quasi-proximity is from Salbany [146], Definition 4.2.2. Proposition 2.21 is comparable to Theorem 3.6 of Murdeshwar and Naimpally [113].

§2.23. Proposition 2.23 and the terminology locally symmetric quasi-proximity are suggested by Theorem 3.17 of Murdeshwar and Naimpally [113]. 2.23 is comparable to Theorem 3.17 of Murdeshwar and Naimpally [113].

§2.24. Mrówka [112] defines a normal metric to be a metric for which the distance between disjoint closed sets is positive. This definition is extended to syntopogenous spaces by Sieber and Pervin [155] and subsequently applied to proximity spaces by Pervin [131].

§2.28. Proposition 2.28 is due to Stoltenberg [163], Theorem 3.1.

§2.29. Corollary 2.29 is from Niemytzki [124], Theorem 3. In this and other papers by Russian authors quasi-metric spaces are called Δ-spaces.

§2.31. Proposition 2.31 is given by Chittenden [25], Section 15.

§2.32. The theorem is from Fletcher [45]. The proof of the theorem is an adaptation of Frink [55], Theorem 3.

§2.33. A list of characterizations of those metrizable spaces for which the fine uniformity has a countable base is given by Rainwater [136], Theorem 2.

§2.34. Lindgren and Fletcher [101], Proposition 4.2.

§§2.36-2.38. The terminology R_1 space is due to Davis [34]. Theorem 2.36 and Examples 2.37 and 2.38 are taken from Lindgren [96] and [97] and Votaw [170]. The product theorem for hereditarily compact spaces, cited in §2.38, is given by Stone [164], Theorem 9.

CHAPTER 3

§3.1. It is natural to define a Cauchy filter as a filter with the property that for each $U \in \mathcal{U}$ there is an $F \in \mathcal{F}$ such that $F \times F \subset U$. Example 3.3 shows that with this definition a $\mathcal{T}(\mathcal{U})$-convergent filter would not necessarily be such a "Cauchy" filter. For this reason Sieber and Pervin [154] choose the definition of Cauchy filter as given in the text. Their definition is motivated, in part, by a corresponding definition for sequences given by Cohen [28].

§3.13. The terminology precompactness is misleading but well established.

§3.15. Murdeshwar and Naimpally [113], Theorems 4.17 and 4.19.

§3.16. Fletcher and Lindgren [48], Lemma, p. 626, and Theorem 4.3.

§3.17. Barnhill and Fletcher [9], Theorem 3.3, and Fletcher and Lindgren [48], Proposition 4.5. The implication (e) => (a) is a result of Kasahara [82], Theorem 1.

§3.18. Iséki and Kasahara [77], Theorem 2, and Arens and Dugundji [6], Theorem 2.4.

§3.22. Fletcher and Lindgren [48], Proposition 4.7. The proof of Proposition 3.22 uses Kasahara [82], Theorem 2.

§3.23. Alexandroff and Urysohn [3], Theorem 7.1.

Problem C. Every Hausdorff can be embedded as a closed nowhere dense subspace of a minimal Hausdorff space; Strecker and Wattel [166], Theorem 3. The observation that an ultrafilter $\mathcal{F}$ is an $\mathcal{FT}$-Cauchy ultrafilter if and only if every closure-preserving subcollection of $\mathcal{F}$ has a cluster point is an unpublished observation of Junnila.

§3.26. The proposition is an unpublished result of Junnila.

§3.27. The first example of a space that admits no convergence complete quasi-uniformity is given by Huffman, Hicks and Carlson [72].

§§3.28-3.38. The term bicomplete is the same as Császár's doubly complete [33] and Salbany's complete [146]. Our presentation of the bicompletion of a quasi-uniform space is modelled upon the presentations of the completion of a uniform space given by Bourbaki [13] and Bushaw [21]. An alternate construction of the bicompletion is given in [99].

§§3.39-3.43. Carlson and Hicks [22], Theorem 4 show that if $(X,\mathcal{U})$ is a locally symmetric transitive T_1 space, then $(\hat{X},\hat{\mathcal{U}})$ is a completion of $(X,\mathcal{U})$. Theorem 3.43 is due to Ward [172], Theorem 6.1. Ward indicates that his construction of $(\hat{X},\hat{\mathcal{U}})$ is motivated by an examination of the Katětov extension of a Hausdorff space [83].

CHAPTER 4

The primary source for this chapter is Nachbin [117].

§§4.1-4.5. A partial order $\leq$ on X such that G($\leq$) is a closed subset of X × X is sometimes called a continuous partial order [107].

Lemma 4.1 is given by Nachbin [117], p. 29; Theorem 4.2 is [117], Theorem 1; Propositions 4.3 and 4.4 are [117], Propositions 4 and 5; Theorem 4.5 is [117], Corollary to Theorem 4; Corollary 4.5 is [117], Theorem 5.

§§4.6-4.8. In [117], Nachbin presents a theory that parallels these results in which preorders are used in place of partial orders.

Several characterizations of uniform ordered spaces are compiled by Choe [26], §4.

§4.9. Blatter [11], Lemma 2.3.

§§4.12-4.13. The compactification nX is defined by Nachbin in [115].

The primary sources for §§4.14-4.27 are Singal and Lal [156], Blatter and Seever [12], and Redfield [137] and [138].

§4.17. Blatter and Seever [12], Theorem 5.16.

§4.19. Redfield [137], Proposition 5.16.

§4.20. The existence of the quasi-uniformity $\mathcal{V}$ may be deduced using Redfield [137], Proposition 2.2 and Nachbin [117], Proposition 11. The proof given here is from Lindgren and Fletcher [100], Theorem 3.6.

§4.21. Nachbin [116], p. 774.

Problem E. The topological problem corresponding to Problem E is solved by Doss [35] and Gál [57]. Blatter [11], Corollary 2.9 solves Problem E for linear orders.

§4.22. Singal and Lal [156], Theorem 4 and Blatter and Seever [12], Theorem 5.16.

§4.23. Singal and Lal [156], Theorem 7 and Blatter and Seever [12], Theorem 5.16.

§§4.24-4.25. Blatter and Seever [12], Theorem 5.25i and Theorem 5.25ii. The corresponding result for proximity spaces and compactifications is Smirnov [160], Theorem 11.

§4.27. Strict compactifications are called Nachbin compactifications by Blatter and Seever [12], Definition 5.17. Proposition 4.27 is due to Blatter and Seever [12], Theorem 5.23.

§4.28. Excellent references for GO spaces are provided by Čech [23] and Lutzer [103]. Proposition 4.28 is comparable to Cech [23], Theorem 17A.23.

§4.30. Lindgren and Fletcher [100], Lemma 3.1.

§4.31. The equivalence of (a) and (c) is given by Lindgren and Fletcher [100], Theorem 3.2; the equivalence of (b) and (c) is given by Blatter [11], Proposition 2.4.

§§4.32-4.33. Lindgren and Fletcher [100], Theorems 3.7 and 3.9.

§4.36. The equivalence of (a) and (c) is proved by Kaufman [84], observation on p. 39.

Problem G. An example of a LOTS X for which υX is not a LOTS is given by Gillman and Henriksen [60], Example 10.11.

Problem H. If $(X,\mathcal{T},\leq)$ is a LOTS and every uniformity compatible with $\mathcal{T}$ is convex, then by Theorem 4.36, X is pseudocompact. Moreover, by Theorem 8 of Kaufman [84], X has a smallest compactification, and so by Theorem XIV of Samuel [147], X is locally compact and has an orderable one-point compactification. Conversely if $(X,\mathcal{T},\leq)$ is a locally compact pseudocompact LOTS that has an orderable one-point compactification, then by Lemma 7 of Kaufman [84], every compactification is orderable and every uniformity compatible with $\mathcal{T}$ is convex.

CHAPTER 5

§5.2. Proposition 5.2 and its corollary are taken from Lindgren and Fletcher [101], Propositions 3.2 and 3.4 and the corollary to Proposition 3.3.

§5.5. Orthocompact spaces are introduced by Sion and Willmott [157], Definition 13.10, where they are called spaces with property Q. The terms orthocompact and metacompact are both due to Arens. According to Arens, these terms are suggested by the conventional terminology for the symmetries of a benzene ring. Orthocompactness has been studied in some detail by Scott [148], [149], [150], and [151].

§5.6. The concept of subinfinite rank is given by Lindgren and Nyikos [102]. The equivalence of (a) and (b) of Theorem 5.6 is taken from Fletcher and Lindgren [47], Theorem 2.2. The equivalence of (a) and (c) is from Scott [148], Proposition 1.2.5. The equivalence of (a) and (e) is from Lindgren and Nyikos [102], Corollary 2.7.

§5.7. Fletcher and Lindgren [47], Theorem 2.1.

§5.8. Junnila (unpublished).

§5.9. Fletcher and Lindgren [49], Proposition 3.1.

§§5.10-5.11. Preorthocompactness is defined by Junnila [78], who obtains the characterization of preorthocompactness of Corollary 5.11. Cushioned refinements are discussed by Michael [108].

§5.13. The concept of an open-finite cover is attributed to Arhangel'skii by Peregudov [128]. The proof of Lemma 5.13 is based on Peregudov [128], Theorem 3. Proposition 5.13 is an observation of van Douwen [36], p. 109.

§5.14. The construction of the space $\mathcal{G}(X)$ is given by Scott [148], Section 8.6. The examples attributed to Bing are given in [10]. The concept of a transitive space, which is mentioned in §5.14, is defined in §6.10.

§5.15. Semi-metacompact spaces are introduced by Peregudov [128], who calls these spaces semi-paracompact spaces. Nearly metacompact spaces are introduced by Heath and Lindgren [67]. The equivalences (a) $\Leftrightarrow$ (b) and (a) $\Leftrightarrow$ (c) of Theorem 5.15 are due to Heath and Lindgren [67], Lemma 1 and Peregudov [128], Theorem 3.

§5.16. This example is announced by Heath and Lindgren [68]. The existence of the sequence $\langle x_{\alpha_n} \rangle$ follows from Moore [111], Theorem 6, page 3.

§5.17. It is observed by Peregudov [128], Assertion 8 that the Niemytzki plane is not semi-metacompact. Phillips is the first to observe that the Niemytzki plane is orthocompact [133], Example 3.9. The proof that the Niemytzki plane is orthocompact is taken from [132], Example 6.16.

§5.18. The concept of a uniform base is introduced by Alexandroff [1] and [2], p. 40. It is observed by Heath and Lindgren [67] that the Sorgenfrey plane is not semi-metacompact so that, in contrast to the corresponding base property ([128], Assertion 4F), semi-metacompactness is not productive.

The result that $\mathbb{N}^{\omega_1}$ is not normal is given by Stone [165], Theorem 3. Corollary 5.18 makes use of Stone's result that a product of metrizable spaces is normal if and only if all but finitely many factors are compact [165], Theorem 4.

§5.19. Theorem 2.1 of Noble [126] is the analogue of Theorem 5.19 for normal spaces. The lemma is taken from Scott [151], Lemmas 1.3 and 1.6; parts (a) and (c) of Theorem 5.19 are comparable to results of Scott [151], Theorems 2.1 and 2.5. Van Douwen proves an interesting result related to Theorem 5.19(b): No uncountable product of infinite discrete spaces is countably orthocompact [36], p. 55.

§5.21. Almost 2-fully normal spaces are first considered by Cohen [27] who shows that every almost 2-fully normal space is collectionwise normal and characterizes the almost 2-fully normal spaces as those spaces for which the collection of all neighborhoods of the diagonal forms a base for the fine uniformity. The terminology almost 2-fully normality is due to Mansfield [105] who studies these spaces as well as the 2-fully normal spaces of Goldman [62]. Theorem 5.21 is due to Junnila [78], Theorem 2.2.10. The example attributed to R. H. Bing by Cohen [27], p. 292 of a collectionwise normal space that is not 2-fully normal is not preorthocompact. Thus this example supports the conjecture that an almost 2-fully normal space is preorthocompact if and only if it is 2-fully normal.

§5.22. The conjecture of Kelley is stated in Kelley [85], p. 208. Corson [30], Theorem 4 proves that the space F_0, defined by A. H. Stone [165], is a nonparacompact almost 2-fully normal space that admits a complete uniformity. Thus Kelley's conjecture does not hold.

§5.23. Theorem 5.23 is announced by Fleischman in [42] and [43]. The proof given here is due to Junnila (unpublished).

§5.24. Burke [20], Example 1.

§5.25. See also Example 7.11.

§5.26. Scott [151], Theorem 5.1.

§5.27. Junnila [78], Proposition 2.3.9. Besides the orthocompact θ-refinable spaces, there is another class of spaces whose closed images are orthocompact. A base $\mathcal{B}$ for a space X is an *orthobase* [102] provided that for each subcollection $\mathcal{B}' \subset \mathcal{B}$, either $\cap\mathcal{B}'$ is open or $\mathcal{B}'$ is a base for the neighborhood filter of some point. Junnila [78], Proposition 2.3.12 shows that the closed continuous image of a space with an orthobase is orthocompact.

§5.28. It follows from Corson and Isbell [31], Theorem 1, that if X is a paracompact Hausdorff space, then C(X) is the fine uniformity if and only if X is a boundedly paracompact Lindelöf space. Hence if X is Hilbert space, the fine uniformity of X is not C(X). The implication (a) $\Leftrightarrow$ (b) of Theorem 5.28 is due to Corson [29], Theorem 2.

§5.30. Weak orthocompactness is defined by Scott [149]. Theorem 5.30 is due to Scott [149], Lemma 2.4.

§5.32. The concept of a weakly Cauchy filter is defined by Corson [29]. Proposition 5.32 is given by Fletcher and Lindgren [50], Theorem 2.2.

§5.33. The term "uniformly locally compact" is used first by Weil [173]. Theorem 5.33 is proved by Fletcher and Lindgren [50], Theorem 2.4.

§5.34. Example 5.34 is a modification, due to Junnila, of an example described by Heath [65] and Kofner [87], Example 2.

§5.35. Example 5.35 is from a private communication to the authors by Junnila.

§5.36. Tamano's Theorem is given in [167].

§§5.37-5.40. Lemma 5.38, as well as the concept of a W-refinement is due to Worrell [177]. The proof of Lemma 5.38 given here is based upon the proof of Theorem 1.1.2 given by Junnila [78]. The remaining definitions and results of these sections are due to Junnila [78] and [80].

CHAPTER 6

§6.1. Monna [110], Theorems 11 and 14.

§6.4. Conditions (a), (b) and (c) of Theorem 6.4 are well-known characterizations of strong zero-dimensionality, Engelking [39] pp. 443-457. Conditions (d), (e) and (f) are part of the folklore.

§6.5. Proposition 6.5 is given by Vedenissoff [168].

§6.8. Theorem 6.8 is due to Monna [110], Theorems 11 and 13.

§6.9. A normal space satisfying condition (a) is given by Wage [171], and the famous counterexample of Roy [141] and [142] satisfies condition (b). The mapping theorem cited in (e) is given by Engelking [39], Exercise 6.2.H.

§6.10. The concept of a transitive space is defined by Fletcher and Lindgren [48].

§6.12. A famous example of a normal P-space that is not metacompact is the Dowker space of Rudin [143], and a simpler example is Example 9L of [61].

§6.14. Proposition 6.14, Junnila [78], Proposition 3.3.2.

§6.15. Junnila [78], Theorem 3.3.4.

§6.18. Proposition 6.18, Junnila [78], p. 71

§6.19. Junnila [78], p. 71.

§6.20. Proposition 6.20, Junnila [78], Proposition 3.3.2.

§6.21. The concept of a semistratifiable space is due to Kofner [86] and Creede [32]. These spaces are called pseudostratifiable by Kofner [86], but the term, semistratifiable, has gained currency. The concept of a G_δ-diagonal sequence is due to Ceder [24], Lemma 5.4. The results of this section are from Junnila [79], Section 4. In a talk at the 1979 meeting of the Virginia Academy of Science, the authors define a topological space X to be *n-pretransitive* (*n-transitive*) provided that for each neighbornet U of X, U^n is a normal neighbornet (U^n contains a transitive neighbornet). In this terminology, results of this section establish that every GO space is 3-transitive and every preorthocompact semistratifiable space is 3-pretransitive.

§6.22. Junnila [78], Proposition 3.1.7. Corollary 6.22 is generalized by Kofner [92], who shows that every space with an orthobase is a transitive space. The definition of an orthobase is given in the notes to §5.27.

§6.23. Junnila [79], Lemma 4.5 and Example 4.6.

§6.24. Junnila [78], Proposition 3.3.8.

§6.25. Junnila (unpublished).

§6.26. The lemmas of this section are given by Hicks and Sharma [70], Theorems 2 and 1. Proposition 6.26 is [70], Corollary 3. Corollary 6.26 is [70], Corollary 2.

The proof of the second lemma uses the following result of Hewitt [69], Corollary to Theorem 4.8: Every first countable T_0 space that has no isolated points can be expressed as the union of an infinite collection of pairwise disjoint dense subsets. The proof of the proposition uses the result of Ryll-Nardzewski and Telgársky [144], Theorem 2 that the weight of a compact scattered Hausdorff space equals its cardinality.

An uncountable subset S of the real line is said to be a *Δ-set* provided that for each decreasing family $\{A_n | n \in \mathbb{N}\}$ of subsets of S with $\bigcap_{n=1}^{\infty} A_n = \emptyset$ there exists a family $\{G_n | n \in \mathbb{N}\}$ of open subsets of S such that $\bigcap_{n=1}^{\infty} G_n = \emptyset$ and for each $n \in \mathbb{N}$, $A_n \subset G_n$. This definition is attributed to van Douwen and Reed by Przymusiński [135], p. 334.

In a private communication, Junnila points out that if S is an uncountable subset of the real line, every neighbornet of S (with its subspace topology) is a normal neighbornet if and only if S is a Δ-set. Since it is known that the existence of a Δ-set is independent of ZFC [44], it cannot even be decided in ZFC whether there is any uncountable subspace of the real line in which every neighbornet is normal.

§6.27. Fox [52].

§§6.29-6.31. The results of these sections are those of Kofner [90]. The Engelking-Lutzer Line is given in [40], Example 4.3.

Problem M. The conclusion that $\{0,1\}^{\lambda}$ is not transitive follows from Engelking [39], Theorem 3.2.2.

§6.32. The Kofner plane is defined by Kofner in [87]. Some of its properties are established in [88], Proposition 1.

§6.34. The result that every first countable Hausdorff space is a closed continuous image of a scattered Hausdorff space follows from Kannan and Rajagopalan [81], Theorem 1.16. Scattered spaces are studied in detail by Semadeni [152].

CHAPTER 7

§7.1. The concept of a σ-interior-preserving collection is introduced by Ceder [24], Theorem 4.2. Theorem 7.1 is proved by Kofner [87], Proposition 1. The corollary is due to Norman [127], Theorem 2 and Sion and Zelmer [158], Theorem 2.2. The stronger result that every T_1 space with a σ-interior-preserving base is quasi-metrizable is attributed to Dolchinov in [119], Theorem 1.

§7.2. Wilson [176], Theorem 3, shows that every second countable T_1 space is quasi-metrizable.

§7.4. Engelking and Lutzer [40], Example 4.2. The proof that a LOTS with a G_δ-diagonal is metrizable appears in Lutzer [104], Theorem p. 557. The result of [104] contrasts with the example of Gruenhage [63] of a LOTS with a point-countable base that is not even quasi-metrizable. Assuming the existence of a Souslin line, however, Heath [66] shows that there is a quasi-metrizable Souslin line with a point-countable base. Such a space is necessarily nonmetrizable and so has no G_δ-diagonal.

§7.5. Stoltenberg [163], Theorem 2.6.

§7.6. Stoltenberg [163], p.67 and Nedev [120], Corollary 7.6.

§7.7. Kofner [87], Example 1.

§§7.10-7.12. Kofner [91], Remark and Examples 1 and 2.

§7.13. Kofner [89], Section 2.

§7.15. The terminology, local uniformity, is taken from Williams [175]. Theorem 7.15 is a result of Fox [51], Theorem 9. Corollary 7.15 is a result of Niemytzki [123], p. 511.

§7.16. The proof of Theorem 7.16 is taken from Williams [175], Theorem 2.9 and Proposition 2.10.

§7.17. Fox [51], Proposition 1. The corollary is motivated by results of Hung [73].

§7.18. Lindgren and Fletcher [98], Theorem 2.1. Condition (a) is due to Hodel [71], condition (c) is due to Martin [106] and condition (d) is due to Sabella [145]. The results for γ-spaces that correspond to the results for quasi-metric spaces given in Propositions 7.8 and 7.9 and Theorems 7.13 and 7.14 may all be obtained by modifying the proofs of these propositions, and proofs for several of these results have appeared in the literature. Many of these results are given by Veličko [169], who calls a γ-space a QS-space.

§7.19. Corollary 7.19 is a result of Kofner [90], Theorem 5. In the terminology given in the notes to §6.21, Proposition 7.19 may be restated as follows: Every m-pretransitive γ-space is quasi-metrizable.

§7.20. Hodel [71], Proposition 4.2, Junnila [78], Corollary 3.2.4 and Nedev and Choban [121], Corollary 8. The proof of Proposition 7.20 uses Heath's well-known characterization of developable spaces [71].

§7.21. The question whether every γ-space is quasi-metrizable originates with Ribeiro [140] and is discussed in Lindgren and Fletcher [98] and Nedev and Choban [121], part III, p. 8. The remarkable example is due to Fox [53].

§7.22. Junnila [78], Lemma 3.2.2.

§7.23. The term subdevelopment is due to Gerlits [59].

§7.24. Theorem 7.24 is based upon Stoltenberg [163], Theorems 3.1 and 3.2.

§7.25. Example 7.25 is due to Kofner [88]. It is observed in Junnila [78], p. 40 that this example is a nonorthocompact space in which every open cover is a quasi-normal cover.

§7.27. The term $\sigma^{\#}$-space is given by Siwiec and Nagata [159].

§7.29. Theorem 7.29 is based upon both Engelking and Lutzer [40], Theorem 4.4, and Kofner [90], Theorem 5.

§7.30. Example 7.30 is Engelking and Lutzer [40], Example 4.3.

§7.31. Atkins and Slaughter [8], Lemmas 3.6 and 3.7.

Problem T. A related problem is raised in [18], Question 1.10.

§7.32. Example 7.32 is given by Burke [19], Example 1 as an example of a separable Moore space that is not K-semimetrizable.

§7.33. The definition of a Cauchy sequence given in §7.33 appears in Cohen [28].

§7.34. Theorem 7.34 is based upon Mrówka [112], Theorem 2.

§7.35. Lindgren and Fletcher [101], Proposition 4.4.

§7.36. Niemytzki and Tychonoff [125].

§7.37. The concept of a base of countable order is due to Arhangel'skii [7]; the characterization of base of countable order used here is given by Worrell and Wicke [178], Theorem 2. An instructive reference on Aronszajn spaces is Wicke and Worrell [174].

§7.38. Example 7.38 is the renowned Moore space Λ of Pixley and Roy [134], p. 77. Fitzpatrick [41], shows in the proof of Theorem 2 that every nonseparable Moore space that satisfies the countable chain condition has no dense metrizable subspace. We are indebted to T. M. Phillips for pointing out that the proof of Fitzpatrich [41], Theorem 1 can be modified to show that every subspace of a complete Aronszajn space contains a dense metrizable subspace.

§7.40. The proof of Theorem 7.40 uses the following result of Worrell and Wicke [178], Theorem 3: An R_0-space is developable if and only if it is a θ-refinable space whose topology has a base of countable order.

Problem V. In a private communication, Hans-Peter Kunzi gives an example of a complete quasi-metric space that is neither quasi-developable nor a Hausdorff space.

§7.41. The example is due to Gerlits [59], p. 345 and in a different guise to Burke [17] and [18], Example 1.9.

BIBLIOGRAPHY

1. P. S. Aleksandrov, On the metrisation of topological spaces (Russian, English summary), Bull. Acad. Polon. Sci. Sér. Sci. Math. Astr. Phys. 8 (1960), 135-140; MR 22 #5024.

2. P. S. Aleksandrov, Some results in the theory of topological spaces, obtained within the last twenty-five years (Russian), Uspehi Mat. Nauk 15 (1960), 25-95; MR 22 #9947 (Translation = Russian Math. Surveys 15 (1960), 23-80).

3. P. S. Aleksandrov and P. Urysohn, Memoires sur les espaces topologiques compacts, Verh. Akad. Wetensch. Amsterdam 14 (1929), 1-96.

4. E. M. Alfsen and J. E. Fenstad, On the equivalence between proximity structures and totally bounded uniform structures, Math. Scand. 7 (1959), 353-360 (Correction, *Ibid.* 9 (1961), 258)); MR 22 #5958, MR 24 #A1112.

5. E. M. Alfsen and O. Njastad, Proximity and generalized uniformity, Fund. Math. 52 (1963), 235-252; MR 27 #4207a.

6. R. Arens and J. Dugundji, Remark on the concept of compactness, Portugal. Math. 9 (1950), 141-143; MR 12, 434.

7. A. V. Arhangel'skii, Some metrization theorems (Russian), Uspehi Mat. Nauk 18 (1963), 5 (113), 139-145; MR 27 #6242.

8. J. M. Atkins and F. G. Slaughter, Pull-back theorems for closed mappings, (unpublished).

9. C. Barnhill and P. Fletcher, Topological spaces with a unique compatible quasi-uniform structure, Arch. Math. (Basel) 21 (1970), 206-209; MR 42 #6784.

10. R. H. Bing, Metrization of topological spaces, Canad. J. Math 3 (1951), 175-186; MR 13, 264.

11. J. Blatter, Order compactifications of totally ordered topological spaces, J. Approximation Theory 13 (1975), 56-65; MR 50 #11153.

12. J. Blatter and G. L. Seever, Interposition and lattice cones of functions, Trans. Amer. Math. Soc. 222 (1976), 65-96; MR 55 #11013.

13. Nicholas Bourbaki, Elements of mathematics. General topology. Part I, Hermann, Paris, Addison-Wesley, Reading Mass. (1966) MR 34; #5044a.

14. G. C. L. Brümmer, Initial quasi-uniformities, Nederl. Akad. Wetensch. Proc. Ser. A 72 = Indag. Math. 31 (1969), 403-409; MR 41 #2617.

15. G. C. L. Brümmer, Oor die totaal-begrendse refleksie, S. Afr. Math. Soc., Annual Report (1970), 47-53.

16. G. C. L. Brümmer, A categorical study of initiality in uniform topology, Ph.D. Thesis, University of Cape Town, 1971.

17. D. K. Burke, A nondevelopable locally compact Hausdorff space with a G_δ-diagonal, General Topology and Appl. 2 (1972), 287-291; MR 47 #7702.

18. D. K. Burke, Spaces with a G_δ-diagonal, TOPO 72 - general topology and its applications (Proc. Second Pittsburgh International Conf., Carnegie-Mellon Univ. and Univ. of Pittsburgh, Pittsburgh, PA, (1972) 95-100. Lecture Notes in Math., 378, Springer, Berlin, 1974; MR 51 #1755.

19. D. K. Burke, On K-semimetric spaces, Proc. Amer. Math. Soc. 73 (1979), 126-128; MR 80b: 54033.

20. D. K. Burke, Orthocompactness and perfect mappings, Proc. Amer. Math. Soc. 79 (1980), 484-486; MR 81e: 54021.

21. D. Bushaw, Elements of general topology, John Wiley and Sons, Inc., New York, 1963; MR 28 #2515.

22. J. W. Carlson and T. L. Hicks, On completeness in quasi-uniform spaces, J. Math. Anal. Appl. 34 (1971), 618-627; MR 43 #6868.

23. E. Čech, Topological spaces, Academia (Czechoslovak Acad. Sci.), Prague, 1966; MR 35 #2254.

24. J. G. Ceder, Some generalizations of metric spaces, Pacific J. Math. 11 (1961), 105-125; MR 24 #A1707.

25. E. W. Chittenden, The metrization problem, Bull. Amer. Math. Soc. 33 (1927), 13-34.

26. T. H. Choe, Partially ordered topological spaces, An. Acad. brasil. Ciênce 51 (1979), 53-63; MR 80k: 54058.

27. H. J. Cohen, Sur un problème de M. Dieudonné, C. R. Acad. Sci., Paris 234 (1952), 290-292; MR 13, 763.

28. L. Cohen, Uniformity properties in topological spaces satisfying the first denumerability postulate, Duke. Math. J. 3 (1937), 610-615.

29. H. H. Corson, The determination of paracompactness by uniformities, Amer. J. Math. 80 (1958), 185-190; MR 20 #1292.

30. H. H. Corson, Normality in subsets of product spaces, Amer. J. Math. 81 (1959), 785-796; MR 21 #5947.

31. H. H. Corson and J. R. Isbell, Euclidean covers of topological spaces, Quart. J. Math. Oxford Ser. (2) 11 (1960), 34-42; MR 23 #A2194.

32. G. D. Creede, Concerning semi-stratifiable spaces, Pacific J. Math. 32 (1970), 47-54; MR 40 #8006.

33. A. Császár, Foundations of general topology, A Pergamon Press Book, The Macmillan Co., New York, 1963; MR 28 #575.

34. A. S. Davis, Indexed systems of neighborhoods for general topological spaces, Amer. Math. Monthly 68 (1961), 886-893; MR 35 #4869.

35. R. Doss, On uniform spaces with a unique structure, Amer. J. Math. 71 (1949), 19-23; MR 10, 557.

36. E. van Douwen, Covering and separation properties of box products. Surveys in General Topology, Edited by G. M. Reed, Academic Press, New York (1980), 55-129.

37. C. H. Dowker, Mappings of proximity structures, General Topology and its Relations to Modern Analysis and Algebra (Proc. Sympos., Prague, 1961), 139-141. Academic Press, New York; Publ. House, Czech. Acad. Sci., Prague 1962; MR 26 #4312.

38. V. A. Efremovič, The geometry of proximity, I (Russian), Mat. Sb. (N.S.) 31 (73) (1952), 189-200; MR 14, 1106.

39. R. Engelking, General Topology, Monografie Mat., 60, Polish Scientific Publishers, Warsaw, 1977; MR 58 #18316b.

40. R. Engelking and D. Lutzer, Paracompactness in ordered spaces, Fund. Math. 94 (1977), 49-58; MR 55 #1303.

41. B. Fitzpatrick, On dense subspaces of Moore spaces, Proc. Amer. Math. Soc. 16 (1965), 1324-1328; MR 32 #1663.

42. W. M. Fleischman, On fundamental open coverings, Proc. International Conference on Topology and its Applications, Herceg Novi, Yugoslavia, 1968.

43. W. M. Fleischman, On coverings of linearly ordered spaces, Proc. Washington State University Conference on General Topology, (1970), 52-55.

44. W. G. Fleissner, Current research on Q sets, Topology, Vol. I, (4th Colloq. Budapest 1978), 413-431. Colloq. Math. Soc. János Bolyai, Vol. 23. North Holland, Amsterdam, 1980.

45. P. Fletcher, Note on metrization of quasi-uniform spaces, Colloq. Math. 24 (1971/2), 81-82; MR 46 #4490.

46. P. Fletcher, On completeness of quasi-uniform spaces, Arch. Math. (Basel) 22 (1971), 200-204; MR 46 #4489.

47. P. Fletcher and W. F. Lindgren, Transitive quasi-uniformities, J. Math. Anal. Appl. 39 (1972), 397-405; MR 47 #5828.

48. P. Fletcher and W. F. Lindgren, Quasi-uniformities with a transitive base, Pacific J. Math. 43 (1972), 619-631; MR 47 #7701.

49. P. Fletcher and W. F. Lindgren, Orthocompactness and strong Čech completeness in Moore spaces, Duke Math. J. 39 (1972), 753-766 (Correction, *Ibid.* 40 (1973), 959); MR 47 #1029.

50. P. Fletcher and W. F. Lindgren, C-complete quasi-uniform spaces, Arch. Math. (Basel) 30 (1978), 175-180; MR 57 #13863.

51. R. Fox, On metrizability and quasi-metrizability, submitted for publication.

52. R. Fox, Pretransitivity and products of suborderable spaces, Mathematical Centre Tracts, No. 142, 115-118, Mathematisch Centrum, Amsterdam, 1981.

53. R. Fox, Solution of the γ-space problem, Proc. Amer. Math. Soc., to appear.

54. R. Fox, Pretransitivity and the gamma-space conjecture. Submitted for publication.

55. A. H. Frink, Distance functions and the metrization problem, Bull. Amer. Math. Soc. 43 (1937), 133-142.

56. I. S. Gál, Proximity relations and pre-compact structures I, II, Nederl. Akad. Wetensch. Proc. Ser. A 62 = Indag. Math. 21 (1959), 304-326; MR 21 #5944.

57. I. S. Gál, Uniformizable spaces with a unique structure, Pacific J. Math. 9 (1959), 1053-1060; MR 22 #966.

58. T. E. Gantner and R. C. Steinlage, Characterizations of quasi-uniformities, J. London Math. Soc. (2) 5 (1972), 48-52; MR 52 #1638.

59. J. Gerlits, On G_δ p-spaces, Topics in topology (Proc. Colloq. Keszthely (1972)), 341-346. Colloq. Math. Soc. János Bolyai, vol. 8, North Holland, Amsterdam, 1974; MR 50 #14676.

60. L. Gillman and M. Henriksen, Concerning rings of continuous functions, Trans. Amer. Math. Soc. 77 (1954), 340-362; MR 16, 156.

61. L. Gillman and M. Jerison, Rings of continuous functions, Van Nostrand, Princeton, NJ, 1960; MR 22 #6994.

62. A. J. Goldman, A Čech fundamental group, summary of lectures and seminars of the Summer Institute of Set Theoretic Topology, Madison, Wisconsin, American Mathematical Society Summer Institute (1955), 107-109.

63. G. Gruenhage, A note on quasi-metrizability, Canad. J. Math. 29 (1977), 360-366; MR 55 #9040.

64. F. Hausdorff, Mengenlehre, Dover Publications. New York, NY, 1944; MR 7, 419. (A photographic reproduction of the third edition [de Gruyter, Berlin-Leipzig 1935]).

65. R. W. Heath, A postscript to a note on quasi-metric spaces, Notices Amer. Math. Soc., 19 (1972), A-338.

66. R. W. Heath, A construction of a quasi-metric Souslin space with a point-countable base, Set-theoretic topology, (Papers, Inst. Medicine and Math., Ohio Univ., Athens, Ohio, 1975-1976), 219-224, Academic Press, New York, 1977; MR 56 #6631.

67. R. W. Heath and W. F. Lindgren, On generating non-orthocompact spaces, Set-theoretic topology, (Papers, Inst. Medicine and Math., Ohio Univ., Athens, Ohio, 1975-1976), 225-237, Academic Press, New York, 1977: MR 56 #3792.

68. R. W. Heath and W. F. Lindgren, A note on semiparacompact spaces, Notices Amer. Math. Soc. 24 (1977), A-636.

69. E. Hewitt, A problem of set-theoretic topology, Duke Math. J. 10 (1943), 309-333; MR 5, 46.

70. T. L. Hicks and P. Sharma, Properties of z-spaces, Topology Proc. 4 (1979), 109-113; MR 81j: 54044.

71. R. E. Hodel, Spaces defined by sequences of open covers which guarantee that certain sequences have cluster points, Duke Math. J. 39 (1972), 253-263; MR 45 #2657.

72. S. M. Huffman, T. L. Hicks, and J. W. Carlson, Complete quasi-uniform spaces, Canad. Math. Bull. 23 (1980), 497-498.

73. H. H. Hung, Some metrization theorems, Proc. Amer. Math. Soc. 54 (1976), 363-367; MR 53 #9159.

74. W. N. Hunsaker and W. F. Lindgren, Construction of quasi-uniformities, Math. Ann. 188 (1970), 39-42; MR 42 #1057.

75. W. N. Hunsaker and W. F. Lindgren, Upper semi-continuous functions and quasi-proximity classes, Bul. Inst. Politehn. Iasi (N.S.) 19(23) (1973) sect. I, 23-30; MR 48 #12472.

76. J. R. Isbell, Uniform spaces, Mathematical Surveys. No. 12, Amer. Math. Soc., Providence, R.I., 1964; MR 30 #561.

77. K. Iséki and S. Kasahara, On pseudo-compact and countably compact spaces, Proc. Japan Acad. 33 (1957), 100-102; MR 19, 668.

78. H. J. K. Junnila, Covering properties and quasi-uniformities of topological spaces, Ph.D. Thesis, Virginia Polytechnic Institute and State University, Blacksburg (1978).

79. H. J. K. Junnila, Neighbornets, Pacific J. Math. 76 (1978), 83-108; MR 58 #2734.

80. H. J. K. Junnila, Metacompactness, paracompactness, and interior-preserving open covers, Trans. Amer. Math. Soc. 249 (1979), 373-385; MR 80b:54024.

81. V. Kannan and M. Rajagopalan, Scattered spaces II, Illinois J. Math 21, (1977), 735-751; MR 57 #13830.

82. S. Kasahara, Boundedness of semicontinuous finite real functions, Proc. Japan Acad. 33 (1957), 183-186; MR 19, 668.

83. M. Katětov, Über H-abgeschlossene und bikompakte Räume, Časopis Pěst. Mat. Fys. 69 (1940), 36-49; MR 1, 317.

84. R. P. Kaufman, Ordered sets and compact spaces, Colloq. Math. 17 (1967), 35-39; MR 35 #3634.

85. J. L. Kelley, General Topology, Van Nostrand, New York, 1955; MR 16, 1136.

86. Ja. A. Kofner, Pseudostratifiable spaces (Russian), Fund. Math. 70 (1971), 25-47; MR 43 #3988.

87. Ja. A. Kofner, Δ-metrizable spaces (Russian), Mat. Zametki 13 (1973), 277-287; MR 48 #3014. (Translation = Math. Notes 13 (1973), 168-174).

88. Ja. A. Kofner, Quasi-metrizable spaces, Pacific J. Math. 88 (1980), 81-89.

89. Ja. A. Kofner, Closed mappings and quasi-metric spaces, Proc. Amer. Math. Soc. 80 (1980), 333-336.

90. Ja. A. Kofner, Transitivity and the γ-space conjecture in ordered spaces, Proc. Amer. Math. Soc. 81 (1981), 629-635.

91. Ja. A. Kofner, Open compact mappings, Moore spaces and orthocompactness, Rocky Mountain J. Math. (to appear).

92. Ja. A. Kofner, Transitivity and orthobases, to appear.

93. V. S. Krishnan, A note on semiuniform spaces, J. Madras Univ. Sect. B 25 (1955), 123-124; MR 17, 66.

94. P. Lambrinos, Quasi-uniform continuity, Bull. Austral. Math. Soc. 8 (1973), 389-392.

95. S. Leader, Spectral structures and uniform continuity, Fund. Math. 60 (1967), 105-115; MR 35 #974.

96. W. F. Lindgren, Topological spaces with unique quasi-uniform structure, Arch. Math. (Basel) 22 (1971), 417-419; MR 47 #5830.

97. W. F. Lindgren, Topological spaces with a unique compatible quasi-uniformity, Canad. Math. Bull. 14 (1971), 369-372; MR 46 #9942.

98. W. F. Lindgren and P. Fletcher, Locally quasi-uniform spaces with countable bases, Duke Math. J. 41 (1974), 231-240; MR 49 #6173.

99. W. F. Lindgren and P. Fletcher, A construction of the pair completion of a quasi-uniform space, Canad. Math. Bull. 21 (1978), 53-59; MR 58 #7562.

100. W. F. Lindgren and P. Fletcher, A theory of uniformities for generalized ordered spaces, Canad. J. Math. 31 (1979), 35-44; MR 80b: 54037.

101. W. F. Lindgren and P. Fletcher, Equinormal quasi-uniformities and quasi-metrics, Glas. Mat. Ser. III 13 (33), (1978), 111-125; MR 58 #18359.

102. W. F. Lindgren and P. J. Nyikos, Spaces with bases satisfying certain order and intersection properties, Pacific J. Math. 66 (1976), 455-476; MR 56 #3794.

103. D. Lutzer, On generalized ordered spaces, Dissertationes Math. Rozprawy Mat. 89 (1971); MR 48 #3018.

104. D. Lutzer, A metrization theorem for linearly orderable spaces, Proc. Amer. Math. Soc. 22 (1969), 557-558; MR 40 #2012.

105. M. J. Mansfield, Some generalizations of full normality, Trans. Amer. Math. Soc. 86 (1957), 489-505; MR 20 #273.

106. H. W. Martin II, Metrization and submetrization of topological spaces, Ph.D. Thesis, University of Pittsburgh, (1973).

107. S. D. McCartan, Separation axioms for topological ordered spaces, Proc. Cambridge Philos. Soc. 64 (1968), 965-973; MR 37 #5853.

108. E. A. Michael, Yet another note on paracompact spaces, Proc. Amer. Math. Soc. 10 (1959), 309-314; MR 21 #4406.

109. E. A. Michael, The product of a normal space and a metric space need not be normal, Bull. Amer. Math. Soc. 69 (1963), 375-376; MR 27 #2956.

110. A. F. Monna, Remarques sur les métriques non-archimédiennes I, II, Nederl. Akad. Wetensch. Proc. 53 (1950), 470-481 and 625-637 = Indag. Math. 12 (1950), 122-133 and 179-191; MR 12, 41.

111. R. L. Moore, Foundations of point set theory, Amer. Math. Soc. Publ. No. 13, revised edition, 1962; MR 27 #709.

112. S. G. Mrówka, On normal metrics, Amer. Math. Monthly 72 (1965), 998-1001; MR 32 #1677.

113. M. G. Murdeshwar and S. A. Naimpally, Quasi-uniform topological spaces, P. Noordhoff Ltd., Groningen, (1966); MR 35 #2267.

114. L. Nachbin, Sur les espaces topologiques ordonnés, C.R. Acad. Sci., Paris 226 (1948), 381-382; MR 9, 367.

115. L. Nachbin, Sur les espaces uniformisables ordonnés, C.R. Acad. Sci., Paris 226 (1948), 547; MR 9, 367.

116. L. Nachbin, Sur les espaces uniformes ordonnés, C.R. Acad. Sci., Paris 226 (1948), 774-775; MR 9, 455.

117. L. Nachbin, Topology and Order, Translated from the Portuguese by Lulu Bechtolsheim. Van Nostrand Mathematical Studies, No. 4., D. Van Nostrand Co., Inc., Princeton, N.J. (1965), (Reprinted by Robert E. Kreiger Publishing Co., Huntington, NY, 1976.) MR 36 #2125.

118. S. A. Naimpally and B. D. Warrack, Proximity spaces, Cambridge Tracts

in Mathematics and Mathematical Physics, No. 59, Cambridge University Press, London-New York, 1970; MR 43 #3992.

119. S. I. Nedev, Generalized-metrizable spaces (Russian), C.R. Acad. Bulgare Sci. 20 (1967), 513-516; MR 35 #7292.

120. S. I. Nedev, Continuous and semicontinuous o-metrics (Russian), Dokl. Akad. Nauk SSSR 193 (1970) 531-534, (Translation = Soviet Math. Dokl. 11 (1970), 975-978); MR 43 #8046.

121. S. I. Nedev and M. M. Choban, On the theory of o-metrizable spaces I, II, III (Russian, English summary), Vestnik Moskov. Univ. Ser. I Mat. Meh. 27 (1972), no. 1, 8-15, no. 2, 10-17, no. 3, 10-15 (Translation = Moscow Univ. Math. Bull. 27 (1972), no. 1, 5-9, no. 2, 65-70, no. 3,

122. R. Nielsen and C. Sloyer, Quasi-uniformizability, Math. Ann. 182 (1969), 273-274; MR 40 #3504.

123. V. Niemytzki, On the third axiom of metric space, Trans. Amer. Math. Soc. 29 (1927), 507-513.

124. V. Niemytzki, Über die Axiome des metrischen Raumes, Math. Ann. 104 (1931), 666-671.

125. V. Niemytzki and A. Tychonoff, Beweis des Satz, dass ein metrisierbarer Raum dann und nur dann Kompact ist, wenn er in jeder Metric vollständig ist, Fund. Math. 12 (1928), 118-120.

126. N. Noble, Products with closed projections, II, Trans. Amer. Math. Soc. 160 (1971), 169-183; MR 44 #979.

127. L. J. Norman, A sufficient condition for quasi-metrizability of a topological space, Portugal. Math. 26 (1967), 207-211; MR 40 #1991.

128. S. A. Peregudov, Certain properties of families of open sets and coverings (Russian), Vestnik Moskov. Univ. Ser. I Mat. Meh. 31 (1976), 25-33, (Translation = Moscow Univ. Math. Bull. 31 (1976), 19-25); MR 25 # 6600.

129. W. J. Pervin, Quasi-uniformization of topological spaces, Math. Ann. 147 (1962), 316-317; MR 25 #3506b.

130. W. J. Pervin, Quasi-proximities for topological spaces, Math. Ann. 150 (1963), 325-326; MR 27 #1917.

131. W. J. Pervin, Equinormal proximity spaces, Nederl. Akad. Wetensch. Proc. Ser. A 67 = Indag. Math. 26 (1964), 152-154; MR 28 #4514.

132. T. M. Phillips, Completeness in topological spaces, Ph.D. Thesis, University of Oklahoma, Norman (1974).

133. T. M. Phillips, Completeness in Aronszajn spaces, Studies in topology (Proc. Conf., Univ. North Carolina, Charlotte, N.C., 1974; dedicated to Math. Sect. Polish Acad. Sci.), 457-465, Academic Press, New York, 1975; MR 50 #14683.

134. C. Pixley and P. Roy, Uncompletable Moore spaces, Proceedings of the Auburn Topology Conference (Auburn Univ., Auburn, Ala., 1969; dedicated to F. Burton Jones on the occasion of his 60th birthday), 75-85, Auburn Univ., Auburn, Ala., 1969.

135. T. C. Przymusiński, Normality and separability of Moore spaces, Set-theoretic topology, (Papers Inst. Medicine and Math., Ohio Univ., Athens, Ohio, 1975-1976), 325-337, Academic Press, New York, 1977; MR 56 #6617.

136. J. Rainwater, Spaces whose finest uniformity is metric, Pacific J. Math. 9 (1959), 567-570; MR 21 #5180.

137. R. H. Redfield, Uniformly convex totally ordered sets, Proc. Amer. Math. Soc. 51 (1975), 289-294; MR 52 #15396.

138. R. H. Redfield, Ordering uniform completions of partially ordered sets, Canad. J. Math. 26 (1974), 644-664; MR 50 #14693.

139. I. L. Reilly, On generating quasi-uniformities, Math. Ann. 189 (1970), 317-318; MR 43 #2663.

140. H. Ribeiro, Sur les espaces à métrique faible, Portugal. Math. 4 (1943), 21-40 (Correction, *Ibid* 4 (1943), 65-68); MR 5, 149, MR 5, 272.

141. P. Roy, Failure of equivalence of dimension concepts for metric spaces, Bull. Amer. Math. Soc. 68 (1962), 609-613; MR 25 #5495.

142. P. Roy, Nonequality of dimensions for metric spaces, Trans. Amer. Math. Soc. 134 (1968), 117-132; MR 37 #3544.

143. M. E. Rudin, A normal space X for which X × I is not normal, Fund. Math. 73 (1971/72), 179-186; MR 45 #2660.

144. C. Ryll-Nardzewski and R. Telgársky, On the scattered compactification, Bull. Acad. Pol. Sci. Sér. Sci. Math. Astronom. Phys. 18 (1970) 233-234; MR 41 #7635.

145. R. R. Sabella, Convergence properties of neighboring sequences, Proc. Amer. Math. Soc. 38 (1973), 405-409; MR 47 #1036.

146. S. Salbany, Bitopological spaces, compactifications and completions, Mathematical Monographs of the University of Cape Town, No. 1, Department of Mathematics, University of Cape Town, 1974; MR 54 #13869.

147. P. Samuel, Ultrafilters and compactifications of uniform spaces, Trans. Amer. Math. Soc. 64 (1948), 100-132; MR 10, 54.

148. B. M. Scott, The behavior of orthocompactness in products with emphasis on certain analogies with that of normality under similar circumstances, Ph.D. Thesis, Universtiy of Wisconsin, Madison (1975).

149. B. M. Scott, Toward a product theory for orthocompactness, Studies in topology (Proc. Conf. Univ. North Carolina, Charlotte, N.C., 1974; dedicated to Math. Sect. Polish. Acad. Sci.), 517-537, Academic Press, New York, 1975; MR 51 #9024.

150. B. M. Scott, Orthocompactness is normality in finite products of locally compact LOTS's, Set-theoretic topology, (Papers, Inst. Medicine and Math., Ohio Univ., Athens, Ohio, 1975-1976), 339-348, Academic Press, New York, 1977; MR 55 #13380.

151. B. M. Scott, More about orthocompactness, Topology Proc. 5 (1980), 155-184.

152. Z. Semadeni, Sur les ensembles clairsemés, Rozprawy Mat. 19 (1959); MR 21 #6571.

153. N. A. Shanin, On separation in topological spaces, C.R. (Doklady) Acad. Sci. URSS (N.S.) 38 (1943), 110-113; MR 5, 64.

154. J. L. Sieber and W. J. Pervin, Completeness in quasi-uniform spaces, Math. Ann. 158 (1965), 79-81; MR 30 #2449.

155. J. L. Sieber and W. J. Pervin, Separation axioms for synotopogenous spaces, Nederl. Akad. Wetensch. Proc. Ser. A 66 = Indag. Math. 25 (1963), 755-760; MR 28 #3399.

156. M. K. Singal and S. Lal, Proximity ordered spaces, J. London Math. Soc. (2) 14 (1976), 393-404; MR 55 #13384.

157. M. Sion and R. C. Willmott, Hausdorff measures on abstract spaces, Trans. Amer. Math. Soc. 123 (1966), 275-309; MR 34 #297.

158. M. Sion and G. Zelmer, On quasi-metrizability, Canad. J. Math. 19 (1967), 1243-1249; MR 36 #4522.

159. F. Siwiec and J. Nagata, A note on nets and metrization, Proc. Japan Acad., 44 (1968), 623-627; MR 39 #3450.

160. Yu. Smirnov, On proximity spaces (Russian), Math. Sb. (N.S.) 31 (73) (1952), 543-574 (Amer. Math. Soc. Transl. Ser. 2, 38 (1964), 5-35); MR 14, 1107.

161. R. H. Sorgenfrey, On the topological product of paracompact spaces, Bull. Amer. Math. Soc. 53 (1947), 631-632; MR 8, 594.

162. E. F. Steiner, The relation between quasi-proximities and topological spaces, Math. Ann. 155 (1964), 194-195; MR 29 #581.

163. R. A. Stoltenberg, On quasi-metric spaces, Duke Math. J. 36 (1969), 65-71; MR 38 #3824.

164. A. H. Stone, Hereditarily compact spaces, Amer. J. Math. 82 (1960), 900-916; MR 22 #11370.

165. A. H. Stone, Paracompactness and product spaces, Bull. Amer. Math. Soc. 54 (1948), 977-982; MR 10, 204.

166. G. Strecker and W. Wattel, On semi-regular and minimal Hausdorff embeddings, Nederl. Akad. Wetensch. Proc. Ser. A 70 = Indag. Math. 29 (1967), 234-237; MR 35 #2261.

167. H. Tamano, On paracompactness, Pacific J. Math. 10 (1960), 1043-1047; MR 23 #A2186.

168. N. Vedenissoff, Remarks on the dimension of topological spaces (Russian, French summary), Uchenye Zapiski Moskov. Gos. Univ. Mathematika 30 (1939), 131-149; MR 2, 69.

169. N. V. Veličko, Quasiuniformly sequential spaces (Russian), C.R. Acad. Bulgare Sci. 25 (1972), 589-591; MR 46 #4454.

170. C. Votaw, Uniqueness of compatible quasi-uniformities, Canad. Math. Bull. 15 (1972), 575-583; MR 47 #1031.

171. M. L. Wage, The dimension of product spaces, Proc. Nat. Acad. Sci. U.S.A. 75 (1978), 4671-4672; MR 80a: 54064.

172. A. J. Ward, A generalization of almost-compactness, with an associated generalization of completeness, Czechoslovak Math. J. 25(100) (1975), 514-530; MR 52 #11851.

173. A. Weil, Sur les espaces à structure uniforme et sur la topologie générale, Act. Sci. Ind. 551, Paris, 1937.

174. H. H. Wicke and J. M. Worrell, Jr., Topological completeness of first countable Hausdorff spaces I, Fund. Math. 75 (1972), 209-222; MR 46 #8189.

175. J. Williams, Locally uniform spaces, Trans. Amer. Math. Soc. 168 (1972), 435-469; MR 45 #5950.

176. W. A. Wilson, On quasi-metric spaces, Amer. J. Math. 53 (1931), 675-684.

177. J. M. Worrell, Jr., A characterization of metacompact spaces, Portugal. Math. 25 (1966), 171-174; MR 38 #675.

178. J. M. Worrell, Jr. and H. H. Wicke, Characterizations of developable topological spaces, Canad. J. Math. 17 (1965), 820-830; MR 32 #427.

LIST OF SYMBOLS

SPECIAL SYMBOLS

a (open spectrum), 2.12*

cl_d (cl_i), 4.1

$\mathcal{C}^F$, 5.28

C(X) (C*(X)), 2.9

C(X) (C*(X)), 2.9

d(Y) (i(Y)), 4.1

δ, 1.22

δ^*, 1.38

$\delta_{\mathcal{U}}$, 1.28

$\mathcal{E}$, 1.1

$\mathcal{FINE}$, 2.3

$\mathcal{FT}$, 2.7

$\mathcal{G}$(X), 5.14

$\leq$, 4.30

<<, 1.24-1.25

$\mathcal{LF}$, 2.7

$\mathcal{M}$, 1.1

$\mathcal{P}$ ($\delta_{\mathcal{P}}$), 2.2

$\mathcal{PF}$, 2.7

$\pi(\delta)$, 1.28

$\mathcal{Q}$, 1.1

Q(δ) (QB(δ)), 1.56

Q($\mathcal{U}$) (QB($\mathcal{U}$)), 1.56

$\mathcal{SC}$, 2.10

T(A,B), 1.33

$\mathcal{U}^*$, 1.3

$\mathcal{U}_A$, 2.6

$\mathcal{U}_\delta$, 1.33

$\mathcal{U}_\omega$, 1.37

$(\tilde{X},\tilde{\mathcal{U}})$ (bicompletion), 3.33

$\mathcal{Z}$, 1.1

OTHER SYMBOLS

$\mathbb{N}$, (natural numbers)

$\mathbb{R}$, (real numbers)

Z, (integers)

f_2, ($f_2(x,y) = (f(x),f(y))$)

$\Delta = U^0$, (diagonal)

$U \circ V = \{(x,z) \mid$ there is a y so that $(x,y) \in U$ and $(y,z) \in V\}$

$U^n = U \circ U^{n-1}$ ($U^1 = U$)

U^∞, 1.6

V, (join)

*References are to sections

$<x_n>$, (sequence whose nth term is x_n)

fil($\mathcal{B}$), (filter for which $\mathcal{B}$ is a subbase)

$f(\mathcal{F}) = \text{fil}(\{f(F) \mid F \in \mathcal{F}\})$

$f^{-1}(\mathcal{F}) = \text{fil}(\{f^{-1}(F) \mid F \in \mathcal{F}\})$

$f(\mathcal{C}) = \{f(C) \mid C \in \mathcal{C}\}$

$\eta(x)$, (neighborhood filter of x)

$\eta^*(x)$, 3.28

$(\leftarrow, a) = \{x \mid x < a\}$

card($\mathcal{C}$), (cardinality of $\mathcal{C}$)

$S_d(x,\epsilon)$, 1.5

$St(x,\mathcal{C}) = \cup\{C \in \mathcal{C} \mid x \in C\}$

$d(x,y)$, 1.5

$d^{-1}(x,y)$, 1.5

$\mathcal{T}(d)$, 1.5

$\mathcal{T}(\delta)$, 1.27

$\mathcal{T}(\mathcal{U})$, 1.4

$(a, \rightarrow) = \{x \mid x > a\}$

INDEX

A

Admits:
  T admits δ, 11
  T admits $\mathcal{U}$, 3
Alexandroff-Urysohn Metrization Theorem, 40
Almost 2-fully normal space, 93, 110-111
Aronszajn space, 178

B

Base:
  Cauchy filter base, 47
  of countable order, 178
  for a quasi-uniformity, 1-2
  uniform base, 107-108, 148, 159, 179
βX (*see* compactification)
Bicomplete quasi-uniform space, 60-66
Bicompletion, 63-65, 91-92, 95

C

C, 31-32, 93-94, 115-116, 127
C*, 31-32, 93-94, 127
C(X), 31-32
C(X,↑), 74, 77-78, 83, 86
C*(X), 31-32
C*(X,↑), 74, 76, 78-80, 84, 86, 89
Canonical embedding, 65-66, 83, 88
Cauchy filter, 47-52, 59-60, 68-69, 176-177
  weakly Cauchy filter, 118
Cauchy sequence, 176
Čech-Stone compactification (*see* compactification)
$cl_d(A)$, 74
$cl_i(A)$, 74
Closed mapping:
  of an orthocompact θ-refinable space, 114-115
  of a quasi-metrizable space, 160
  and strongly zero-dimensional spaces, 130
Closure-preserving collection, 29, 59, 117, 139-140
  σ-closure-preserving collection, 172, 175
Coarser quasi-proximity, 15
  quasi-uniformity, 2
Coarsest proximity, 19-20
Coarsest quasi-proximity, 15, 19-20
Coarsest quasi-uniformity, 2, 19-20
Coarsest uniformity, 19-20, 126
Cocushioned:
  cocushioned in, 103
  cocushioned pairbase, 163
  σ-cocushioned pairbase, 163
  cocushioned refinement, 103
Cofinite topology, 40, 44-45, 173
Compact map, 156
Compact open map, 113, 157-179
Compact ordered space, 76-81, 85-86
  uniquely determined, 86
Compact space:
  vs. equinormal, 38
  hereditarily compact, 43-45
  if and only if, 58, 116
  vs. Lebesgue quasi-uniformities, 97-98
  vs. locally compact, 20
  and normal neighbornets, 141
  and point symmetry, 39-40
  vs. preorthocompact, 109

[Compact space:]
products of, 117
and total boundedness, 65-66
and transitive space, 146
unique uniformity, 9
Compactification, 66-67, 69-71
order compactification, 79
equivalent, 79-81, 87-88
Nachbin, 80, 92
if and only if, 81
is strict, 90
vs. βX, 93-94
strict, 89
Stone-Čech, 80, 94, 120-121, 124
Compatible, 3, 12
Completely regular space, 31, 36, 154
Complete structures:
complete Aronszajn space, 178-180
complete development, 180
complete metric, 177-178
complete quasi-metric (*see* quasi-metric)
complete quasi-uniformity, 50-53, 58-60, 118-119, 176
complete unformity, 110, 119
Completion of a quasi-uniform space, 67-69, 71, 180
vs. regular quasi-metric completion, 179
uniform completion, 65, 67
CoNagata space (*see* γ-space)
Condensed onto, 172
Conjugate quasi-metric, 3
Conjugate quasi-uniformity, 1
Continuous partial order, 187
Convergence complete quasi-uniformity, 50-51, 59-60, 101-102
Convex quasi-uniformity, 84-85, 90, 92-95
Countably compact space:
characterizations of, 54
vs. compact, 40, 58, 178
vs. equinormal, 38
has at most one quasi-metric subtopology, 173
products of, 108-109
$SC \subset PF$, 33
Countably metacompact space:
and γ-refinable is preorthocompact, 168
vs. countably orthocompact, 105
vs. orthocompact, 102, 109
Cube, 77
Cushioned in, 103
Cushioned refinement, 103

D

Decreasing function, 74
Decreasing set, 74-77
Δ-set, 192
Δ-space, 185
Determines, 81-87, 89, 95
Developable space, 39
not condensed onto a T_2 γ-space, 176
vs. hereditarily weakly orthocompact, 120
if and only if, 39-40, 169
not nearly metacompact, 106
orthocompact,
if and only if hereditarily orthocompact, 171
if and only if non-Archimedeanly quasi-metrizable 171
preorthocompact, 137
if and only if quasi-metrizable, 171
vs. quasi-metrizable space, 119-120, 169-171
not semi-metacompact, 107
Development, 39
Directed cover, 115-119, 122
Discrete proximity, 15
Discrete uniformity, 2-3
Dominates, 79-81, 87-88

E

Embedding, 7, 21
order embedding, 74, 78, 83, 89-91
Embedding Lemma, 78, 80
Embedding Theorem, 78
Engelking-Lutzer Line, 146, 175
Entourage, 1
Equinormal:
metric, 41-42, 177
quasi-metric, 41-42, 177
quasi-proximity, 37-38
quasi-uniformity, 37-38, 98
Equivalent order compactifications (*see* compactification)
Equivalent subbases, 2
Evaluation map, 78, 80
Extension:
of maps, 61, 81
of quasi-uniformities, 69-70

F

Far from, 10
Fine quasi-uniformity, 28 (*see also* transitive space)
  belongs to δ_P, 33
  has a countable base if and only if, 42
  not fine transitive, 147
  vs. *LF*, 153
  and perfect maps, 51
  vs. *PF*, 137, 139, 153
  products and subspaces of, 34-35, 55
Fine transitive quasi-uniformity, 30 (*see also* transitive space)
  belongs to δ_P, 33
  complete, 59-60
  convergence complete, 60, 101
  equals *FINE*, 130
  equals *PF*, 137, 139
  and orthocompactness, 100-101
  and perfect maps, 51
  precompact if and only if, 58
  products and subspaces of, 34-35, 55
  totally bounded if and only if, 57-58
  and weak orthocompactness, 116
Finer quasi-proximity, 15
Finer quasi-uniformity, 2
Finest proximity, 18, 38
Finest uniformity, 127-128, 130
  each Lebesgue quasi-uniformity contains, 98
  of a GO space is convex, 93
  has a countable base, 41
  LF contains, 99
  in a quasi-proximity class, 26
Finite-to-one open map:
  preserves (pre)orthocompactness, 113
  preserves quasi-metrizability, 156
Finite topology, 35, 43-44, 57-58
F_σ-discrete space, 132-133, 139-140
F_0, 110-111

G

γ-refinable, 168
  and countably metacompact is preorthocompact, 168
[γ-refinable]
  in a developable space, 171
γ-sequence, 165
γ-space, 165 (*see also* local quasi-uniformity)
  vs. developable space, 166
  if and only if, 163-164
  not quasi-metrizable, 166-167
  vs. quasi-metrizable, 162, 165
  vs. $\sigma^\#$-space, 174
G_δ-diagonal, 136, 140-141, 169, 175
  and orthocompact is $\sigma^\#$-space, 175
  and θ-refinable is $\sigma^\#$-space, 175
  and quasi-metrizable spaces, 153-154
G_δ-diagonal sequence, 136
G_δ-set, 57, 139
Generalized ordered space (GO space), 90-95
  characterization of, 91-92
  and γ-spaces, 166
  is hereditarily orthocompact, 111
  metrizable, 93
  neighbornets of, 143-146
  pseudocompact if and only if, 93-94
  is transitive, 145
  vs. uniform ordered space, 92
G(X), 105, 131

H

Hereditarily compact (*see* compact)
Hereditarily (pre)orthocompact (*see* (pre)orthocompact)

I

Increasing function, 74-75, 79-82
Increasing set, 74-78
Induced topology, 3-4
Infimum, 2
Interior-preserving collection (*see also* quasi-metric, non-Archimedean), 29, 43, 54, 57, 102, 151, 153
  vs. closure-preserving, 29
  and convergence completeness, 60
  and ω_1, 161
  and orthocompactness, 100
  vs. $P = FT$, 31, 45
  and σ-orthocompactness, 102
  and transitive quasi-uniformities, 29-31
  and weak orthocompactness, 116

J

Junnila's Theorem, 120, 123

K

Kofner plane, 147-149, 153, 155, 164, 173-174

L

Lebesgue quasi-uniformity (*see also* quasi-uniform cover), 97-101, 167
  contains each neighborhood of Δ, 99
  is convergence complete, 101
  is equinormal, 98
  $\mathcal{FT}$ is, 100
  in Hilbert space, 115
  vs. orthocompact, 100
  and subdevelopable, 169
Lindelöf degree, 109
Lindelöf space, 57, 115-116, 128, 130
Linearly ordered space (LOTS), 90-92, 95
  nonmetrizable, 154
  vs. quasi-metrizable space, 174
Locally compact space, 19-20, 40, 70, 118-119, 126, 132, 141
  uniformly locally compact, 119
Locally convex, 92
Locally finite covering quasi-uniformity, 30
  belongs to $\delta_{\mathcal{P}}$, 33
  contains finest uniformity, 99
  vs. $\mathcal{FINE}$, 153
  vs. paracompact, 116
  and perfect maps, 51
  precompact if and only if, 55, 58
  products and subspaces, 34, 55, 118
  totally bounded if and only if, 55, 58
Locally symmetric quasi-metric, 40-41, 162
Locally symmetric quasi-proximity, 36-39
Locally symmetric quasi-uniformity, 37-38, 48, 50, 70, 98, 162
Locally transitive space, 134
Local quasi-uniformity, 161-162
Local uniformity, 162
Lower semi-continuous function, 24, 54, 70
Lower topology, 4

M

Metacompact space:
  $\mathcal{G}(X)$ is preorthocompact if and only if, 105
  if and only if, 105-106, 116, 120-124
  vs. orthocompact space, 111, 112
  products of, 117-118
  vs. quasi-metrizable, 161
  and semistratifiable, 136-137
  vs. uniform base, 108
Metrizable space:
  conditions implying, 40-42, 155, 162-163
  examples of non, 154
  open cover consisting of, 161
Michael line, 1, 4, 12-13, 56, 144, 153
  has no completion, 180
  finite product of is transitive, 135
  each neighbornet is normal, 141
  M^n, 142
  M^{ω_0}, 143, 166
Minimal $\mathcal{U}$*-Cauchy filter, 63-65
MOBI, 158
Monotone collection, 100, 111
Moore space, 39 (*see* developable space)

N

Nagata-Smirnov Metrization Theorem, 163
Near, 10
Nearly metacompact, 105
  example of non-, 106
Neighborhood:
  δ-neighborhood, 10
  $\mathcal{U}$-neighborhood, 3
  of the diagonal, 8-9, 99
  semi-neighborhood of the diagonal, 183
Neighbornet, 4-5
  closed neighbornet, 4
  normal neighbornet, 5, 28, 135-144, 192
  normal sequence of neighbornets, 5, 160

[Neighbornet]
open neighbornets, 4
symmetric neighbornet, 4-5, 99
transitive neighbornet, 4, 132-138, 141, 143-147
unsymmetric neighbornet, 4
Niemytzki plane, 107, 153
Niemytzki-Tychonoff Theorem, 178
Non-Archimedeanly metrziable 129
Non-Archimedeanly quasi-metrizable (*see* quasi-metrizable space)
Normally ordered space:
vs. compact topological ordered space, 76
if and only if, 74-76, 78-79
not uniform ordered, 79
vs. uniform ordered space, 78
n-pretransitive, 191, 193
n-transitive, 191

O

ω_1, 59, 161, 177
Open-finite cover, 104-105, 109
Open-finite base, 106, 108, 111, 113
Open map:
of a preorthocompact space, 113
and strongly zero-dimensional spaces, 130
of a transitive space, 149
of a quasi-metrizable space, 156-158
Open spectrum, 33, 56
point-finite, 33-34, 53-54, 56-57
Order compactification (*see* compactification)
Order embedding (*see* embedding)
Order homeomorphism, 74
Ordered cube, 77-79
Orthobase, 190-191
Orthocompact space, 100 (*see also* preorthocompact space)
characterizations of, 100
closed continuous image of, 114-115
examples of, 138, 147-148
examples of non-, 60, 119, 131, 170, 189
finite-to-one maps preserve, 113
FT is convergence complete, 101
and G_δ-diagonal is a $\sigma^\#$-space, 175

[Orthocompact space]
hereditarily orthocompact, 102, 104, 111, 171
Kofner plane is, 148
compact open maps do not preserve, 158
vs. paracompact, 111
vs. realcompact, 111, 149
and semistratifiable is transitive, 137
vs. σ-orthocompact, 102, 104
not transitive, 147-149
weakly orthocompact space, 116-120

P

Pairbase, 163
Paracompactness:
characterizations of, 103, 116, 167
vs. orthocompactness, 99, 103, 112
products of, 108, 117
and Tamano's Theorem, 120
Partial refinement, 114
Perfect map:
of a (pre)orthocompact space, 112-113
of a quasi-metrizable space, 160
and quasi-uniformities in $\delta(P)$, 51
of a weakly orthocompact space, 117
Pervin quasi-proximity, 28, 33-37, 43-45
Pervin quasi-uniformity, 28
in a compact space, 116
is a covering quasi-uniformity, 30
= FT if and only if, 31
= $PF(= SC)$ if and only if, 57
is the only admissible quasi-uniformity if and only if, 42-45
is totally bounded, 27-28, 58
perfect maps, 51
products and subspaces, 34-35, 55
subbase for, 27, 32
for the upper topology on $\mathbb{R}$, 28, 82, 92
Pixley-Roy space, 179
Point-finite covering quasi-uniformity, 30
belongs to δ_P, 33
of a countably compact space, 33, 54, 58
vs. $FINE$, 137, 139, 153
vs. metacompact, 116
and perfect maps, 51
products and subspaces, 34, 55-56
totally bounded if and only if, 57-58

Point-finite open spectrum (*see* open spectrum)
Point-separating cover, 172
Point-star refinement, 121
Point-symmetric:
  quasi-metric, 39-40, 169-171
  quasi-proximity, 36-39
  quasi-uniformity, 36, 38, 69, 98
Precise refinement, 100
Precompact space (*see also* totally bounded space), 51-55, 58, 152
Preorthocompact space (*see also* orthocompact space)
  and almost 2-fully normal spaces, 110
  characterization of, 103-104
  and compact open maps, 113
  developable, 137
  examples of non-, 108, 111-113, 117, 119, 131
  F_σ-discrete, 139-140
  vs. γ-refinable, 168
  hereditarily preorthocompact, 104
  vs. metacompact, 103-106
  not orthocompact, 170
  not preserved by compact open maps, 113
  not preserved by perfect maps, 112
  vs. normal, 108
  open finite-to-one maps preserve, 113
  and perfect maps, 112, 117
  of products, 108-109
  vs. σ-orthocompact, 113
  semistratifiable, 136-137, 171
  subspaces of, 104
  $X\times\beta X$, 120-124
Product quasi-proximity, 21-22, 25, 45
Product quasi-uniformity, 7-8, 22, 25, 34-35, 49, 51
Product topological ordered space, 74, 79
Property Q, 188
Proximity, 10
Pseudocompactness, 55, 58, 93-95
Pseudo-open map, 156
P-space, 129-132

Q

$\mathcal{Q}$, 1-2, 4, 8, 12-13, 16, 23-24, 82
qp-continuous function, 21, 23-24, 26, 82
qp-embedding, 21, 87
qp-equivalent, 12
qp-invariant, 37, 81
qp-isomorphism, 21
QS-space, 193
Quasi-metric, 3
  complete if and only if, 176
  continuous, 154
  equinormal is complete, 177
  non-Archimedean, 129
  precompact not totally bounded, 152
Quasi-metrizable space, 4
  characterized in a transitive space, 153
  closed mappings of, 160
  compact open mappings of, 157-159
  completely quasi-metrizable,
    vs. complete Aronszajn space, 178
    vs. complete Moore space, 180
    vs. developable space, 181
    not quasi-developable, 194
    is subdevelopable, 179
  conditions implying quasi-metrizability, 151-153, 156, 160-161, 165-166, 169-171, 174
  countable products of, 156
  countably compact, 40, 178
  countably metacompact, 168
  examples of non-, 119-120, 157-159, 166-167, 192
  vs. every neighbornet is normal, 140
  vs. every open cover is quasi-normal, 168
  finite-to-one open maps preserve, 156
  vs. γ-space, 162, 165-166
  vs. G_δ-diagonal, 153-154
  vs. metrizable, 40
  not orthocompact, 113, 170
  not preorthocompact, 113
  not transitive, 155
  non-Archimedeanly quasi-metrizable space, 129, 151-153, 166, 169-172, 174
  perfect maps preserve, 160
  point symmetric if and only if, 169-171
  precompact vs. totally bounded, 152

[Quasi-metrizable space]
  pseudo-open mappings of, 156
  vs. σ-interior preserving base, 151, 155
  totally bounded, 152
Quasi-metrization problem, 153
Quasi-normal cover (*see also* Lebesgue quasi-uniformity), 97, 119, 131, 137, 168-169
Quasi-proximity, 10
  associated quasi-proximity, 87-89
  class, 12-15, 23-26
  induced by $\mathcal{U}$, 12
  product, 21-22
  topology induced by δ, 11
Quasi-pseudo-metric, 4, 28
Quasi-uniform cover (*see also* Lebesgue quasi-uniformity), 97, 115-120
Quasi-uniformly continuous function, 6-9, 12-13, 21, 23-24, 26, 34, 48-49, 61-62
Quasi-uniform space, 1
Quasi-uniformity, 1
  generated by d, 3
  induced by $\mathcal{U}$, 7
Quasi-unimorphism, 6

R

R_0-space, 6, 36-39, 43
R_1-space, 43
Relative quasi-uniformity, 7, 34-35, 49-51, 55, 63-64

S

Scattered space, 141, 149
Second countable space, 152
Semi-continuous quasi-uniformity:
  contains C(X), 116
  of a countably compact space, 54
  and perfect maps, 51
  precompact if and only if, 54, 58
  products and subspaces, 34, 55-57
  subbase for, 32
  is transitive, 33
Semi-metacompact space, 105-108
Semistratifiable space, 135-139, 166, 171, 172
Separate, 77
σ-cocushioned pairbase (*see* cocushioned)
σ-interior preserving base, 151, 153, 171
σ-locally finite base, 153, 163
σ-orthocompact space, 102, 104, 113
σ-point-finite base, 152-153
$\sigma^{\#}$-space, 172-176
Sorgenfrey line:
  determined by P^{-1}, 82
  has no completion, 180
  nonnormal neighbornet of, 144
  order compactification of, 86, 92
  $PF \neq FINE$, 153
  precompact subspace of, 152
  topology of, 4, 28
  vs. quasi-uniform covers, 118
  Z, 1-2, 13, 82
Sorgenfrey plane, 118, 153
Souslin line, 192
Stone-Čech compactification (*see* compactification)
Strict order compactification, 89
Strong inclusion, 11-12, 14-17, 19-20
Strongly zero-dimensional space (*see* zero-dimensional space)
Subbase for a quasi-uniformity, 1
Subdevelopable space, 169, 179-180
Subinfinite rank, 100
Subtopology, 172-173, 175
Supremum quasi-proximity, 15-17, 22, 66
Supremum quasi-uniformity, 2, 12, 16-17, 26, 28, 66
Syntopogenous structure, 184-185

T

Tamano's Theorem, 120
θ-refinable space, 114-115, 134, 180
T_1-space, 6, 67-69
Topological ordered space, 73
Topological spaces (*particular spaces listed in this index*):
  Engelking-Lutzer line, F_0, Kofner plane, Michael line, Niemytzki plane, ω_1 (space of countable

[Topological spaces]
ordinals), Pixley-Roy space, Sorgenfrey line, Sorgenfrey plane
Totally bounded quasi-uniformity, 12-17
- compactifications of, 69-71
- with a countable base, 152
- if and only if, 31, 52, 65
- *LF*, 55, 58
- *P* is, 27-28
- *P* is finest totally bounded quasi-uniformity, 28
- *PF*, 57-58
- *SC*, 57-58
- quasi-uniform continuity preserves, 23
- vs. uniformly discrete sets are finite, 22-23

Transitive neighbornet (*see* neighbornet)
Transitive quasi-uniformity, 27, 33, 35
Transitive space, 45, 130-135
- closed continuous image of, 135
- vs. compact, 146
- countable union of, 132-133
- examples of non-, 147, 170
- F_σ-discrete space is, 133
- GO-space is, 145
- locally transitive space, 134
- metrizable space is, 137
- orthocompact if and only if, 131
- orthocompact semistratifiable space is, 137
- not preorthocompact, 131
- products of, 149
- P-space is, 131
- vs. $\sigma^\#$-space, 173
- subspaces of, 132, 146
- topological sum of, 134

Transitive uniformity, 125-130
2-fully normal, 110
T_0-space, 6, 64-66

U

Uniform base, 107-108, 148
Uniform cover, 97, 115-116
Uniformity, 1, 167

[Uniformity]
- coarsest, 20
- and compactifications, 66, 69-70
- completion of, 65
- and continuous maps, 19
- convex, 84-85
- with a countable base, 26
- fine, 99
- local, 162
- metric, 97
- and the Niemytzki-Tychnoff Theorem, 178
- precompact if and only if totally bounded, 52
- transitive, 125-130
- $\mathcal{U}_\delta$, 14
- and uniform covers, 98

Uniformly discrete subspace, 22-24
Uniformly locally compact space, 119
Uniform ordered space, 78-80, 84-90
- vs. compact ordered space, 78
- vs. GO space, 92, 94-95
- if and only if, 80, 84
- products of, 79

Unique proximity of a compact Hausdorff space, 19
Unique uniformity of a compact Hausdorff space, 9
$\mathcal{U}_\omega$, 15, 17, 25-26, 31, 66, 81-82, 126
Upper semi-continuous function, 154-155
Upper topology for $\mathbb{R}$, 28, 82, 92
Urysohn's Lemma, 24, 75, 88-89

W

Weakly Cauchy filter, 118-119
Weakly orthocompact space, 116-120
Weakly regular space, 183
W-refinement, 121

Z

Zero-dimensional space, 125-130
- basic properties of, 130
- compactification, 126
- not quasi-metrizable, 119, 157-158
- not weakly orthocompact, 119
- strongly zero-dimensional space, 127-130